12752358   cds.

GROUNDWATER

Second Edition

GROUNDWATER

Second Edition

ROBERT BOWEN

Professor of Geology
Institute of Geology, University of Cologne
Federal Republic of Germany

WATER RESOURCES
CENTER ARCHIVES

FEB -- 1999

UNIVERSITY OF CALIFORNIA
BERKELEY

ELSEVIER APPLIED SCIENCE PUBLISHERS
LONDON and NEW YORK

ELSEVIER APPLIED SCIENCE PUBLISHERS LTD
Crown House, Linton Road, Barking, Essex IG11 8JU, England

Sole Distributor in the USA and Canada
ELSEVIER SCIENCE PUBLISHING CO., INC.
52 Vanderbilt Avenue, New York, NY 10017, USA

First edition 1980
Second edition 1986

WITH 63 TABLES AND 55 ILLUSTRATIONS

© ELSEVIER APPLIED SCIENCE PUBLISHERS LTD 1986

British Library Cataloguing in Publication Data

Bowen, Robert
 Groundwater.—2nd ed.
 1. Water, Underground
 I. Title
 551.49   GB1003.2

Library of Congress Cataloging-in-Publication Data

Bowen, Robert
 Groundwater.

 Bibliography: p.
 Includes index.
 1. Water, Underground. I. Title.
 GB1003.2.B69   1986   553.7′9   85-25389

ISBN 0-85334-414-0

Special regulations for readers in the USA

This publication has been registered with the Copyright Clearance Center Inc. (CCC), Salem, Massachusetts. Information can be obtained from the CCC about conditions under photocopies of parts of this publication may be made in the USA. All other copyright questions, including photocopying outside of the USA, should be referred to the publisher.

All rights reserved. No part of this publication may be reproduced, stored in a retrieval system, or transmitted in any form or by any means, electronic, mechanical, photocopying, recording, or otherwise, without the prior written permission of the publisher.

Photoset in Malta by Interprint Limited
Printed in Great Britain by Galliard (Printers) Ltd, Great Yarmouth

Preface

In his Olympian Odes, Pliny called water 'the best of all things' and, next to air, it is the most essential human requirement. Water constitutes the hydrosphere of the Earth, a planetary feature unique in the solar system, which shows characteristic physico-chemical effects associated with its circulation in the hydrologic cycle.

Nearly all water occurs in the vast sink of the oceans, which is being polluted by man-made toxicants on an ever-increasing scale; unfortunately, this process is happening in fresh waters also. Among these fresh waters are groundwaters, which originate mainly from precipitation and subsequent infiltration through soils and surface rocks of continents into saturated subterranean water-bearing bodies termed aquifers.

The continuous increase in population in many countries and the accelerating urbanization so evident in most developing lands accompany a steadily rising demand for these groundwaters. This book examines the diverse aspects of groundwater, beginning with an historical survey: groundwater has been used for millennia, and is mentioned in the Bible. Thus, at Nineveh about 700 B.C., King Sennacherib ordered fresh spring water, a concentrated discharge of groundwater, to be conveyed to his palace along a waterproofed canal, 80 km long, to replace muddy water from the Tigris River. The Athens of the fifth century B.C. was supplied with groundwater from wells and tunnels, as was Rome, where an estimated $38\,000\,m^3$ was consumed every day by the million or so residents.

Hydrometeorology and infiltration, the mechanism whereby surface water penetrates below ground, are then discussed. Voids in which water may exist, and through which it may be transmitted, are investigated; then analyses of the movement of groundwater and water well hydraulics are followed by a description of possible environmental effects. An account is given of geothermal waters and the quality of groundwater, threatened in many places today by viral, bacterial, radioactive, arsenical, lead, aluminium, nitrate and other pollutants, is considered in some detail. Later chapters are devoted to such specialized topics as saline water intrusion, the artificial recharge of groundwaters, and isotope hydrology in investigations of groundwater, as well as the geophysical technology employed on and below the ground surface in order to obtain information regarding locations, thicknesses, compositions, permeabilities, transmissivities, yields and other parameters of aquifers, including water quality and movement. Finally, groundwater in construction is considered, together with groundwater modelling and groundwater management. Groundwater also affects landscape, especially in karstlands, where appropriate features can be utilized in interpreting the geomorphological history of a region. However, groundwater as a geomorphological agent is only now being widely studied and there is insufficient information available to justify the matter being discussed in this book. For those interested, there is a recent interesting treatment which was published in 1984 and edited by R. G. LaFleur.[1]

There are several appendixes, including dictionaries of water and isotope terms together with a review of the legal aspects of groundwater. Indexes cover authors cited, subjects and locations mentioned in the book.

A first edition of this book was published in 1980, but the present second edition is considerably enlarged and completely re-written; it has been expanded and updated with many more illustrations and new material.

The literature of groundwater is growing at such a rate that limitations of space restrict references to only a representative number. The problem of choosing these references may be illustrated by the fact that GEOREF lists 4802 groundwater papers since 1929, the year to which it currently goes back; of these, 1699 appeared between 1980 and July 1985, the time of writing. GEOARCHIVE (again, at the time of writing) records no less than 12 195 papers with groundwater as descriptor since 1974, the inception year for data at present, and 4262 of these were published since 1980. In addition, there are 2422 papers listed which refer to aquifers and there have been 22 with this word in their titles since 1984. However, the references cited in this book provide satisfactorily for general requirements, offering a reasonably complete survey of fundamental principles and their field applications.

For greater, more specialized, coverage of various topics in groundwater studies, databanks of the above-mentioned type may be consulted, together with the many relevant research and applied journals such as the *Journal of Hydrology, Water Resources Research* and *Ground Water*, publications such as the United States Geological Survey Water Supply Papers, and those produced by the United States Environmental Protection Agency and various agencies of the United Nations Organization such as UNESCO, UNICEF, FAO and the International Atomic Energy Agency which periodically issues symposia on isotope hydrology, data on stable isotopes in precipitation and other valuable sources of information.

ROBERT BOWEN
Cologne

REFERENCES

1. LaFleur, R. G., Ed., 1984. *Groundwater as a Geomorphic Agent.* Allen and Unwin, Boston, 390 pp.

Contents

Preface v

Chapter 1. GROUNDWATER AND HYDROLOGY: AN HISTORICAL SURVEY 1
1.1. Aspects of Water 1
1.2. Hydrology and Groundwater in History. . . . 6

Chapter 2. HYDROMETEOROLOGY AND INFILTRATION 25
2.1. Meteorological Data 25
2.2. Infiltration and Percolation Below Ground . . . 38

Chapter 3. VOIDS IN ROCKS 49
3.1. Soils and Rocks 49
3.2. Subterranean Vertical Distribution of Water Input. . 55
3.3. Groundwater Occurrence. 62

Chapter 4. THE MOVEMENT OF GROUNDWATER . . 78
4.1. Darcy's Law 78
4.2. Anistropic Aquifers 87
4.3. Rates and Directions of Flow in Groundwater . . 89
4.4. Vertical Movements of Groundwater 101

Chapter 5. GROUNDWATER HYDRAULICS . . . 104
5.1. Steady Flow 104
5.2. Unsteady Flow 112

5.3.	Unsteady Radial Flow	117
	5.3.1. Unconfined aquifers	117
	5.3.2. Leaky aquifers	118
5.4.	The Method of Images	120
5.5.	Interfering Wells and Those Only Partially Penetrating	124
5.6.	Well Losses and Specific Capacity	126

Chapter 6. WATER WELLS 131

6.1.	Introduction	131
6.2.	Construction Technology	132
6.3.	Development of Wells	142
6.4.	The Equipment for Pumping	143
6.5.	The Protection and Restoration of Wells	147
6.6.	Non-vertical Wells	149

Chapter 7. ENVIRONMENTAL EFFECTS ON GROUNDWATER 152

7.1.	Groundwater Levels	152
7.2.	Evapotranspirative Fluctuations	154
7.3.	Meteorological Fluctuations	156
7.4.	Tidal Fluctuations	158
7.5.	Population Centres	159
7.6.	Seismicity	160
7.7.	Subsidence	161

Chapter 8. GEOTHERMAL WATERS 166

8.1.	Occurrence	166
8.2.	Chemical and Isotopic Thermometry	167
8.3.	Quantitative Chemical Geothermometers	167
8.4.	Qualitative Chemical Geothermometers	170
8.5.	Isotope Thermometers	171
8.6.	Hydrothermal Reactions and Reservoir Temperature Estimates	172
8.7.	Subterranean Mixing and Boiling Phenomena	177
8.8.	Geothermal Fields	180
8.9.	Thermal Waters from Springs and Wells	181
8.10.	Reservoir Engineering	182
8.11.	Geochemical Evidence of Higher Temperature Elsewhere	184
8.12.	Hyperthermal Geothermal Fields	185

Chapter 9. GROUNDWATER QUALITY 188
9.1. Groundwater Salinity 188
9.2. Quality Analyses 194
9.3. Water in Irrigation 203
9.4. Subterranean Geochemical Variations 207
9.5. Temperature 207
9.6. Pollution 208
9.7. Recent Developments Regarding Pollutants . . . 211

Chapter 10. SALINE INTRUSION 214
10.1. Occurrence 214
10.2. The Ghyben–Herzberg Relation 215
10.3. The Fresh Water/Saline Water Interface . . . 219
10.4. Oceanic Islands 220
10.5. Karstic Terrains 222
10.6. Controlling Saline Water Intrusion 222
10.7. Origin and Movement of Saline Contamination . . 224

Chapter 11. ARTIFICIAL RECHARGE 227
11.1. Increasing Natural Groundwater Supply . . . 227
11.2. Project Objectives 228
11.3 The Methods of Recharge 228
11.4. Unplanned Recharge 235
11.5. Re-using Municipal Wastewater 236
11.6. Recharge Mounds 237
11.7. Long Island, New York, USA 237
11.8. Recharge Induction 239
11.9. Artificial Recharge Activities in Europe . . . 240
11.10. Major Problems in Artificial Recharging . . . 241
11.11. Heat Storage 242

Chapter 12. ISOTOPE HYDROLOGY AND
GROUNDWATER. 245
12.1. Isotope Hydrology. 245
12.2. Environmental Isotope Hydrology 245
12.3. Artificial Isotope Hydrology 268

Chapter 13. GEOPHYSICAL INVESTIGATIONS OF
GROUNDWATER — I. SURFACE METHODS. 275
13.1. Above-surface Methods. 275
13.2. On-surface Methods 279

CONTENTS

Chapter 14. GEOPHYSICAL INVESTIGATIONS OF GROUNDWATER — II. SUBSURFACE METHODS 288
14.1. Geological Methods 288
14.2. Geophysical Methods 290
    14.2.1. Resistivity Logging. 292
    14.2.2. Spontaneous Potential Logging . . . 294
    14.2.3. Radioactive (Nuclear) Logging . . . 295
        14.2.3.1. Natural Gamma Logging . . 295
        14.2.3.2. Gamma–Gamma Logging . . 296
        14.2.3.3. Neutron Logging 297
        14.2.3.4. Neutron–Gamma Logging . . 298
    14.2.4. Temperature Logging 300
    14.2.5. Caliper Logging 300
    14.2.6. Fluid-conductivity Logging 300
    14.2.7. Fluid-velocity Logging 301
    14.2.8. Television Logging. 301
    14.2.9. Casing Logging 301
    14.2.10. Acoustic Logging 301
    14.2.11. Environmental Isotope Logging . . . 302

Chapter 15. GROUNDWATER IN CONSTRUCTION . . 305
15.1. Groundwater as a Hazard 305
15.2. Building Foundations 306
15.3. Tunnels 309
15.4. Settlement 310
15.5. Dams 312
15.6. Excavation Beneath Rotterdam 316
15.7. Permafrost 317

Chapter 16. GROUNDWATER MODELLING . . . 322
16.1. Models for Groundwater Characteristics . . . 322
16.2. Porous Media Models 322
16.3. Non-electrical Analogue Models 323
    16.3.1. Viscous Fluid Models 324
    16.3.2. Membrane Models 325
    16.3.3. Moiré Pattern Models 326
    16.3.4. Blotting Paper Models 326
    16.3.5. Thermal Models 326

16.4.	Electrical Analogue Models		327
	16.4.1.	Continuous Systems	327
	16.4.2.	Discontinuous Systems	328
16.5.	Digital Computer Models		329

Chapter 17. GROUNDWATER MANAGEMENT . . . 336
- 17.1. Groundwater Basins 336
- 17.2. Extraction of Groundwater 336
- 17.3. The Equation of Hydrologic Equilibrium . . . 338
- 17.4. The Investigation of Groundwater Basins . . . 338
- 17.5. Safe Yields 340
  - 17.5.1. Mining Yield 340
  - 17.5.2. Perennial Yield 341
- 17.6. Salinity 343
- 17.7. Conjunctive Usage 344
- 17.8. Case Histories 345
  - 17.8.1. High Plains, New Mexico/Texas, USA . . 345
  - 17.8.2. Los Angeles Coastal Plain, California, USA . 346
  - 17.8.3. Indus River Valley, Pakistan 346
  - 17.8.4. Libya 347
  - 17.8.5. USSR 348
  - 17.8.6. Federal Republic of Germany . . . 348
    - 17.8.6.1. The Upper, Unconfined Aquifer . 352
    - 17.8.6.2. The Lower, Confined Aquifer . 353
    - 17.8.6.3. Watershed Movement . . . 353
    - 17.8.6.4. Surface Streams 353
    - 17.8.6.5. Remedies 354
  - 17.8.7. North Africa and the Arabian Peninsula . 355
- 17.9. Water Balances 356
  - 17.9.1. Modry Dul 357
  - 17.9.2. Lake Chala 357
  - 17.9.3. Turkey 357
  - 17.9.4. Other Cases 357
  - 17.9.5. Global Water Balance 358

APPENDIXES 361
1. Dictionary of Water Terms 361
2. Dictionary of Isotope Terms 384
3. Analysis of Water Quality with Reference to its Chemistry 392

4. Some Representative Porosities and Permeabilities for Various Geologic Materials 395
5. Groundwater in Law 396
6. Organizations Connected with Groundwater Matters . . 402
7. Measurements 406

Author Index. 411

Subject Index 418

CHAPTER 1

Groundwater and Hydrology: An Historical Survey

'Everything flows.' HERACLITUS, 540–480 B.C.

1.1. ASPECTS OF WATER

Water comprises hydrogen and oxygen, the common isotopic species being $^1H_2{}^{16}O$. In the water molecule, two hydrogen atoms are attached by covalent bonding to each atom of oxygen. Such molecules are arranged randomly in the liquid state so that angles between constituent atoms of neighbouring molecules vary continuously. Each oxygen atom can attract more than two hydrogen atoms and each of the latter can be surrounded by several oxygen atoms. Breaking of hydrogen bonds precedes a rolling flow of molecules and, as this process occurs easily, water possesses a low viscosity. In the solid state, two hydrogen atoms are bonded to an oxygen atom at an angle of 105° and the overall pattern is hexagonal, as may be seen in an ice crystal. Each group of molecules in the crystal has a central one surrounded by four others at the corner of a tetrahedron, all being joined by hydrogen bonds. Interacting attractive forces produce an inwardly directed pressure and, when the temperature reaches 0°C, collapse occurs through thermal agitation of the molecules. In an ice crystal, the distance between the centre of one oxygen atom and the next is 2·72 Å, increasing to 2·9 Å during melting.

As well as in its common species, water may be found in others. In 1934, H. C. Urey discovered deuterium, a second isotope of hydrogen with an atomic weight of 2; this may combine with oxygen to form 'heavy water', D_2O ($^2H_2{}^{16}O$). Deuterium is found in nature, but it is rare at a mere 200 ppm in natural waters. It is interesting that the properties of heavy water differ from those of ordinary water. It has a slightly higher boiling point (101·4°C), and it freezes at a higher temperature (3·8°C). In addition, its viscosity is higher. Perhaps its most surprising attribute is that, as it is inert in normal metabolic processes, rats die of thirst if given it and plants will not grow in it. A third isotope of hydrogen, ^{3}H or T as it sometimes written, is weakly radioactive and important in hydrology. It forms in the upper atmosphere through cosmic ray bombardment and other processes, thereafter precipitating in rain or snow. Its half-life is just over 12 years so that it disappears from the environment after 50 years. This characteristic can be used to 'date' groundwater.

Oxygen also has a series of isotopes of which the stable oxygen-18, ^{18}O, is valuable in hydrology despite its rarity in natural waters; see Table 1.2, Section 1.2.

A. M. Buswell and W. H. Rodebush[1] summarized the isotopic species of water; their list includes 33 in all, although most are hardly found in nature and some might not be considered real varieties of water at all. That discipline which deals with the properties, occurrence, distribution and movement of water on and under the surface of the Earth constitutes hydrology. O. E. Meinzer[2] in 1942 defined it as the science concerned with the occurrence of water in the Earth, its physical and chemical reactions with the rest of the Earth and its relation to life on Earth. Twenty years later, the US Federal Council of Science and Technology for Scientific Hydrology stated through its Ad Hoc Panel that 'hydrology is the science that treats of the waters of the Earth, their occurrence, circulation and distribution, their chemical and physical properties and their reaction with their environment including their relation to living things. The domain of hydrology embraces the full life history of water on the Earth'.[3] Clearly, there must be a hydrology of the past as well as of the present. Its objectives were described by K. J. Gregory in 1983 as including 'reconstruction of the components of the hydrological cycle, of the water balance, and of sediment budgets for the time before continuous hydrological records ... embracing an understanding of the way in which changes in the hydrological cycle occurred and establishing how they differed from the contemporary hydrological picture', ambitious undertakings indeed. He edited a perspective in which such matters are discussed, including climatic change, paleomagnetism and archeology.[4]

The perpetual movement of water from sea to atmosphere to land and back to the sea is termed the hydrologic cycle, a system powered by solar and planetary forces. The sun provides energy for the evaporation of seawater, the planetary gravitational field and Coriolis force are controls on winds, and the gravitational fields of the sun and moon affect the balance of forces according to E. E. Adderley[5] and E. G. Bowen.[6]

Water evaporating, mainly from the oceans but also to a lesser extent from the continents, is carried into the atmosphere and may drift for great distances before returning to the Earth's surface as precipitation of one kind or another. After falling, this precipitation undergoes one or more of the following processes:

On land
(i) interception by vegetation which prevents its reaching the ground;
(ii) collection on the Earth's surface where soils, rocks, etc., are impermeable;
(iii) seepage into permeable materials;
(iv) contribution to streams which flow to lower ground and finally may reach oceans;

On the oceans
(v) recombination with their surface waters.

It is just possible that sporadic phenomena such as meteorite showers or solar flares or sunspots which cause changes in the electromagnetic field of the Earth may influence the hydrologic cycle. Some attempts have been made to correlate precipitation with meteorite showers and also to correlate river hydrographs with numbers of sunspots, but these are peripheral to most hydrologic studies and, in any case, they have no practical effect on the availability of water.

When precipitation stops, water on the planetary surface may re-evaporate or, if brought up through plant roots, transpirate. In vegetated areas, these two processes are termed evapotranspiration. Some water percolates into the ground and may reach the water table, which is a natural level of free groundwater.

Groundwater is water which occupies all voids in a geological bed, thus saturating it to comprise the zone of saturation. This is to be distinguished from an unsaturated zone of aeration which may overlie it and in which the voids are filled with water and air.

Unsaturated zones extend to the surface of the ground, as may saturated ones, for example when the water table emerges to produce springs.

Figure 1.1 shows the hydrologic cycle. Associated with this is a geochemical cycle discussed by G. H. Davis *et al*.[7] and shown in Fig. 1.2. Essentially, the geochemical cycle has the following characteristics. Evaporating water from the oceans carries a small amount of dissolved mineral matter away. Nitrogen compounds and molecules of oxygen and

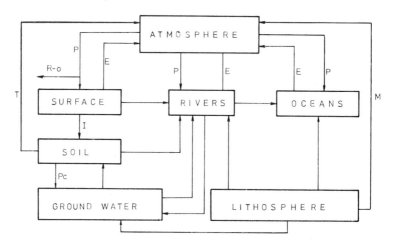

Fig. 1.1. The hydrologic cycle. R-o, runoff; T, transpiration; P, precipitation; E, evaporation; I, infiltration; Pc, percolation; M, volcanic or magmatic water.

carbon dioxide are all dissolved in water precipitated during storms. Carbon dioxide, a product of organic decomposition, is also dissolved as water percolates through upper layers of the soils. Dilute carbonic acid facilitates chemical reactions with mineral fragments which release bicarbonates and carbonates (which may dissolve). Other soluble minerals or salts are dissolved by incoming water. Once water reaches the geological matrix many reactions may occur, e.g. less soluble compounds may precipitate as solubility limits are attained or bacterial reduction of sulphates in solution may take place. In the end, water either returns to the atmosphere by evaporation (leaving mineral matter behind in the soil) or it returns to the sea as groundwater discharge or as streamflow carrying a mineral content.

The hydrologic and geochemical cycles are highly variable and indeed subcycles have been noted within them. For instance, water may reach vegetation and evaporate from it directly or water may reach the ground and be held in soils by capillary and molecular forces until it evaporates. In the first case the cycle is by-passed, and in the second the water cannot be included in surface or groundwater discharge.

R. G. Kazmann mentioned that, of the 30 in (762 mm) average depth of water falling annually on the United States, 21–22 in (533–559 mm) are vaporized and only 8–9 in (203–229 mm) appear as clearly defined liquid water on or below the ground surface.[8] Naturally, water

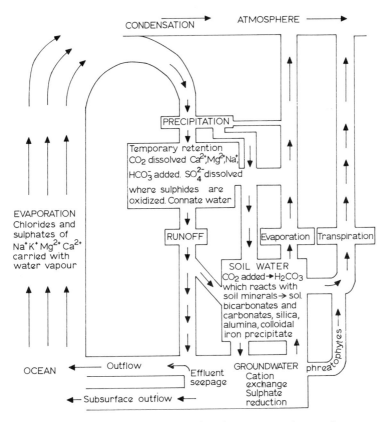

Fig. 1.2. The geochemical cycle of surface water and groundwater.

not found in the runoff phase of the cycle goes to sustain plant life and this includes crops which Man plants and harvests.

The quantities of water passing through the hydrologic cycle in a given period of time can be expressed in the hydrologic (continuity) equation:

$$I - O = \pm \Delta S \tag{1.1}$$

I being the total inflow (precipitation, infiltration, vadose and groundwater), O the total outflow (evapotranspiration, surface and subsurface runoff) and ΔS the change in storage in the various forms of retention and interception. Details are conditioned by considerations of geography and geomorphology. Data from Germany may be compared with those cited above from the USA in order to emphasize this point. The data shown in Table 1.1 were provided by S. Clodius[9] for West Germany, and by K.-F. Busch and L. Luckner[10] for East Germany. West German records were for the period 1891–1930 and the bracketed percentages refer to proportions of the precipitation.

TABLE 1.1
QUANTITIES OF WATER (mm/y) IN THE HYDROLOGIC CYCLE[9,10]

	West Germany	East Germany
Precipitation	803	628
Evaporation	410 (51%)	479 (76·3%)
Outflow	393 (49%)	149 (23·7%)
Subsurface (groundwater)	112 (14%)	70 (11%)

The results from the United States of America cited above reflect the great variability of that enormous landmass. The lower figures obtained from East Germany are a measure of its continental climate, quite different from that of West Germany.

Subsurface groundwater movements result from recharge entering and moving down gradient, i.e. from higher to lower hydraulic head. Gradients usually arise according to the types of rock involved and also the structural characteristics of rock masses.

1.2. HYDROLOGY AND GROUNDWATER IN HISTORY

As a science, hydrology may be said to have commenced in the seventeenth century: it was preceded by hydraulic engineering in the shape of maintenance of water supplies, channel improvement, levee construction for protection against floods, and canal construction for crop irrigation. Such activities had been practised for at least 5000 years and involved complex operations of which the results have been exceeded only in the past century (mainly due to advances both in earth moving machinery and communications).

Early civilizations in the valleys of the Tigris and Euphrates rivers and also along the Nile valley, all extremely arid regions, depended on irrigation, of which the earliest examples come from predynastic Egypt before the Old Kingdom (3400 B.C.). In fact, the Pharaonic basin irrigation system goes back at least as far as the first, legendary Pharaoh, Menes, who captured the Nile waters and harnessed the floods of that river. The method was to divide the flood plain into a series of areas enclosed by dikes, ranging in size from 100 to 40 000 acres (40·5 to 16 188 hectares). When floods occurred, the basins so constructed became filled with water which was trapped by closing the dikes. After saturation of the land, the standing water was drained off to lower basins or even to the river and planting became possible. Every year, the flood water deposited organic silt which enriched the soil as well as elevating it. It is believed that in the last 1300 years or so, silt deposited in the Nile Valley has raised the land about 2·4 m. This estimate is based on complete records since the Arab conquest and, by extrapolation, it is probable that the flood plain of Pharaonic times was approximately 6 m lower that it is now. The ancient Egyptians had to devise a means of measuring and recording the height of the Nile at many localities so as to compare the daily river rise with data from previous years. This was in order that they could determine probable flood heights and so predict the acreage which might be flooded in any particular reach of the river. If flooding became too severe, overtopping of the dikes would be followed by washing out and the area remaining for irrigation would be reduced. Nile height data enabled the Egyptians to forecast the high-water mark at any downstream point quite accurately and hence select higher land downstream for extra cultivation to replace flood-lost area upstream.

As long ago as the Twelfth Dynasty (about 2000 B.C.) information was being recorded as far upstream as the Second Cataract near the present Egyptian–Sudanese border: in fact, rock-engraved readings have been discovered. No doubt, such data could have been dispatched to the Pharaoh and preparations for the high water set in hand before it arrived. It is interesting to observe that the Tennessee Valley Authority today maintains river gauges in the upper reaches of the Tennessee Valley and the data from these are transmitted to a computing centre which enables predictions and downstream preparations to be carried out. Actually, the task is more difficult than was the case with the Egyptians because the Nile is more regular, and hence more predictable, than the Tennessee River.

As M. S. Drower has indicated, it is perfectly possible that reclamation of marshland took place in the lowlands of the Persian (Arabian) Gulf and along the lower Euphrates River far earlier than 3000 B.C.,[11] and G. Sarton has reported that drainage and irrigation canals then in use have left traces clearly visible from the air today.[12] This is a region which is more complicated and less predictable than the Nile; also, the Tigris–Euphrates system waters remove about five times the sediment concentration of the Egyptian river so that the canals and distributaries must have been choked with silt quickly. Their maintenance must have been a considerable problem and one made more difficult by the higher mineralization of the waters of the twin rivers of Mesopotamia which could cause excess concentration of salt in the soil.

Flooding was a perennial, practically unpredictable hazard in the region and, when peaking on both rivers occurred simultaneously, there were catastrophic consequences. This may well be echoed in the Biblical account of Noah and the Great Flood, and in the construction of earthen levees and walls around towns and cities, though these may have been meant to protect not only against floods, but also against attackers. Kazmann drew attention to the fact that the Persians, who came from northern highlands, constructed walled cities only after their conquest of Babylon.[8]

Hammurabi's rise to power in Mesopotamia about 1760 B.C. entailed the incorporation into his Code of laws in regard to irrigation. These stipulated, for instance, that every district had to maintain its own canals properly; also, every individual was required to keep his part of the elaborate ditch and dike system in order. If he did not, and a neighbour's land was flooded as a result of this negligence, a very substantial compensation had to be paid.

Although baked clay tablets exist showing some details of tax assessment records, the map referring to the relevant property does not define the hydraulics of the irrigation and drainage system adequately. Perhaps during high water periods water was permitted to enter canals, from which it went into channels for irrigation purposes. These might have been drainage ditches flanked by low earthen embankments and the associated fields must have been small, possibly less than 10 acres (4 hectares) per irrigated zone. Of course, with time, silt deposition would have infilled them and the canals would have had to be progressively elevated, one consequence of which would have been reduction in the area irrigated. This is because, without pumping equipment, the natural height of the river itself would have had to provide the required motive power to force water over the land. As well as for such applications, hydrologic knowledge may have been employed even then for military purposes: there is the story that, in his assault on Babylon, the Persian King Cyrus the Great diverted the Euphrates River so as to use its former bed as a route for his troops to enter the city (which, in fact, he captured easily through surprise).

It is clear that the study of water is far from new: systematic irrigation, drainage of land, establishment of water rights and obligations as well as navigational considerations began very early in Man's history. Unfortunately, there was only a low technological level and therefore a science of hydrology could not develop, all work of hydrologic nature proceeding more or less by rule of thumb. However, these early achievements were very remarkable; Mesopotamian irrigation works endured until their destruction by the Mongols, i.e. over a period of over 4000 years.

There was possibly a parallel development of hydrologic engineering in the Yellow River region of China at a similarly early time, but the much greater humidity of Europe obviated any such need there. Nevertheless, Rome had a city system of water supply at the start of the Christian era which was described by Julius Frontinus, the water commissioner from 97 to 103 A.D. His responsibilities included the operation and maintenance of the eight aqueducts which supplied the city. It is interesting that there was no provision for water storage, but the fact that the aqueduct water sources comprised five springs and several streams tended towards uniformity of flow. There is no possible doubt that the Romans had no real hydrologic knowledge or, for that matter, no understanding of basic hydraulics either.

The first recognition of the hydrologic cycle is ascribed to Leonardo da Vinci (1452–1519) and Bernard Palissy (1509–1589), who independently realized that water circulates from the sea to the atmosphere, thence to the land and eventually back to the sea.

Groundwater has been used since antiquity, as A. K. Biswas mentioned, and the Old Testament contains many references to it as well as to springs and wells.[13] It was obtained from dug wells, and from horizontal wells known as qanats. This word derived from a Semitic word meaning 'to dig' and there are variants on it which include karez, foggara and falaj (discussed by G. B. Cressey[14]). Enormous numbers of qanats still exist and cover a swath from the arid regions of southwestern Asia and Afghanistan through North Africa. Generally, a gently inclined tunnel extends through alluvial materials and leads water by gravity flow from under the water table at the upper end to a surface outlet on the ground with an associated irrigation canal. A vast number of qanats occur in Iran, about 22 000 in all, and they supply three-quarters of all the water used in that country. P. Beaumont in 1971 referred to the longest one, which is near Zarand; it is 29 km in length, has a source well depth of 96 m, and there are no less than 966 shafts along it.[15] The total volume of excavated material is estimated to have been over 75 000 m^3.

Normally, the wells involved are shallower, but some have been found with depths exceeding 250 m. Qanats are usually much shorter than the one mentioned above, 5 km being an average figure. Their discharges vary with seasonal water table fluctuations and rarely exceed 100 m^3/h. Density on the ground can be high; for instance aerial photographs of the Varamin Plain, about 40 km southeast of Tehran, indicate 266 qanats in an area of 1300 km^2.

Use of groundwater preceded by far any clear understanding of its origin, much less of its movement, and there were several theories propounded by the Greeks, including Aristotle, Thales, Plato and Homer, and the Romans, among them Seneca, Pliny and Vitruvius. They believed that seawater, conducted through subterranean channels below mountains, became purified and raised to the surface as springs, Vitruvius, however, was perceptive enough to suggest that rain could percolate through rocks and emerge basally as streams. This was a great improvement on Aristotle's belief that air probably enters dark caverns under mountains, then condenses into water which contributes to springs.

Greek ideas persisted through the Middle Ages until Bernard Palissy again proposed Vitruvius's infiltration theory in 1580. He was ignored. More influential were Johannes Kepler (1571–1630) and René Descartes (1596–1650).

Kepler was a German astronomer who compared the Earth to a gigantic animal which imbibes oceanic water, digests it and discharges groundwater and springs.

Descartes, a French philosopher, reiterated the Greek idea regarding conduction of seawater through underground channels beneath mountains and supplemented it with processes of vaporization and condensation. Unfortunately, being famous does not always mean being right. In the seventeenth century, the first accumulation of observational and quantitative data was made and true scientific theories were erected on them. The most important Europeans to contribute to these were Pierre Perrault (1611–1680), Edme Mariotte (1620–1684) and Edmund Halley (1656–1742).

Perrault was a lawyer who held important governmental financial and administrative posts. He also wrote *De l'Origine des Fontaines* (published in 1674), dedicating it to Christian Huygens, the Dutch mathematician, astronomer and physicist then living in France. Perrault measured rainfall over three years and estimated runoff from the upper Seine River basin. In 1674, the precipitation in the basin exceeded river discharge by a factor of six, which showed clearly that the Greeks were wrong to think that spring water could not be derived from rainfall because the rainfall was insufficient in amount. He also did experiments on evaporation of water and other liquids, establishing too the rough limits of capillarity in sands.

O. E. Meinzer wrote of Mariotte, a French physicist, that he 'probably deserves more than any other man the distinction of being regarded as the founder of groundwater hydrology, perhaps I should say of the entire science of hydrology'.[16] The Frenchman used floats to ascertain the velocity of water in the Seine and, in order to demonstrate that the source of groundwater could be precipitation, compared seepage into the cellar of his Paris Observatory with the depth of rainfall in the vicinity.

Edmund Halley, an English astronomer, reported in 1693 on measurements of evaporation, indicating that evaporation from the sea was quite sufficient to account for all springs and the flow of streams. He evaporated salt solutions initially at the concentration of seawater, and concluded that the evaporation from the Mediterranean was enough to balance the amount of water entering through rivers draining into this sea. Additionally, he studied the effects of altitude on the rate of precipitation. It must be mentioned that Mariotte used Perrault's data to develop a theory for groundwater replenishment through precipitation and infiltration, an extraordinarily modern concept for his time. But the real significance of all this work was its practicality, which finally ended Greek speculation and paved the way for hydrology to become a real science. The foundations of modern geology were laid in the eighteenth century and contributed greatly to further understanding of the occurrence and movement of groundwater. C. Herschel[17] translated the books of Frontinus on the water supply of Rome and mentioned that 'to appreciate (his) position with regard to a proper knowledge of the velocity of efflux and generally of the velocity of running water, it is instructive to follow the development of the art from his time until we arrive at the formula $V = \sqrt{2gh}$ now known to every beginner in hydraulic science, and the very foundation stone of that science as it is known at the present day. This formula and the numerical values it gives to velocities of efflux were not discovered until about the year 1738, when Daniel Bernoulli and John Bernoulli, his father, each published a different mathematical demonstration of this law.

'Castelli (1577–1644), a Benedictine monk, the pupil of Galileo, first showed the quantity of efflux in a given time depended by law on, or was a function of, the depth of water in a bowl — that is, was a function of the head. It was his pupil Torricelli (1608–1647), the inventor of the barometer, the grandson, in a professional sense, of Galileo, who first proved, in 1644, or only two years after Galileo's death, that the velocities of efflux are as the square roots of the head ... Huygens (1629–1695), the inventor of pendulum clocks, first found the numerical value of the acceleration of gravity, commonly represented by the letter "g", in 1673. Sixty-five more years had to elapse until the genius of the two Bernoullis, father and son, in 1738, finally laid the foundation of modern, determinate hydraulics'.

Many others contributed to the development of hydraulic engineering and hydrology. Among the most noteworthy were B. F. Belidor (1697–1761), who proposed a siphon theory for ebb and flow of springs, and J. d'Alembert (1717–1776), who aided in the creation of the mathematical foundations for hydraulic theory.

In less than a quarter of a century after the work of the Bernoullis, P. Frisi published his *Treatise on Rivers and Torrents*, which included quantitative data on stream flow. He was an excellent engineer who did not have much faith in applying mathematics to hydrodynamic problems. This is probably because statistics and probability theory did not appear until 50 years later. Hence his statement[18] that 'all hydraulic problems are beyond the reach of geometry and calculus. The difficulty... is increased in proportion to the number of conditions (i.e. variables)... thus mechanical problems become so much more complicated as the number of bodies whose motions are sought and which act in any way on each other is augmented ... in a fluid mass (moving) in a tube or canal the number of bodies acting together is infinite (and so) it follows that to determine the motion of each body depending on an infinity of equations (is) beyond the powers of algebra to reach'. Frisi would be quite correct today, in that it is impossible to compute, under turbulent flow conditions, the path and velocity of a particular water molecule. The point is that it is not necessary to do this at all. All that is required is to find out the overall mass movement of water molecules, and this can be discovered in the field as well as being computed. In a stream, for instance, it is possible to multiply the velocity of a stream by its cross-sectional area — instantaneous velocities are measured at several points. The total flow can be understood even though turbulent flow is involved. It is theoretically feasible to determine the movement of individual molecules directly by the use of radioactive tracers, but, as noted, there is no need to do this.

Until the last century, the development of human knowledge of the surface water–groundwater interrelation as well as of groundwater movement was slow. Partly, this was because well sinking is slow and expensive. In addition, getting water out of wells required power sources and the pumps available were both large and inefficient. However, attempts to de-water mines stimulated pump construction, although the immediate objective — to get rid of water — dominated activities in those days. Only when people started to use the water, rather than remove it, was the principle of groundwater movement discovered and understood.

This utilization of groundwater accelerated during the Industrial Revolution when wells with ever smaller diameters were drilled, often by the cable-tool method, and artesian conditions were frequently ecountered — no doubt to the delight of the owner who thus avoided having to buy a pump. An adverse factor was that only hollowed-out trees were available as casing and so alternative surface water sources, normally available, were preferred. Consequently, wells were either shallow and of large diameter, penetrating the water-bearing bed (aquifer) which was non-artesian, not yielding flowing water, or they were deep, of small diameter (say 0·6 m) and expensive, especially when it was quite possible that they would not actually penetrate an artesian aquifer with water confined under pressure in it, water which would flow freely out at the surface. As a result, groundwater was not utilized in quantity in the nineteenth century, except from springs. However, with the population continually rising and hence ever greater human, industrial and animal pollution being produced, it became necessary to develop groundwater resources as well as to construct water-filtration plants. The main stimulus was the desire to control epidemics caused by water-borne diseases by freeing water from pathogenic micro-organisms. At that time, there were daily outbreaks of diseases such as dysentery, cholera and typhoid in one large city or another. Nowadays, their incidence has decreased and it is no longer possible to obtain information regarding the sanitary quality of private or public water supplies by studying them from an epidemiological standpoint. Work such as the construction of the first slow-sand filter for Poughkeepsie, New York, in 1872 made a sizeable contribution. However, the problem is by no means completely solved. Pathogenic bacteria and viruses are still found in the unsaturated zone and derive from septic tanks, leaky sewage lines, sanitary landfills, waste oxidation ponds and land applications of waste water. As well as these allochthonic pathogenic micro-organisms (parasitic bacteria and exterotoxine-producing bacteria) which enter the unsaturated zone and may reach the groundwater, there are autochthonic soil and groundwater micro-organisms. The latter will flourish under favourable ecological conditions and attain high population densities, according to P. Hirsch and E. Rades-Rohkohl in 1983.[19] Allochthonic constituents usually become eliminated in the subsurface environment. However, under oligotrophic conditions, they can survive with no substantial decrease in the germ number during the first few weeks. In fact, it might even rise. Afterwards, the elimination of bacteria and viruses can be described by an exponential function, thus:[20]

$$C_{(t)} = C_0 \exp[-\lambda(t-t_0)] \qquad (1.2)$$

($t \leqslant t_0$ and $t_0 \leqslant 20$ days), C_0, $C_{(t)}$ being the initial concentration and the concentration at time t, and λ an elimination constant [$(\ln 2)/\tau_{1/2}$] is half-life (mostly between 1 and 20 days).

Bacteria and viruses can survive for months in sewage, septic tanks and waste deposits. Species of *Salmonella* persist for anything up to ten weeks and, in appropriate soils, may endure for as long as three months. H. Althaus *et al.* noted that coliform bacteria can last for no less than seven months.[21] A problem in control is that survival times differ from species to species, also from one subsurface environment to another. As well as from living organisms, inorganic pollution may come from organic substances, mineral oil products and products of the chemical industry. Gas components may migrate into pores in the unsaturated zone or even diffuse from solution plumes in aquifers. Subterranean dissolved oil constituents may undergo rather rapid microbial degradation, but sufficient dissolved molecular oxygen must be available. If this is not the case, a different mechanism comes into play and anaerobic processes entail recourse to the oxygen of the nitrates and sulphates. Consequently, a reduced groundwater type is formed which may convert back to the oxidized type to the extent to which the contaminated water mixes with oxygen-bearing groundwater. For details, F. Schwille's 1976 paper may be consulted.[22]

Turning again to water pumping from deep small-diameter wells, the problems were solved only by the 1880s. One approach was air-lift pumping, in which a stream of air was forced into a tube ending deep in the well where it mixed with the water in the well and formed a mixture of lower density than the original water. This mixture rose to the surface as a result of the pressure of the surrounding aquifer water. In the Frasch process, the system is adapted to pump liquid sulphur from certain wells. In the late 1890s, centrifugal pumps sufficiently small to enter a 0·75 m hole were constructed and rotated by shafts carrying steam or electricity from the ground surface.

In the United States, groundwater resources were exploited very fast indeed. The first deep well in Chicago, for instance, was drilled in 1864, and by 1890 there were hundreds of such wells producing about 50 million litres of water daily. Although air-lift pumps had been introduced by then, they were rapidly replaced by the much more efficient turbine. It is not surprising that, by 1915, the water level in Chicago was over 45 m below ground and over 200 million litres were being extracted per day. Around 1960, the water level had declined a further 120 m, approximately 350–400 million litres being extracted daily.

These developments led to systematic studies of groundwater aimed at evaluating available resources and facilitating new discoveries as well as improved technology. Among the earliest American workers was T. C. Chamberlain, who published a paper on requisites and conditions for artesian flow in 1885 through the US Geological Survey and, in 1873–1879, conducted investigations into artesian conditions in the state of Wisconsin.

A towering figure in nineteenth century hydrology is the French hydraulic engineer Henri Darcy (1803–1858), who studied the movement of water through sand and, in a treatise of 1856, defined a relation now termed Darcy's law which governs groundwater flow in most alluvial and sedimentary formations. Subsequent European contributions which clarified further the hydraulics of groundwater development were made by J. Boussinesq, J. Dupuit, A. Thiem, P. Forscheimer and G. A. Daubrée. In the twentieth century many workers were active; among the most important were G. Thiem, C. V. Theis and C. E. Jacob.

In America, O. E. Meinzer, N. H. Darton, F. Leverett and W. H. Norton are noteworthy and work on the hydraulic, quantitative aspects of hydrogeology was effected by A. Hazen, F. H. King and C. S. Slichter, all predecessors of Theis. In Europe, as well as Adolph and Günther Thiem, there were K. Keilhack, E. Prinz, E. Martel, A. Herzberg, W. Baydon-Ghyben and J. Pennink. Incidentally, it may be of interest that the term 'hydrogeology' was first used by Lamarck in 1802, although in a different sense from today. The modern application began with an Englishman, Lucas by name, in 1879. There is also the word 'geohydrology', which is frequently used. D. K. Todd in 1980[23] stated that this is identical in connotation with 'groundwater hydrology'. As B. Hölting indicated in 1980, there has been much discussion à propos the two, mainly directed at which one is better to use scientifically, hydrogeology or geohydrology.[24] Hölting wrote that 'heute hat sich allgemein durchgesetzt, Hydrogeologie also Oberbegriff zu verwenden für die Wissenschaft die sich mit der Erforschung der Grundwassers befasst. Die Geohydrologie ist Teil der Hydrogeologie und erforscht die Hydrologie des unterirdischen Wassers'. This agrees with Todd and entails regarding geohydrology as a subsidiary part of hydrogeology, a restricted meaning.

It must be made clear that the history of hydrology and groundwater is an aspect of the overall historical development of science and engineering which took place since the middle of the last century. The relation is demonstrable. For instance, deep wells could be drilled and lined only after the introduction of the Bessemer process in 1855 and the open-hearth process in 1860 provided mass-produced steel and steel casing. J. E. Brantly gave an excellent account of well drilling.[25] In total, it is possible to assert that, from the Chicago water wells in 1864 and also Drake's oil well of 1859, the modern era in hydrology commenced. Adequate support technology was available and the products of wells were much in demand. Up to then, deep wells were rare and usually drilled for salt. Afterwards, they became common and offered a wide range of products.

A great stimulus to measurements of surface water and studies of its natural occurrence was the need for data connected with canal construction and water supply in the early part of the nineteenth century; nevertheless there was no requirement for vast control, conservation and distribution projects of the type of those in antique and modern Mesopotamia, Egypt and Sri Lanka. Developments in the USA were both rapid and important. In 1879, the US Geological Survey was founded and its first Director, Major J. W. Powell, was appointed in 1881. He obtained Federal Government funding for stream gauging six years later and the actual field work started in 1888. In 1871 T. G. Ellis constructed a rugged stream gauging current meter which was utilized by the Corps of Engineers.

In 1903, the Bureau of Reclamation was established and completed construction of its first large-scale project, the Elephant Butte Dam on the Rio Grande, in 1916, after which there was more dam building as well as water impounding and irrigation of desert lands. Activity slackened later and hence, whereas from 1903 to 1910 about three million acres were covered by water projects, from 1910 to 1930 only a million or so acres were reclaimed using impounded water.

The literature was sparse until the Second World War. In fact, only three books appeared in the USA, those of D. W. Mead in 1904 and

1919[26,27] and A. F. Meyer's in 1917;[28] these were general texts on hydrology which were not superseded until O. E. Meinzer et al. published a volume in 1942 which was sponsored by the National Research Council and published in the McGraw-Hill 'Physics of the Earth' series.[2]

After the War, data accumulation accelerated tremendously and information on rates of evaporation and precipitation, the occurrences of the various aquifer types, water levels in wells, surface water mineralization, stream stages and discharge, was built up at a staggering rate with which it is extremely hard even for the specialist to keep up. Perhaps most interesting was the introduction of isotopic techniques to groundwater, surface and atmospheric water studies. Isotope hydrology is a relatively recent field of scientific inquiry which deals with significant parameters such as evaporation rates, age, velocities and directions of flow of groundwaters, stream flow velocities and so on.

For groundwater, one of the most important queries relates to the safe yield or, where 'mining' of a source is intended, the total yield. Here, appropriate aquifer characteristics such as interconnections, porosities, transmissivity and dispersivity can be investigated.

Isotopic hydrology is divisible into two main branches. Firstly, there is environmental isotope hydrology which uses isotopic variations established in waters by natural fractionation processes. Isotopic tracing in this approach cannot be controlled by Man, but can be observed and interpreted to solve a number of hydrological problems on the basis of general knowledge of isotopic variations in nature. Regional hydrological problems can be studied using environmental isotopes if natural conditions establish measurable variations in isotope concentrations of different waters (as is usually the case). Secondly, artificial isotope hydrology can be applied to hydrologic investigations. This employs radioactive isotopes which are injected at a well-defined point of a system being investigated. Subsequently, the isotopic concentration is monitored so as to determine its evolution with time. Highly detailed information can be obtained, but its validity is restricted to an area around the point of injection. The conditions present when the injection is made are also extremely important. However, if measurements are made at a number of points and at many different times, a reasonably accurate description of the hydrological system being analyzed can be provided.

The main environmental isotopes used in hydrology are listed in Table 1.2. However, many others, e.g. those of helium, nitrogen, chlorine,

TABLE 1.2
ENVIRONMENTAL ISOTOPES USED IN HYDROLOGY

Isotope	Relative abundance in nature (%)	Decay	Half-life	Maximum energy (keV)
^{1}H	99·985	Stable		
^{2}H(D)	0·015	Stable		
^{3}H(T)	10^{-15} (pre-1952), 10^{-12}	—	12·26 y	18·1
^{12}C	98·89	Stable		
^{13}C	1·11	Stable		
^{14}C	$1·2 \times 10^{-10}$ (in modern carbon before 1950)	—	5730 y	156
^{16}O	99·76	Stable		
^{17}O	0·04	Stable		
^{18}O	0·2	Stable		

iodine, uranium, thorium, radium and radon, are used as well, to a greater or lesser extent.

Some important radioisotopes are listed in Table 1.3.

In spite of the progress made, there is still much to do, especially in the less developed countries and particularly regarding the evaluation of the peripheral effects of hydrological projects. This was recognized as long ago as 1961 when the then-President of the Hydrology Section of the AGU, W. C. Ackermann, gave his opinion that the science is 'still ... primitive' and, as evidence, cited the 1960 publication *Design of small dams* of the Bureau of Reclamation.[29] Mentioning that this uses the 'latest available scientific and engineering methods', Ackermann quoted from the section on 'Estimating runoff from rainfall', thus:

'Two methods may be used to determine rainfall excess: by assuming a constant average retention rate throughout the storm period, and, by assuming a retention varying with time. Because the use of varying retention rate requires a complicated method of computation, and because present knowledge of the exact shape of the infiltration curve is rather limited, it is often preferable to assume an average retention (sometimes referred to as infiltration index) with an estimate of initial loss being made if antecedent conditions are relatively dry'. His comment was: 'Mind you, I am not quoting this to be critical of the Bureau of Reclamation — I believe this is the best we can do'. He then gave another instance, this from Illinois where he was, for many years, the Chief of the State Water Survey Division. There, the Agency made an

TABLE 1.3
RADIOISOTOPES USED IN HYDROLOGY

Isotope	Chemical form	Half-life	Decay mode	In drinking water Max. permissible concn. (Ci/ml)	Min. concn. detectable (Ci/ml)
^{3}H(T)	H_2O	12·26 y	β^-	3×10^{-3} (10^6 T.U.)	10^{-6} (300 T.U.) without enrichment
^{24}Na	Na_2CO_3	15 h	β^-	2×10^{-4}	10^{-8}
^{51}Cr	Cr–EDTA, $CrCl_3$	27·8 d	EC	2×10^{-3}	8×10^{-7}
^{58}Co	CoEDTA $K_3[Co(CN)_6]$	71 d	EC (85%) β^+ (15%)	10^{-4}	6×10^{-8}
^{82}Br	NH_4Br	35·7 h	β^-	3×10^{-4}	2×10^{-8}
^{110}Ag	$K[Ag(CN)_2]$	249 d	β^- (98%) IT (2%)	3×10^{-5}	3×10^{-8}
^{131}I	KI	8·05 d	γ	2×10^{-6}	8×10^{-8}
^{198}Au	$AuCl_3$	64·8 h	β^-	5×10^{-5}	10^{-7}

exhaustive study of rainfall intensity, frequency and duration. Ackermann referred to this being 'welcomed by a number of agencies who will use it in the formula $Q = CIA$, for estimating peak discharges. This is an ancient method and we are still guessing at the value of $C(I)$.

In the first edition of this book in 1980, the introduction included a table showing the distribution and residence times of various constituents of the planetary water balance. Groundwater down to 4000 m is described as: Volume, $8 \cdot 35 \times 10^6 \, km^3$; percentage of total, $0 \cdot 59$; residence time, 5000 years. These, with other data, are derived from R. L. Nace[30] and H. W. Menard and S. M. Smith.[31] It is perhaps useful to add information on the situation in America taken from the Ad Hoc Panel on Hydrology publication.[3] For the continental USA, excluding Alaska and Hawaii, the appropriate groundwater figures are given in Table 1.4.

TABLE 1.4
CIRCULATION AND RESIDENCE TIME OF GROUNDWATER IN CONTINENTAL USA

Source	Volume (m^3)	Animal circulation (m^3/y)	Residence time (y)
Shallow less than 800 mm depth	$6 \cdot 3 \times 10^9$	310×10^9	200
Deep, more than 800 mm depth	$6 \cdot 2 \times 10^9$	$6 \cdot 2 \times 10^9$	10 000

The literature is now assuming monumental proportions. A major source in the USA is the US Geological Survey, which carries out most of the field measurements and investigations. Much of this work is published in cooperation with individual states and the results appear as Circulars, Professional Papers and Water Supply Papers. P. F. Clarke *et al.* in a USGS Circular in 1978, provided a guide to obtaining data from that organization.[32]

After 1935, groundwater measurement records in various key wells were published in Water Supply Papers under the title *Ground-water Levels in the United States*. Up to the War, records for each year appeared in single volumes, but subsequently they have been compiled in six volumes, covering areal sections of the United States (North-western, South-western, North Central, South Central, North-eastern and South-eastern). From 1956, it has been the practice to publish only water level information from a basic network of observation wells located to facilitate accumulation from the minimum number of wells in the most important groundwater zones. Of course, this does not preclude occasional issuing of additional papers on both the geology and the groundwater resources of various localities. These may be found in the Publications of the Geological Survey.

In 1971, the US Geological Survey established the National Water Data Storage and Retrieval System known as WATSTORE, from which can be obtained water data: it is accessible to the public in America through any district office of the Survey. Included are files on water quality and groundwater site inventory. The former provide analyses of the biological, chemical, physical and radiochemical features of 200 different constituents which occur in groundwater. It is also possible to obtain computer-printed tables, graphs, etc.

There are also the information resources of the various state geological and water resources agencies, although there are differences in activity between states, California being among the most productive with its very large water resources agency which is responsible for extensive groundwater investigations. G. J. Giefer and D. K. Todd edited a relevant publication in 1972, followed by another in 1976, when a further paper by Giefer alone outlined sources of information on water resources in journals of civil engineering, geology, geophysics, agriculture, etc., as well as in those of water resources and supply.[33-35] However, the only organization in the USA which recognizes hydrology as a separate science is the AGU, the American Geophysical Union. It published three journals, the *Transactions*, the *Journal of Geophysical Research* and the *Bulletin of Water Resources Research*, which are invaluable. Also, publications of the International Association of Hydrological Sciences, which is an organization within the International Union of Geodesy and Geophysics, and the *Journal of Hydrology* should be consulted for information.

In addition, the National Water Well Association in the USA publishes *Ground Water* and the *Water Well Journal*, important reading for all involved in the developing and managing of groundwater resources.

In Great Britain, there are many useful sources of data including the journals of the Institution of Civil Engineers, the Institute of Hydrology (Wallingford), the Meteorological Office and, occasionally, the National Environmental Research Council (NERC) as well as conventional geological publications.

There are also some valuable source journals in the Federal Republic of Germany such as *Wasser und Boden, Dtsch. Gewässerkdl. Mitt.*,

Meyniana, Baumachine Bautechnik, Z. Geophys., Phs. Blätter, Bes. Mitt. Dtsch. Gewässerkundl. Jahrb., Z. Kulturtechn. Flurberein., Z. Angew. Geologie, Geologische Rundschau, etc.

In the German Democratic Republic, useful journals include *Wasserwirtschaft* and *Wassertechnik* in East Berlin, and in Austria there are journals such as the *Steir. Beitr. Hydrogeologie*.

Poland produces useful data from time to time, and some may be found in *Archiwum Hydrotechniki* (*Warsaw*).

Of course, the other European countries have similar journals; it is often worthwhile to consult them, as well as the many publications which appear sporadically from UNESCO and the International Atomic Energy Agency (Technical Reports Series, etc.), and other United Nations agencies.

An exhaustive source is provided by the DIALOG Information Retrieval Service, from DIALOG Information Services, Inc., which has offered services since 1972 and now has more than 200 databases available. Those of significance in hydrology are GEOARCHIVE and GEOREF. Some details follow.

GEOARCHIVE, 1969 to date, has 519 000 records with monthly updates. It is a comprehensive geoscience database which indexes over 100 000 references annually and it is based in England (Geosystems, London). To give the reader an idea of what it offers, in May 1985 it listed 12 195 groundwater references for the period 1980 to that date.

GEOREF, 1929 to date North America, 1967 to date worldwide, contains 939 000 records, again with monthly updating. It is slightly more expensive to use than GEOARCHIVE, but can provide a wider spectrum of information. It is based in the USA (American Geological Institute, Falls Church, Va). In both cases, it is cheaper to use offline rather than online services unless immediate data printout is required.

GEOREF accesses data from over 4500 international journals and about 40% of these originate in the USA, 7% come from international organizations and the balance derive from the rest of the world, i.e. outside the USA. DIALOG services are useful because they obviate much manual searching and thereby may be cost-effective. Also, they are available 22 hours every weekday as well as having a Saturday service. DIALOG is based in Palo Alto, Ca.*

---

*DIALOG Information Services, Inc., 3460 Hillview Avenue, Palo Alto, California 94304, USA. In Europe: Learned Information/Dialog, Oxford, England. Also in Australia, Japan and Canada.

REFERENCES

1. BUSWELL, A. M. and RODEBUSH, W. H., 1956. Water. *Scientific American*, **194**, 4, 76–89.
2. MEINZER, O. E., Ed. 1942. *Hydrology*. Physics of the Earth Series, McGraw-Hill, New York, reprinted by Dover Publications, New York, 1949, Vol. 1, No. 12, p. 14.
3. AD HOC PANEL ON HYDROLOGY, 1962. *Scientific Hydrology*. Federal Council for Science and Technology, Washington, D.C., 37 pp.
4. GREGORY, K. J., Ed., 1983. *Background to Palaeohydrology — a Perspective*. John Wiley and Sons, Chichester, UK, 486 pp.
5. ADDERLEY, E. E., 1963. The influence of the moon on atmosphere. *J. Geophys. Res.*, **68**, 5, 1405–8.
6. BOWEN, E. G., 1963. A lunar effect on the incoming meteor rate. *J. Geophys. Res.*, **68**, 5, 1401–4.
7. DAVIS, G. H. et al., 1959. *Ground-water Conditions and Storage Capacity in the San Joaquin Valley, California*. US Geol. Surv. Water Supply Paper 1469, p. 168.
8. KAZMANN, R. G., 1965. *Modern Hydrology*. Harper and Row, New York.
9. CLODIUS, S., 1963. Zum Schema des Wasserkreislaufs. GWF 104, Oldenbourg, München, pp. 755–6.
10. BUSCH. K.-F. and LUCKNER, L., 1974. *Geohydraulik*. Enke, Stuttgart, 442 pp.
11. DROWER, M. S., 1954. Water supply, irrigation and agriculture. In: *History of Technology*, Vol. 1, Chapter 5. Oxford University Press, Oxford, pp. 520–57.
12. SARTON, G., 1952. *A History of Science: Ancient Science Through the Golden Age of Greece*. Cambridge, Mass., Harvard University Press, p. 59.
13. BISWAS, A. K., 1970. *History of Hydrology*. Elsevier, New York, 348 pp.
14. CRESSEY, G. B., 1958. Qanats, karez and foggaras. *Geogr. Rev.*, **48**, 27–44.
15. BEAUMONT, P., 1971. Qanat systems in Iran. *Bull. Intl Assn Sci. Hydrol.*, **16**, 39–50.
16. MEINZER, O. E., 1934. The history and development of ground-water hydrology. *J. Wash. Acad. Sci.*, **24**, 6–32.
17. FRONTINUS, Sextus Julius. *Two books on the water supply of the city of Rome*. Translation by C. Herschel, 1899. Dana Estes and Co., Boston, pp. 216–219.
18. FRISI, P., 1762. *Treatise on Rivers and Torrents*. Translation by J. Garstin, 1860. John Weale, London, p. 57.
19. HIRSCH, P. and RADES-ROHKOHL, E., 1983. Die Zusammensetzung der natürlichen Grundwasser-Mikroflora und Untersuchungen über ihre Wechselbeziehungen mit Fäkalbakterien. Forum 'Mikroorganismen und Viren in Grundwasserleitern'. *Proc. DVGW-Wasserfachliche Aussprachetagung, München, 1–4 March, 1983*.
20. BERG, G., 1967. *Transmission of Viruses by the Water Route*. John Wiley–Interscience Publishers, New York, 447 pp.
21. ALTHAUS, H., JUNG, K. D., MATTHESS, G. and PEKDEGER, A., 1982. Lebensdauer von Bakterien und Viren in Grundwasserleitern. *Umweltbundesamt Materialien 1/82*. Erich Schmidt Verlag, Berlin 190 pp.
22. SCHWILLE, F., 1976. Anthropogenically reduced groundwaters. *Hydrol. Sciences Bull.*, **XXI**, 4, 12.

23. TODD, D. K., 1980. *Groundwater Hydrology*, 2nd edn. John Wiley, New York, 535 pp.
24. HÖLTING, B., 1980. *Hydrogeologie*. Enke, Stuttgart, 340 pp.
25. BRANTLY, J. E., 1961. In: *History of Petroleum Industry*, Vol. 5, Chapter 23. Oxford University Press, Oxford.
26. MEAD, D. W., 1904. *Notes on Hydrology*. S. Smith and Co., Chicago.
27. MEAD, D. W., 1919. *Hydrology*. McGraw-Hill, New York.
28. MEYER, A. F., 1917. *The Elements of Hydrology*. John Wiley, New York.
29. ACKERMANN, W. C., 1961. Needed — three wise men. *Trans. Amer. Geophys. Union*, **42**, 1, 5–8.
30. NACE, R. L., 1967. Water resources: a global problem with local roots. *Environmental Sci. Tech.*, **1**, 550–60.
31. MENARD, H. W. and SMITH, S. M., 1966. Hypsometry of ocean basin provinces. *J. Geophys. Res.*, **71**, 18, 4305–25.
32. CLARKE, P. F. et al., 1978. *A Guide to Obtaining Information from the USGS*. US Geological Survey Circular 777, 36 pp.
33. GIEFER, G. J. and TODD, D. K., 1972. Eds: *Water Publications of State Agencies*. Water Information Center, Port Washington, New York, 319 pp.
34. GIEFER, G. J. and TODD, D. K., 1976. *Water Publications of State Agencies, First Supplement, 1971–74*. Water Information Centre, Huntington, New York, 189 pp.
35. GIEFER, G. J., 1976. *Sources of Information in Water Resources*. Water Information Center, Port Washington, New York, 290 pp.

CHAPTER 2

Hydrometeorology and Infiltration

2.1. METEOROLOGICAL DATA

These data are relevant because most groundwater derives from and constitutes part of the hydrologic cycle, including not only surface, but also atmospheric or meteoric water.

The hydrology in any region is mostly determined by its weather and, to some extent, by its topography and geology. In order to examine the hydrology information is collected on relative humidity, temperature, wind velocity and a number of other factors, among which precipitation is important: infiltrating water derives from it and can make a major contribution to the groundwater system in an area.

Precipitation results from evaporation of the oceans, water vapour being absorbed therefrom into air streams which transport it away. Such moisture-bearing air retains the water vapour until it cools below the dewpoint temperature and precipitates as rain or, if it is cold enough, as snow or hail. In fact, precipitation may occur as sleet, fog and dew as well as in the forms mentioned above. Rainfall records are kept, using rain gauges, in almost all countries. Naturally, coverage is greater in developed areas; for instance in the United Kingdom there are 6500 rain gauges, of which the majority provide daily values of rainfall. Standard rain gauges are made of copper and consist of a 5-in diameter copper cylinder with a chamfered upper edge which collects the rain and permits it to drain through a funnel into a graduated glass measuring cylinder for daily examination. Of course, there are prescribed patterns for the standard gauge and also for its installation and operation. Automatic recording rain gauges are useful in countries where the installation sites are remote: they often depend upon the tipping bucket principle. Rainfall enters a typical automatic gauge through a copper funnel with a phospher bronze rim set into a rolled brass cylinder, at the base of which is the interior measuring device. Precipitation is conducted to a specially shaped and plated twin bucket which is mounted on tungsten rods and designed to roll rather than pivot in order to reduce friction and wear. Also, the bucket is gold-plated to provide low viscosity. Each tip is monitored by a switch connected to a module which accepts a data cartridge, both being battery-powered. Wire mesh in the funnel reduces the possibility of foreign objects entering it. Deep funnel types are available which enable the rain gauge to be placed on the ground surface rather than having to be buried and which also minimize 'splash-out'. Insulation jackets and heaters can facilitate the employment of such gauges down to temperatures of $-15°C$ or even lower.

Rain gauges have been used for centuries and records were being collected in India as early as 400 B.C. for use in crop growing and weather forecasting. In Great Britain, regular observations were made as early as 1677 and W. Heberden in 1769 indicated that rain gauge catch seemed to be a function of height above the ground surface.[1] Later, it was suggested that raindrops accrete by condensation as they near the ground.[2] In addition, it was realized that wind influences rain gauge catch, which decreases as the wind velocity increases with elevation.[3] Attempts were made to reduce or, if possible, eliminate this effect and this is how siting in pits began. In fact, H. Koschmieder even asserted in 1934 that pit gauges are the only appropriate method for measuring rainfall without error.[4] However, others tried shields of various types, one of the most widely used of which was that devised by F. E. Nipher in 1878, this being rigid in a fluted trumpet shape which surrounded the gauge and opened upwards.[5] In 1937, J. C. Alter described his flexible wind shield composed of a baffle, open top and bottom, with free swinging leaves surrounding the gauge.[6]

In 1958, L. L. Weiss and W. T. Wilson surveyed rain gauge literature and gave a table which compared average rain gauge catch in shielded gauges with that in unshielded ones.[7] They expressed the catch of shielded gauges as percentages of catch in nearby unshielded ones and gave values ranging from 97–110% for gauges equipped with Nipher-type rigid shields, through 104–111% for gauges equipped with Alter-type flexible shields, to 94–34% for pit gauges. All investigators agree that exposed rain gauges have a 5–10% average error in rainfall catch, and also that the error is not consistent, varying from 0 to 70% depending upon wind velocities. Errors in measuring snowfall are even greater because lighter snowfall is more easily affected by wind than is rain.

Various analytical procedures have been proposed for estimating true precipitation, but none is wholly satisfactory. For instance, E. L. Neff in 1977 gave equations of simple linear regression type relating catch in control gauges with that in proximate surface gauges.[8] The number of rain gauges required in any given area varies according to the local conditions and, for monthly percentage-of-average rainfall estimates, the suggested figures are two for 26 km^2, six for 260 km^2, 12 for 1300 km^2, 15 for 2600 km^2, 20 for 5200 km^2 and 24 for 7800 km^2, according to A. Bleasdale.[9] Hydrologists may be interested in intensity, duration, frequency and areal extent of rainfall, all of which correlate with infiltration and therefore bear on the recharging water which may end by contributing to the groundwater system in any particular region. Generally, the greater the intensity of rainfall, the shorter is the length of time during which it continues. The connection may be expressed thus:

$$i = \frac{a}{t+b} \qquad (2.1)$$

where i is the intensity in mm/h, t the time in hours and a, b are locality constants. Highest intensities are about 40 mm in a minute, 200 mm in 20 minutes and 26 m in a year.

In Great Britain, maximum recorded rainfall approximates to this equation:

$$R = 106 D^{0.46} \qquad (2.2)$$

where R is rainfall (mm) and D is a time period (h).

Intensity–duration–frequency relationships are interesting. Frequencies may be calculated from the following equation:

$$n = 1.25 t (r + 0.1)^{0.282} \qquad (2.3)$$

where n is the number of occurrences in 10 years, r the depth of rain in inches and t the duration of rain in hours. In SI form, this becomes:

$$n = 1.214 \times 10^5 t (P + 2.54)^{0.282} \qquad (2.4)$$

where P is the depth of rain in mm and n, t have unchanged units.

Precipitation does not usually occur uniformly over an area, however, and there are variations both in intensity and in total depth of fall from the centres to the peripheries of storms. D. J. Holland has quantified this[10] and shown that the ratio between point and areal rainfall over areas up to 10 km^2 and for storms lasting between 2 and 120 minutes is provided by:

$$\frac{\bar{P}}{P} = 1 - \frac{0 \cdot 3\sqrt{A}}{t'} \qquad (2.5)$$

where $\bar{P}$ is the average rain depth over the area, P the point rain depth measured at the centre of the area, A the area in km^2 and t' an 'inverse gamma' function of storm time obtained from appropriate data embodied in a correlation curve between storm t and t' (see C. P. Young's 1973 paper[11]). Precipitation depth can be averaged over an area; one way of doing this is due to A. H. Thiessen who, in 1911, defined the zone of influence of each meteorological observation station by drawing lines between pairs of gauges, bisecting the lines with perpendiculars and assuming that all the area enclosed within the boundary formed by these intersecting perpendiculars has received rainfall of the same amount as the enclosed gauge. Another approach is to draw isohyets, contours of equal rainfall depth. Areas between successive isohyets are measured and assigned an average rainfall value. The overall average for the area is derived from weighted averages. Quite possibly, this is the optimum method because the isohyets can be drawn to take into account local effects such as the prevailing wind and irregularities in topography.

For duration, standards are selected, namely two days and 60 minutes. M5, the standard of frequency, is the depth of rainfall with a return period of five years and it can have a set of different durations, e.g. there are one-minute M5 and 25-day M5 values. In NERC's *Flood Studies Report* volumes, analyses of wide arrays of British data are provided in maps, tables and graphs. Methods for estimating return periods are given. Two-day M5 rainfall is mapped in the *FSR* (*Flood Studies Reports*) for the United Kingdom using values from 6000 stations. Sixty-minute M5 rainfall is mapped for the British Isles as a ratio, $r = 60$-minute M5/two-day M5.

Since evaporation can affect the yield from underground water supplies, it is appropriate to consider it briefly here. Water evaporates from all terrestrial surfaces and, if through vegetation, it transpirates — hence the term evapotranspiration. However, whilst evaporation proceeds as long as any input of heat is available, transpiration takes place almost exclusively by day when solar radiation is the stimulus. At night, the pores of plants close and practically no moisture escapes. Also, a distinction must be made between potential and actual evapotranspiration because, if insufficient water is present, the latter may be much less than the former. The most significant meteorological factors which affect evaporative processes are solar radiation, wind, relative humidity and temperature. Evaporation may affect groundwater and it increases as the water table nears the surface of the ground. The rate depends also on the soil structure, which controls the capillary tension above the water table and thus its hydraulic conductivity. Under isothermal conditions, the upward movement is in the liquid phase, so it is not truly evaporative (evaporation is the conversion of water into water vapour), but if a soil has a high surface temperature, it may dessicate and establish upwards movement of water vapour responding to a vapour pressure gradient; see for instance the 1972 paper of C. D. Ripple *et al.*[12] Change of state of molecules of water from liquid to gas needs an input of energy, the latent heat of vaporization, and hence it occurs most actively under direct solar radiation. Consequently, clouds inhibit evaporation. Wind is important because, as water vapour enters the atmosphere, the boundary layer between earth and air becomes saturated. This layer must be removed and continually replaced by drier air if evaporation is to continue. Movement of air in this boundary layer depends upon wind, the velocity of which is therefore significant. Finally, if ambient temperatures of air and ground are high, evaporation is accelerated, and the capacity of air to absorb water vapour rises with temperature.

Estimation of evaporation is feasible by various methods. One is the storage equation approach which consists of compiling a water budget for any particular area, i.e. an account of all water input and output relating to a catchment or drainage basin. Regular, systematic rain gauging provides a good approximation of the quantity of water arriving from the atmosphere. Water which leaves on the surface can be assessed if regular stream gauging enables flow-rating curves to be prepared. The difference between the two must be caused by change in storage in the catchment (either in surface lakes and depressions or in aquifers), or by a difference in subterranean flow into and out of the catchment or by evaporation and transpiration. The hydrologic (continuity) equation of Chapter 1 (eqn (1.1)) may be rewritten, therefore, as:

$$E = P + I \pm U - O \pm S$$

where E is evapotranspiration, P the total precipitation, I the surface inflow, U the subterranean outflow, O the surface outflow and S the change in storage, surface and subsurface.

By making observations over a long period of time, S may be ignored if the beginning and end of the study are effected under the same seasonal conditions. Knowledge of the geology may show that large underground flows are not possible; U may be neglected in such cases. Where S and U can be ruled out, E may be derived, at least to a first approximation.

Empirical formulae exist which can be employed, usually referring to evaporation from an open water surface. The logic behind this is that water must be present if evaporation is to occur, and the assumption that a free water surface is present is reasonable. However, the results show potential and not actual evaporation. Occasionally these coincide, e.g. in reservoirs, but in dealing with land surfaces the availability of water becomes critical. So does water table level and vegetation, as well as the type of soil which is present. Two basic cases exist, one where the air temperature is the same as that of the water surface, the other where these parameters differ. The first is rare, and the appropriate expression is:

$$E_a = C(e_s - e)f(u) \tag{2.6}$$

in which E_a is open water evaporation in unit time (for air and water at the same temperature $t°C$) in mm/day, C an empirical constant, e_s the saturation vapour pressure of the air at $t°C$ in mm mercury, e the actual vapour pressure in the air above in mm mercury and u the wind velocity at a standard height. The generally valid equation which has been obtained is:

$$E_a = 0.35(e_s - e)(0.5 + 0.54 u_2) \tag{2.7}$$

with u_2 denoting wind speed in m/s at a height of 2 m. E_a is in mm/day, as before. Normally, the second case occurs and an appropriate equation is of the type:

$$E_0 = C(e'_s - e)f(u) \tag{2.8}$$

where E_0 is the evaporation of the lake in mm per day, e'_s is the saturation vapour pressure of the boundary layer of air between the air and the water; the temperature of this boundary layer is not that of either the air or the water and it is practically impossible to measure. Among the empirical formulae that have been developed is one that was derived for the Ijsselmeer in the Netherlands and is applicable to these and similar conditions:[13]

$$E_0 = 0.345(e_w - e)(1 + 0.25 u_6) \tag{2.9}$$

where e_w is the saturation vapour pressure at a given temperature (t_w) of the surface water of the lake in mm mercury, e is the actual vapour pressure in mm mercury and u_6 is the wind velocity in m/s at a height of 6 m above the surface.

Turning now to the very well-known Penman theory, the relevant nomenclature is as follows:

E_0 = evaporation from open water (or its equivalent in heat energy);
e_w = saturation vapour pressure of air at a water surface temperature t_w;
e = actual vapour pressure of air at temperature t
  = saturation vapour pressure at dew point t_d;
e_s = saturation vapour pressure of air at a temperature t;
e'_s = saturation vapour pressure of air at boundary layer temperature t'_s;
n/D = cloudiness ratio
  = actual/possible hours of sunshine;
R_A = Angot's value of solar radiation arriving at the atmosphere;
R_C = sun and sky radiation actually observed arriving at the Earth's surface on a clear day;
R_I = net amount of radiation received as surface absorption after reflection;
R_B = radiation from the surface of the Earth.

H. L. Penman, in 1948, proposed both a theory and a related formula for the estimation of evaporation from weather data.[14] Two requirements of the theory have to be fulfilled if continuous evaporation is to take place. These are that a supply of energy is available in order to provide latent heat of vaporization and that a mechanism exists which can remove water vapour once it is produced.

As regards the supply of energy, short-wave radiation reaches the planetary surface during daylight, the actual amount varying according to season, latitude, time of day and cloudiness. Total radiation which can be expected at a point under cloudless conditions and assuming a perfectly transparent atmosphere has been tabulated by Angot and expressed as R_A in Penman theory. The equations given below demonstrate regional differences in values for R_C, n/D being included in the expressions. In southern England, Penman gave:

$$R_C = R_A(0.18 + 0.55 n/D) \qquad (2.10)$$

and quoted Kimball for Virginia, USA, as:

$$R_C = R_A(0.22 + 0.54 n/D) \qquad (2.11)$$

and Prescott for Canberra, Australia, as:

$$R_C = R_A(0.25 + 0.54 n/D) \qquad (2.12)$$

From these, it may be noted that, on cloudless days, roughly 75% of solar radiation reaches the terrestrial surface, compared with about 20% which penetrates when a complete cloud cover exists and $n/D = 0$.

Of course, some of R_C is reflected as short-wave radiation, the quantity being dependent upon the surface reflectivity, i.e. the reflection coefficient, r. Expressing the net amount of radiation absorbed as R_I, then:

$$R_I = R_C(1-r) = R_A(1-r)(0.18 + 0.55 n/D) \tag{2.13}$$

in southern England. Some of the R_I is radiated by the Earth in the form of long-wave radiation, especially during the night when the air is dry and the sky clear. This is the component R_B mentioned above, and net outflow can be expressed empirically as follows:

$$R_B = \sigma T_a^4 (0.47 - 0.077\sqrt{e})(0.20 + 0.80 n/D) \tag{2.14}$$

where σ is the Lummer and Pringsheim constant (117.74×10^{-9} g cal/cm^2 day) and T_a is the absolute temperature of the Earth ($= t^\circ C + 273$), e being the vapour pressure in the air in mm mercury.

Therefore, the net quantity of energy finally remaining at a free water surface ($r - 0.06$) is given by H, where

$$H = R_I - R_B = R_C - rR_E - R_B$$

$$= R_C(1-r) - R_B$$

$$= R_A(0.18 + 0.55 n/D)(1 - 0.06) - R_B$$

and from this,

$$H = R_A(0.18 + 0.55 n/D)(1 - 0.06) - (117.4 \times 10^{-9})$$
$$\times T_a^4(0.47 - 0.077\sqrt{e})(0.20 + 0.80 n/D) \tag{2.15}$$

and this heat is consumed thus:

$$H = E_0 + K + S + C \tag{2.16}$$

E_0 being the heat available for evaporation from open water, K the convective heat transfer from the surface, S the increase in heat content of the water mass (storage) and C the increase in heat content of the environment (negative advected heat). Over some days, and frequently within a single day, both the heat storage and environmental storage are small compared with other changes and consequently, to a small error, $H = E_0 + K$.

Considering now vapour removal, as was seen on p. 30 evaporation may be described by the expression $E_0 = C(e'_s - e)f(u)$, but e'_s cannot be evaluated if the air and water are at different temperatures (as they usually are). Penman assumed that vapour transport and heat transport by eddy diffusion are both controlled by the same mechanism, atmospheric turbulence, one being governed by $(e'_s - e)$ and the other by $(t'_s - t)$, t'_s being the temperature of the boundary layer of air between air and water also mentioned on p. 30. To a near approximation, it is possible to write:

$$\frac{K}{E_0} = \beta = \frac{\gamma(t'_s - t)}{e'_s - e} \qquad (2.17)$$

where γ = psychrometer constant (0·66 if t is in °C, e in mbar). Because $H = E_0 + K = E_0(1 + \beta)$, then

$$R_0 = \frac{H}{1 + \beta} = \frac{H}{1 - \gamma \frac{t'_s - t}{e'_s - e}} \qquad (2.18)$$

From this equation $t'_s - t$ can be eliminated by substitution because $t'_s - t = (e'_s - e_s)/\Delta$ where e_s is the saturation vapour pressure at temperature t, and Δ is the slope of the vapour pressure curve (tan α where α is the angle between this curve and the horizontal); see Fig. 2.1. The reasoning is that t'_s is never too far from t for the straight line approxima-

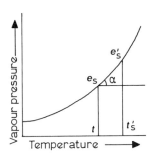

FIG. 2.1. Saturation vapour pressure curve: e_s is the saturation vapour pressure at temperature t, e'_s is the saturation vapour pressure of air at boundary layer temperature t'_s and α is the angle which the curve makes with the horizontal.

tion to be unreasonable. Therefore, it is possible to write:

$$E_0 = \frac{H}{1+\frac{\gamma}{\Delta}\left[\frac{e'_s-e_s}{e'_s-e}\right]}$$ (2.19)

and e_s must be eliminated. Because $e'_s - e_s = (e'_s - e) - (e_s - e)$ and, since $E_a = C(e_s - e)f(u)$ (eqn (2.6)), taking into account that $E_0 = C(e'_s - e)f(u)$ as noted previously in eqn (2.8), it is possible to write:

$$\frac{E_a}{E_0} = \frac{e_s - e}{e'_s - e}$$ (2.20)

where E_a is the evaporation in terms of energy for the hypothetical case of equal temperatures of both air and water.

Recapitulating, by substituting values of $e'_s - e_s = (e'_s - e) - (e_s - e)$ and eqn (2.20) into eqn (2.19), the following expression can be derived:

$$E_0 = \frac{H}{1+\frac{\gamma}{\Delta}\left[\frac{(e'_s-e)-(e_s-e)}{e'_s-e}\right]}$$ (2.21)

and

$$E_0 = \frac{H}{1+\frac{\gamma}{\Delta}\left(1-\frac{E_a}{E_0}\right)}$$ (2.22)

Thus

$$E_0 = \frac{\Delta H - \gamma E_a}{\Delta + \gamma}$$ (2.23)

The parameter Δ can be derived from the saturation vapour pressure curve. Some typical values are as follows:

$t = 0°C$, $\Delta = 0.36$; $t = 10$, $\Delta = 0.61$; $t = 20$, $\Delta = 1.07$; $t = 30$, $\Delta = 1.80$.

Reference to earlier equations for E_a and H demonstrate that E_0 may now be computed from the standard meteorological observations of mean air temperature, relative humidity, wind velocity at a standard height and hours of sunshine. The equation has been found to give good results in many parts of the world, although direct evaporation observations are also desirable.

Penman also reported the results of experiments conducted on turfed and bare soil in order to determine how their evaporation rates (E_T and E_B) compared with that from open water (E_0). He concluded that the rate of evaporation from a freshly wetted soil was about 90% of that from an open water surface exposed to the same weather; thus $E_B/E_0 = 0.90$. In the case of grass surfaces, the comparison was more erratic and the following data were obtained for values of E_T/E_0 for southern England:

November to February	0.6
March to April } September to October }	0.7
May to August	0.8
Entire year	0.75

All are below unity because vegetation is more reflective than open water and, in addition, plant transpiration ceases at night.

In 1946, C. W. Thornthwaite published a paper based upon his experiments in the USA using lysimeters and investigating the correlation between temperature and evapotranspiration.[15] He devised a method which enabled potential evapotranspiration from short, closely planted vegetation with adequate water supply to be estimated. If t_n is the average monthly temperature of the consecutive months of the year in °C (where $n = 1, 2 \ldots 12$) and j is the 'monthly heat index', then:

$$j = \left(\frac{t_n}{5}\right)^{1.514} \tag{2.24}$$

and the 'annual heat index', J, is given by:

$$J = \sum_{1}^{12} j \text{ (for the 12 months)} \tag{2.25}$$

Potential evapotranspiration for any month with average temperature t (°C) is then given as PE_x by:

$$PE_x = 16\left(\frac{10t}{J}\right)^a \text{ mm per month} \tag{2.26}$$

where

$$a = (675 \times 10^{-9})J^3 - (771 \times 10^{-7})J^2 + (179 \times 10^{-4})J + 0.492 \tag{2.27}$$

PE_x is a theoretical standard monthly value based on 30 days with 12 hours of sunshine daily and the real PE for a particular month and an average temperature t (°C) is given by:

$$PE = PE_x \frac{DT}{360} \text{ mm} \qquad (2.28)$$

where D is the number of days in the month and T the average number of hours between sunrise and sunset in the month.

It has been proposed to simplify matters as follows:

$$j = 0.09 t_n^{3/2} \qquad (2.29)$$

$$a = 0.016 J + 0.5 \qquad (2.30)$$

The method is an empirical one, also rather complex, so a nomogram is necessary in order to solve it. Thornthwaite published one which is reproduced in Fig. 2.2. First, the heat index must be obtained and, using

t(°C)	PE_x
26.5	135.0
27.0	139.5
27.5	143.7
28.0	147.8
28.5	151.7
29.0	155.4
29.5	158.9
30.0	162.1
30.5	165.2
31.0	168.0
31.5	170.7
32.0	173.1
32.5	175.3
33.0	177.2
33.5	179.0
34.0	180.5
34.5	181.8
35.0	182.9
35.5	183.7
36.0	184.3
36.5	184.7
37.0	184.9
37.5	185.0
38.0	185.0

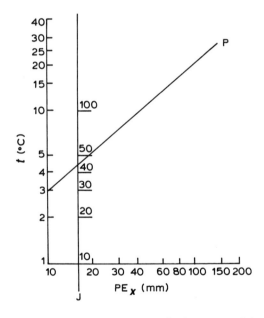

Fig. 2.2. Nomogram for finding potential evapotranspiration PE_x. (modified from Thornthwaite): t°C = mean monthly temperature; J = annual heat index; P is the point of convergence.

the nomogram, the unadjusted value of potential evapotranspiration is derived by drawing a straight line from the J value of the location through the point of convergence at $t = 26.5°C$. For higher values of t, the adjoining Table can be utilized. Twelve values are obtained, one for each of the 12 months. The unadjusted values can be adjusted for day and month length using $PE - PE_x \, DT/360$, and thereafter totalled to provide annual potential evapotranspiration.

The method has been found to give good results whatever the vegetational cover, although different types of vegetation affect the true value at a given locality. The formula is based upon temperature and, of course, this does not directly correspond with the incoming solar radiation because of the 'heat inertia' of land and water. However, transpiration responds directly to solar radiation. Thus, the approach must be used carefully in order to make sure that conditions do not change abruptly in a particular month. If data for many consecutive months are being employed, the cumulative differences can be neglected. Often, both this and the Penman method are utilized.

As mentioned above, evaporation can be measured directly and for this purpose pans are used; in the USA, the standard or Class A pan is circular with a diameter of 1·22 m and a depth of 254 mm. It is filled to a depth of 180 mm and placed on a timber grillage with the base of the pan 150 mm above the ground level. In the UK, the standard pan is 1·83 m square and 610 mm deep. It is filled to a depth of 550 mm and set in the ground so that the rim projects 76 mm above the surrounding ground. Another type, the Peirera pan, is circular like the Class A pan, but deeper and sunk into the ground with a 3-in air space surrounding it. Actually, water in the Class A pan has a greater temperature range than water in the square one, but differs in being homogeneous whereas water in the latter may stratify. A doubling of the wind run can increase evaporation by up to 20%. Since such pans have very small capacities and are very shallow in comparison with lakes or rivers and also are located either very near or on the surface of the land, proportionately larger quantities of advected heat from the atmosphere can be absorbed by their contained water through the sides and the bottom. Some pans are more susceptible to this effect than others. As a result, pan evaporation is too high and an appropriate pan coefficient must be applied. This can vary from 0·65 to more than unity, depending on the dimensions and location of the pan concerned. Usually, the American Weather Bureau Class A pan has a coefficient of about 0·75 and the coefficient of the British standard pan is about 0·92. The relatively insignificant sizes of such pans make results obtained from them very difficult to relate to large open bodies of water, but they still have their uses and should be involved in studies of evaporation.

2.2. INFILTRATION AND PERCOLATION BELOW GROUND

Groundwater is composed of constituents of differing origins, but by far the most important source is the hydrologic cycle, i.e. from surface and atmospheric (meteoric) waters. These must penetrate underground, of course, and so infiltration and percolation have to be examined: but before this is done, three alternative, much less important sources are described briefly.

The first is connate water, water which has not been in contact with the atmosphere for a substantial part of a geological time interval. It is composed of fossil, interstitial water which has migrated from its original location. Connate water may have derived from either oceanic (saline) or freshwater sources and usually it is highly mineralized, hence being capable of polluting groundwater wells after entering an aquifer system.

Secondly, there is magmatic water which comes from magma. If it has separated at great depth, such water may be referred to as plutonic. When it arises from relatively shallow depths, say 4 or 5 km, the water may be termed volcanic. New water which has not previously been a part of the hydrosphere and which may be of either cosmic or magmatic origin is termed juvenile water.

Finally, metamorphic water is the term which is applied to water which has been associated with rocks during their metamorphism.

As B. Hölting mentioned in 1980, it is uncertain how great a contribution such waters make to groundwater, but it is certainly much less than that of recharge through infiltration from above ground.[16] This has been believed for some time and, as mentioned in Chapter 1, E. Mariotte, in the seventeenth century, developed the theory that groundwater arises solely through infiltration, a remarkable achievement in his time. The concept of 'juvenile water' was introduced much later, in fact in 1909, by the Austrian geologist H. E. Suess. On the land surface, precipitation may be followed by evaporation, infiltration or depression storage, always presuming that interception by vegetation or artificial surfaces such as roads does not take place. However, any surplus obeys the law of gravity and flows over the surface to the nearest stream channel. Such streams coalesce into rivers and eventually reach the sea. When rainfall is especially intense, the surplus water (runoff) may exceed the capacity of such channels and flooding results. Infiltration may occur also during this process. As well as entering soil in this way, water may percolate into the saturated ground zone beneath the water table (or phreatic surface). Such water flows slowly through aquifers to river channels or sometimes reaches the sea directly.

Different soil types allow water to infiltrate at different rates, each one having a different infiltration capacity, f, which is measured in mm/h (sometimes in in/h). Rain penetrates gravel or a sandy soil easily and, if the water table is below the surface, there will be no surface runoff even if the rain is heavy. On the other hand, a clayey soil will resist infiltration and hence the surface becomes covered with water rapidly even if the rainfall is very light. The rate of rainfall affects the quantity of water which infiltrates as well as how much will run off.

In 1976, S. H. Nassif and E. M. Wilson published the results of extensive work on infiltration during which they used a weighable laboratory catchment with an area of 25 m² and concluded that, for any soil under constant rainfall, the rate of infiltration decreases in accordance with the following expression:

$$f = f_c + \mu \exp(-Kt) \tag{2.31}$$

which is of a form first used by R. E. Horton;[17,18] f is the rate of infiltration at any time t (mm/h), f_c is the infiltration capacity at large values of t (mm/h), f_0 is the initial infiltration capacity at a time $t=0$ (mm/h), t is the time from the beginning of rainfall (min), $\mu = f_0 - f_c$ and K is a constant for a particular soil and surface (min^{-1}). K is a function of surface texture: it is small if vegetation occurs, smoother surfaces yielding larger values.

The parameters f_0 and f_c are functions of both the type of soil and the cover. A bare sandy or gravelly soil will have high values of both whereas a bare clayey soil will give low values. However, both sets of values will increase if the soils in question are turfed. The infiltration capacity f_c is also a function of slope (up to a limiting value varying between 16 and 24%, after which there is little variation) as well as of both the initial moisture content and the intensity of rainfall. The drier the soil, the larger is f_c and if intensity, i, increases, f_c also does so. In fact, rainfall intensity is more influential on f_c than any other variable. It used to be thought that f_c was a constant for a particular soil, but this may not, in fact, be the case. In 1983, E. M. Wilson stated that the rate of infiltration seems to be controlled to a great extent by the surface pores.[13] A small increase in the hydrostatic head over these pores causes an increase in the flow through the soil surface. Figure 2.3 shows what may be happening. It shows surface soil grains and it may be observed that the governing factor is the head, h, over the smallest cross-section of a pore: h continues to increase with rainfall intensity until some limiting value is attained, at which point runoff precludes further increase. However, it is probable that, in nature, such a point is hardly ever reached. It should be added that other research workers attributed this increase in f_c at higher rainfall intensities to a lack of homogeneity in their experimental catchment watershed.[19] Yet others drew attention to the importance of the superficial layer of a soil.[20]

FIG. 2.3. The hydraulic head on pores between soil grains: $h=$ head over the smallest cross-section; h increases to h' with rainfall intensity to attain a limiting value where further increase is prevented by surface runoff.

The rate of infiltration in any particular soil represents the sum of percolation and water entering storage above the water table; usually, the soil is not saturated, so storage increases for quite a long time, f_c decreasing under steady rainfall intensity for a similar length of time. It is inadvisable, therefore, to attribute average values of rain loss to catchments in order to determine net rain.

Raindrop sizes vary with the intensity and typical figures for dimensions are 0·1 cm with 0·1 cm/h, 0·2 cm with 1·3 cm/h and 0·3 cm with 10·2 cm/h. Raindrops erode soil and also compact it, thus reducing f_c, and the amount of soil in runoff is related to raindrop energies. The median dimensions of raindrops increase with intensity of rainfall (for low and moderate falls) in the relation $D_{50} = aI^b$, where D_{50} is the median diameter in mm, I the intensity (mm/h) and a and b are constants. With high-intensity rainfall, drop dimensions actually diminish, the maximum sizes being of the order of 5–6 mm diameter. The drop sizes influence the velocity of impact. A drop of rain in free fall under gravity accelerates until its weight equals the frictional resistance of air. At this stage, the drop of rain attains its terminal velocity. Because fall distances to maximum velocity are short, nearly all raindrops hit the ground at terminal velocity.

The relationship between rainfall and soil loosening, separation and transportation is expressed as $G = KDV^{1.4}$, where G is the weight of soil splashed in g, K is a constant for the particular soil type involved, D is the diameter of the raindrops in mm and V is the impact velocity (m/s). A compound parameter termed the EI_{30} index has been derived in order to explain soil losses in terms of rainfall. This is the product of kinetic energy E (in foot-tons/acre-inch) and intensity I (in/h), where I_{30} refers specifically to the 30-minute intensity, i.e. the maximum average intensity in any half-hour period during a storm.

Not only can large raindrops compact soils and reduce f_c, they can also wash very fine particles into the voids. The surface may become 'puddled' and then the f_c value drops very sharply. Other factors which may cause compaction include Man or animals or vehicles moving over the surface. High values of f_c are promoted by a dense vegetal cover because the root systems provide means of ingress to the subsoil; the layer of organic material is also useful in this respect, as are burrowing animals. The transpiration of the vegetation aids too because it removes soil moisture and assists infiltration.

Frost heave, leaching of soluble salts and desiccation cracks are other phenomena which tend to increase f_0. On the other hand, trapping of interstitial air decreases f_0. Flow in interstices is laminar in type and so viscosity has a direct effect on resistance to flow. If all things are equal, it may be said that f_c and f_0 have higher values during the warm seasons of the year.

There is no instrument with which infiltration can be measured directly. However, there are methods of investigating infiltration capacity. One is by the use of infiltrometers. A typical infiltrometer is a wide-diameter short tube or some other impervious boundary surrounding an area of soil. Normally, two such rings are utilized concentrically and are flooded to a depth of 5 mm over the surface, this depth being maintained by constant refilling. The inflow of water to the central tube is measured, the outer tube acting to minimize peripheral effects from the outer, dryer soil. Tests of this type provide rather useful comparative data, but do not simulate real conditions. Sprinkler tests, which may be applied to larger areas, simulate rainfall and runoff better. The in-coming water is measured, as also is the runoff, the difference being taken as infiltration. However, this gives values of f which are higher than those of natural conditions. Again, sprinkler tests are most important as a means of comparing different vegetative conditions, soils, etc. Laboratory measurements have been carried out in which input, output and storage change were assessed, for as much as 7 tonnes of soil in a layer 200 mm thick.[17] In this way, absolute figures for infiltration may have been obtained, but again natural conditions were not adequately simulated — for instance, there is atmospheric pressure below the soil layer in the laboratory, which is not the case in nature. Some research workers tried to improve sprinklers by selecting small, 'homogeneous' drainage basin and measuring input, output and storage. Both R. E. Horton[18] and L. K. Sherman[21] outlined techniques applicable to such basins.

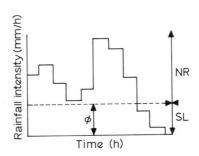

Fig. 2.4. Infiltration loss by ϕ-index: NR = net rain, i.e. quantity of runoff, SL = storm loss (infiltration mainly).

Another approach is by the ϕ index, an index of infiltration defined as the average rainfall intensity above which the volume of infiltration becomes equal to the volume of runoff. Figure 2.4 shows a rain storm plotted on a time base in terms of average hourly intensity. The zone above the dashed line represents the runoff, measured in mm, over a catchment area. The blank zone below this line is measured rainfall not appearing as runoff and hence represents all losses including surface detention and evaporation as well as infiltration. This last parameter constitutes by far the largest loss in a great many catchments. Consequently, although the approach does not recognize the variation of f with time, it is widely employed as a method for rapid estimation of probable runoff from large catchments in the case of particular rain storms.

Finally, the f_{av} method may be applied; this is an extension of the ϕ index by which an attempt is made to allow for depression storage and brief rainless periods during storms as well as eliminating those rain periods in which the rainfall intensity is less than the assumed infiltration capacity. An approximate position for an f_{av} line on a graph of intensity (mm/h) against time (h) parallel to the time axis is chosen and may be moved vertically until various losses are balanced and the demand of runoff values satisfied.

As a result of infiltration, water enters soils and the existing soil moisture conditions are significant. The methods so far discussed relate to the average limiting values for the infiltration capacity and refer to extremely small-scale areas which can be covered by infiltrometer experiments, so they do not permit information to be obtained regarding the actual quantity of rainfall which is absorbed by the soil. Of course, it can be stated that the amount involved will depend upon the wetness of the soil at the start of the rainfall. This initial soil moisture greatly affects the infiltration capacity and also the runoff during the first stages of storms. Clearly, therefore, some means of assessing initial soil moisture is necessary. There are at least two ways of doing this, one American and the other British.

The antecedent precipitation index used in the USA is based upon the assumption that soil moisture is depleted at a rate which is proportional to the quantity of storage in the soil; therefore there is a logarithmic relationship of the following type:

$$I_t = I_0 k^t \tag{2.32}$$

where I_0 is the initial value of the index in mm, I_t is the index value t days later and k is a recession constant having a value around 0·92 (it can range between 0·85 and 0·98).

Where t is unity, then a day's value *is* k times that of the previous day. If precipitation occurs, the value of the index is increased by an indeterminate amount (because some of the rain may have disappeared from the catchment as surface runoff). The amount added to the index should be the basin recharge alone, but the difference made by including all the precipitation is usually small. There is a progressive daily diminution in this parameter due to evapotranspiration which varies seasonally; consequently, k also varies seasonally.

In the UK, the estimated soil moisture deficit approach is employed. Evapotranspiration continually removes soil moisture, but precipitation replaces it; hence measurements of these two processes give a method for estimating the parameter which is sometimes referred to as smd. No equation is required. Where evaporative loss exceeds the input from precipitation, the vegetation utilizes accumulated soil moisture to continue transpiring, so the smd increases. The result is that transpiration becomes increasingly difficult. The UK Meteorological Office estimates smd twice monthly, assuming that each measuring station represents an area with the following characteristics: half of it is covered by vegetation with short roots which can draw up to 3 in of moisture before evapotranspiration starts to decline below potential, 30% is covered by long-rooted vegetation which can draw up to 8 in of moisture, and 20% is riparian (the water table is near the surface and evapotranspiration is unrestricted). In the case where the difference between rainfall and evapotranspiration is 3 in, the subsequent evaporation data which are calculated take into account different rates and areas so that a weighted figure for the entire catchment is computed.

In fact, smd estimates are compiled from observed rainfall and estimated evaporation at a network of 176 stations with an appropriate map for the entire country being published within a day of the receipt of values at each of the stations. Estimated smd is kept up to date daily or weekly by additions and subtractions of precipitation and evapotranspiration until the next map is published. Such maps demonstrate that the likelihood of flooding taking place varies greatly from one region to another. Estimated smd is utilized to predict the proportion of runoff resulting from particular storms exactly as the antecedent precipitation index is used in the United States. The main problem with the latter is the enormous amount of work required to obtain appropriate relationships between storms and runoff over such a vast area.

As noted above, the initial soil moisture content in an area affects both infiltration capacity and runoff during the initial stages of storms. A relevant factor is the catchment wetness index, CWI. A period of five days preceding a storm is taken as providing information on the history of recent rainfall, where:

$$\text{CWI} = 125 + \text{AP15} - \text{smd} \tag{2.33}$$

$$\text{API} = (0.5)^{1/2} P_{d-1} + 0.5 P_{d-2} + (0.5)^2 P_{d-3}$$
$$+ (0.5)^3 P_{d-4} + (0.5)^4 P_{d-5} \tag{2.34}$$

where 0.5 is a decay index, API being the antecedent precipitation index and P the daily rainfall, suffixes indicating the relevant day.

One of the most important aspects of a catchment is the soil cover, which has a considerable influence on runoff. Soils will be discussed in greater detail in Chapter 3, but here it may be pointed out that soil maps, obtainable in most countries, can be used as a basis for the calculation of a soil index term for employment in equations recommended for flood studies.

The 'winter rain acceptance potential' is significant and represents the reverse of the runoff potential. It is influenced by the position of the groundwater level as well as by perviousness and slope. There are five classes of winter rain shown on British soil maps (Table 2.1). The appropriate soil index may be calculated from the proportions of a catchment area covered by each of these categories.

TABLE 2.1
WINTER RAIN CLASSIFICATION

Class	Winter rain acceptance	Runoff
1	Very high	Very low
2	High	Low
3	Moderate	Moderate
4	Low	High
5	Very low	Very high

Direct measurement of soil moisture can be carried out using nuclear equipment. The principle involved is that fast neutrons are slowed down in soils by elastic collisions with hydrogen nuclei, which are proportional in number to the water content. A probe contains a neutron source and a detector for slow neutrons. The relevant reaction is $^9Be(^4He, ^1n)^{12}C + 5.75$ MeV and the alpha emitter may be ^{226}Ra, ^{227}Ac, ^{210}Po, ^{241}Am or any other which is suitable. Yields vary by an order of magnitude from 10^6 n/s Ci. Either scintillation or gas proportional counters may be employed in the detector.

An example is the Wallingford soil moisture probe, developed by the UK Institute of Hydrology in association with the Atomic Energy Authority. It is so designed that it can be used in the field in all types of terrain and weather. The probe is 740 mm long and can be inserted into an aluminium access tube which is permanently installed in the ground. There is a shield for it together with a suspension cable and a meter. Epithermal neutron detection is achieved by conversion of events into electrical pulses which are displayed visually. Obviously the wetter the soil, the more collisions occur and hence the greater is the number of slow neutrons recorded as pulses. There is a so-called 'sphere of influence' which varies in radius from ≥ 300 mm under zero moisture conditions, decreasing with increasing levels of moisture to a minimum around 150 mm. The indicated moisture value represents the mean value for this sphere of influence or importance. The weighted mean of a set of readings taken down the profile at 100 mm intervals provides the total moisture storage in any given section of the profile. Of course, a number of aluminium tubes should be installed in a catchment to give adequate data and the neutron probe can be transported from site to site so as to measure *in situ* several soil moisture contents. This gives a better overall picture of the situation.

Certain errors can arise, however. One relates to calibration of the probe; ideally, this should be done at every single site, usually an impracticable proposition. Otherwise, the instrument is calibrated in the laboratory in a number of 'infinite' models uniformly filled with a soil of known moisture content for a constant dry density. In 1983, P. Schudel compared the accuracy of a neutron probe against soil-water storage changes measured over two years in weighable loess-soil monoliths.[22] To evaluate the precision of the neutron probe measurements, the relative deviation of measured changes in water content was defined as:

$$\Delta X = (\Delta W_N - \Delta W_L)/\Delta W_L \tag{2.35}$$

where W_N is the water content determined by the neutron probe and W_L that determined by weighing. The mean value of ΔX gave a systematic deviation of neutron meter measurements from those of the lysimeters of $\sim 25\%$. Much of the error probably arises from the calibration of the neutron probe with surface soil cores, bearing in mind also how greatly water content changes with depth. The weighing or gravimetric technique simply involves weighing a soil sample, removing its water content by oven-drying and subsequently re-weighing the sample.

In the region of vadose water, there is a negative pressure head of water; this may be termed a positive suction or tension. It can be measured using a tensiometer installed in a column of soil, the instrument being a U-tube with a side arm which enters the soil through a porous cup. The local value for this parameter is measured by Δh, the depression in water level in the U-tube, i.e. the suction head. Instruments of this type function in a range from atmospheric pressure (about 1000 cm of water) to about 200 cm of water (800 cm water tension), but, unfortunately, calibration data for soil suction and water content show that the relationship between the two is not a simple one. The structure of the soil and its degree of compaction as well as the effects of wetting or drying have an influence on the results. There are a number of additional approaches based upon the removal of soil water from soil samples by evaporation, leaching or chemical reaction, after which the amount removed can be measured. These are variants on the gravimetric technique.

REFERENCES

1. HEBERDEN, W., 1769. Of the different quantities of rain which appear to fall at different heights over the same spot of ground. *Phil. Trans.*, **59**, 359–62.
2. PHILLIPS, J., 1833. Report of experiments on the quantities of rain falling at different elevations above the surface of the ground. *Brit. Assn Adv. Sci. Rept. Trans. Sect.*, **3**, 403–12.
3. ABBE, C., 1889. Determination of the amount of rainfall. *Amer. Met. J.*, **6**, 6, 241.
4. KOSCHMIEDER, H., 1934. Methods and results of definite rain measurements. *Mont. Weather Rev.*, **62**, 5–7.
5. NIPHER, F. E., 1878. On the determination of true rainfall in elevated gages. *Amer. Assn Adv. Sci.*, **27**, 103–8.
6. ALTER, J. C., 1937. Shielded storage precipitation gages. *Mont. Weather Rev.*, **65**, 262–5.
7. WEISS, L. L. and WILSON, W. T., 1958. Precipitation gage shields. *Extrait C.R. Rapp. Assem. Gen. Toronto, Ont*, *I*, 462–84.
8. NEFF, E. L., 1977. How much rain does a rain gage gage? *J. Hydrol.*, **35**, 213–20.
9. BLEASDALE, A., 1965. Rain gauge networks development and design with special reference to the United Kingdom. *Intl Assn Sci. Hydrol., Symp. on Design of Hydrol. Networks, Quebec.*
10. HOLLAND, D. J., 1967. Rain intensity–frequency relationships in Britain. *Brit. Rainfall*, HMSO, London.
11. YOUNG, C. P., 1973. Estimated rainfall for drainage calculations. *LR 595*, Road Research Laboratory, HMSO, London.
12. RIPPLE, C. D. *et al*, 1976. Estimating steady-state evaporation rates from bare soils under conditions of high water table. *U.S. Geol. Surv. Water Supply Paper 2019-A*, 39 pp.
13. WILSON, E. M., 1983. *Engineering Hydrology*. The Macmillan Press Ltd, London.
14. PENMAN, H. L., 1948. Natural evaporation from open water, bare soil and grass. *Proc. Roy. Soc.*, **A 193**, 120.
15. THORNTHWAITE, C. W., 1946. The moisture factor in climate. *Trans. Amer. Soc. Civ. Eng.*, **27**, 1, 41.

16. HÖLTING, B., 1980. *Hydrogeologie.* Enke, Stuttgart, 340 pp.
17. NASSIF, S. H. and WILSON, E. M., 1976. The influence of slope and rain intensity on runoff and infiltration. *Bull. Intl Assn Sci. Hydrol.*, **20**, 4.
18. HORTON, R. E., 1933. The role of infiltration in the hydrologic cycle. *Trans. Amer. Geophys. Union*, **14**, 443–60.
19. BOUCHARDEAU, A. and RODIER, J., 1960. Nouvelle méthode de détermination de la capacité d'absorption en terrains perméables. *La Houille Blanche, A*, 531–6.
20. SOR, K. and BERTRAND, A. R., 1962. Effects of rainfall energy on the permeability of soils. *Proc. Amer. Soc. Soil Sci.*, **26**, 3.
21. SHERMAN, L. K., 1943. Comparison of F-curves derived by the methods of Sharp and Holtan and of Sherman and Mayer. *Trans. Amer. Geophys. Union*, **24**, 465.
22. SCHUDEL, P., 1983. The accuracy of measurements of soil-water content made with a neutron-moisture meter calibrated gravimetrically in the field. *J. Hydrol.*, **62**, 355–61.

CHAPTER 3

Voids in Rocks

3.1. SOILS AND ROCKS

If soils and rocks contain pores or voids, these may admit infiltration into the zone of aeration and eventually contribute to the groundwater. In consolidated rocks, water may find ingress through structural features such as faults or joints if they are not blocked. Also, it may be stored in these features under certain conditions.

The classification of soils is essential because of their great variability in properties such as compressibility, shear strength, porosity and permeability. Porosity and void ratio are interconnected parameters which can be expressed by:

$$n = \frac{V_v}{V_t} = \frac{e}{1+e}$$

$$e = \frac{V_v}{V_s} = \frac{n}{1-n}$$

where n is the porosity, e the void ratio, V_t the total volume, V_v the volume of voids and V_s the volume of solids.

Porosity may be defined as the percentage ratio of the volume of voids to the total volume so that:

$$n = 100 \left(\frac{W-D}{W-S} \right)$$

where W is the saturated weight, D is the dry weight and S is the weight of a saturated sample suspended in water.

Permeability is the capacity of a rock to transmit water. In an aquifer, there is a variable resistance to subterranean flow of water through pores of which permeability may be regarded as a measure. To some extent it is a function of porosity, but the grain size, distribution, orientation, arrangement and shape of particles all influence it.

Quantitatively, both porosity and permeability have a wide range. Thus, in a granular mass comprising perfect uniform spheres, the loosest possible packing would have $n = 47 \cdot 6\%$, as compared with $n = 26\%$ for the densest possible packing. For spherical grains, the volume of each grain is given by $V = d^3/6$, where d is the diameter of the sphere. Of course, in nature, such an arrangement does not occur and because of the variability of grain sizes, the pore spaces are very variable. Original interstices in both sedimentary and igneous rocks arose from the phenomena responsible for the relevant geological formation and they may occur in both sedimentary and igneous rocks. Subsequently, additional interstices may have developed, for example as a result of solution processes or through the activities of plants and animals. Water is held in them by various forces: for instance, in capillary openings it may be through surface tension, whilst in sub-capillary ones adhesive forces may operate. Effective porosity is a term used for the amount of interconnected pore space which is available for fluid flow, i.e. the parameter e noted earlier.

Some representative values of porosity for various types of material may be cited from the 1967 paper of D. A. Morris and A. I. Johnson[1] (Table 3.1).

G. E. Manger in 1963 demonstrated that, in sedimentary rocks subject to compaction, porosity decreases with depth of burial.[2] The relationship

TABLE 3.1
REPRESENTATIVE POROSITY VALUES

Material	Porosity (%)
Coarse, repacked gravel	28
Coarse sand	39
Silt	46
Clay	42
Fine-grained sandstone	33
Limestone	30
Dune sand	45
Loess	49
Tuff	41
Basalt	17
Weathered granite	45

has the form: $n_z = n_0 \exp(az)$, where n_z is the porosity at a depth z, n_0 is the porosity at the surface and a is a constant.

Intrinsic permeability, k, is a property of the medium which is quite independent of fluid properties. It can be expressed as:

$$k = \frac{K\mu}{\rho g}$$

where μ is dynamic viscosity, ρ is fluid density, g is acceleration due to gravity and K is hydraulic conductivity; k has units of area so small that the US Geological Survey gives them as $10^{-12} m^2$.

In practical work on groundwater, water being the prevailing medium, K is utilized. Any medium has a unit hydraulic conductivity if it will transmit, in unit time, a unit volume of groundwater at the prevailing kinematic viscosity (i.e. dynamic viscosity divided by fluid density) through a cross-section of unit area, measured at right angles to the direction of flow, under a unit hydraulic gradient; it has units of velocity. Some representative values of hydraulic conductivity are cited in Table 3.2.[1]

TABLE 3.2
REPRESENTATIVE HYDRAULIC CONDUCTIVITY VALUES[1]

Material	Hydraulic conductivity (m/day)	Type of measurement
Coarse, repacked gravel	150	Repacked
Coarse, repacked sand	45	Repacked
Silt	0·08	Horizontal hydraulic
Clay	0·0002	Horizontal hydraulic
Fine-grained sandstone	0·2	Vertical hydraulic
Limestone	0·94	Vertical hydraulic
Dune sand	20	Vertical hydraulic
Loess	0·08	Vertical hydraulic
Tuff	0·2	Vertical hydraulic
Basalt	0·01	Vertical hydraulic
Weathered granite	1·4	Vertical hydraulic

The classification of soils is very important and various approaches have been made. Some form of mechanical analysis (granulometry) is required. The wartime Airfield Classification was modified in 1952 by A. Casagrande of Harvard, who first devised it, together with the US Department of the Interior and the Corps of Engineers to produce the

Unified Soil Classification from which are excluded grains exceeding 3 in diameter, i.e. cobbles (3–12 in diameter) and boulders (more than 12 in diameter). In this scheme, there are two divisions, namely coarse and fine grains. The dividing line between these is the (US) No. 200 sieve size, i.e. 0·074 mm. The categories are listed in Table 3.3. Whilst gravels may

TABLE 3.3
UNIFIED SOIL CLASSIFICATION

Class	Size range
G (Gravel)	3 in to No. 4 sieve (3/16 in)
Coarse	3 in to 3/4 in
Fine	3/4 in to No. 4 sieve
S (Sand)	No. 4 sieve to No. 200 sieve
Coarse	No. 4 sieve to No. 10 sieve
Medium	No. 10 sieve to No. 40 sieve
Fine	No. 40 sieve to No. 200 sieve
M (Silt)	
C (Clay)	

be measured directly, sieve methods are applied to finer materials down to the silts, but below these dimensions it becomes necessary to utilize electron microscopy.

The Wentworth scale is a well-known classification which has been in use for many years; it, too, takes into account grain sizes. However, the commonest schema in the USA and elsewhere is perhaps that of Krumbein who, in 1934, introduced the phi-symbol, ϕ, which many sedimentologists feel gives the best results. Phi represented the negative logarithm to base 2 of the various particle sizes expressed in mm, so that:

$$\phi = -\log_2 \xi$$

where ξ is the grain diameter in mm. On this scale, grains of 1 mm diameter have a phi-value of zero, those of 0·5 mm a value of 1 and those of 0·25 mm a value of 2, and so on. Grains with diameters less than 1 mm have positive phi-values, while those with greater diameters have negative phi-values. Figure 3.1 compares the Wentworth and phi-scales and presents other important data.

There is a great distinction between clays and the other soil materials. Clays are cohesive and the other materials are not. Also, clays are acidic aluminosilicic, with a structure which reflects layerings of various thicknesses; they act like Bingham bodies and are thixotropic. Soils may

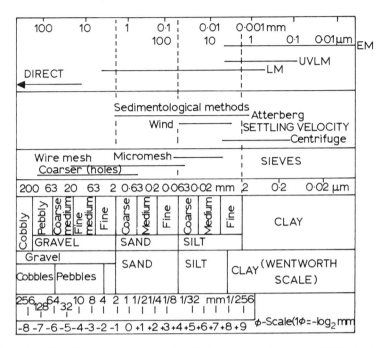

FIG. 3.1. Classification scales for soils. EM, electron microscope; UVLM, ultraviolet light microscope; LM, light microscope.

be well graded or poorly graded and, as well as grains and air, they often contain water. The quantity of water present produces three main consistencies: liquid, in which the soil is either in suspension or behaves as a viscous fluid; plastic, in which the soil can be deformed without elastic rebound; or solid, in which the soil cracks on deformation or rebounds elastically. These three states of soil consistency are usually described with reference to that fraction of a soil which is smaller than the No. 40 sieve size, the upper limit of the fine-sand component. For this particular soil fraction, the water content, as a percentage of dry weight, at which the soil passes from the liquid into the plastic state is called the liquid limit. The water content of the soil at the boundary between the plastic and the solid state is called the plastic limit. The difference between the liquid and plastic limits constitutes the plasticity index, which corresponds to the range of water contents within which the soil is plastic. These parameters comprise the Atterberg limits and can be used to distinguish between clays (plastic) and silts (non-plastic or only slightly plastic).

Another method of describing the texture of a particular soil is by means of a soil-textural triangle. Based upon information from the US Department of Agriculture Soil Survey Manual of 1951,[3] the texture of the soil is defined by the relative proportions of sand, silt and clay present in the particle-size analysis; for example, a soil composed of 20% clay, 40% sand and 40% silt would be designated a loam. There are differences in categorization between the various schemes for classification so that, in some, the dividing line between silt and clay is placed at 0·005 mm (5 m) and in others at 0·002 mm. D. A. Morris and A. I. Johnson in 1967 produced a soil classification based on particle size (Table 3.4).[1]

TABLE 3.4
SOIL CLASSIFICATION BASED ON PARTICLE SIZE

Material	Particle size (mm)
Clay	0·004
Silt	0·004–0·062
Very fine sand	0·062–0·125
Fine sand	0·125–0·25
Medium sand	0·25–0·5
Coarse sand	0·5–1·0
Very coarse sand	1·0–2·0
Very fine gravel	2·0–4·0
Fine gravel	4·0–8·0
Medium gravel	8·0–16·0
Coarse gravel	16·0–32·0
Very coarse gravel	32·0–64·0

As well as sieving for determination of size distribution with particles coarser than 0·05 mm, rates of settlement for smaller particles in suspension may be measured and the results plotted on a particle-size distribution graph. The effective particle size is the 10%-finer-than value (d_{10}) and the distribution of particles is characterized by the uniformity coefficient, U, given by:

$$U = d_{60}/d_{10}$$

where d_{60} is the 60%-finer-than value. A uniform material has a low uniformity coefficient, whereas a well-graded material possesses a high uniformity coefficient.

3.2. SUBTERRANEAN VERTICAL DISTRIBUTION OF WATER INPUT

Allusion was made to the zones of aeration (unsaturation) and saturation in Chapter 1, Section 1.2, and pollution problems in zones of aeration were mentioned later in the same section. The zone of aeration comprises interstices which are occupied partly by water and partly by air, the water in it being termed vadose water (from the Latin *vadosus*, shallow). The zone may be subdivided into an upper soil water zone discussed above, an intermediate vadose zone and a lower capillary zone. The saturated zone extends from the upper surface of saturation down to underlying, impermeable rock. If there are no overlying, impermeable beds, then the water table (or phreatic surface, derived from the Greek *phrear, -atos*, a well) forms the upper surface of the zone of saturation, being the surface at atmospheric pressure which manifests as that level at which water stands in a well penetrating an aquifer. Because of capillary attraction, saturation in fact extends slightly above the water table, but there the water is held at less than atmospheric pressure. As noted earlier, water in the zone of saturation is referred to as groundwater, although the term phreatic water is also used. Figure 3.2 illustrates groundwater in relation to the subsurface water profile.

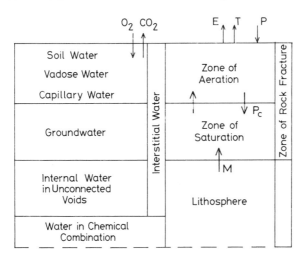

FIG. 3.2. Groundwater in relation to the subsurface water profile: E = evaporation; T = transpiration; P = precipitation; P_c = percolation; M = magmatic (volcanic) water.

It is proposed to examine both the zone of aeration and that of saturation in more detail now.

In the zone of aeration, the uppermost part comprises the soil-water zone, in which water is below saturation level, except when excess water falls on the ground through unusually high rainfall or irrigation, a temporary phenomenon. The soil-water zone extends from the ground down through the major root region, its thickness varying according to soil type and vegetation. The quantity of water present in the soil-water zone depends to a great extent on how recently the soil has been exposed to moisture. In hot and arid conditions, a water–vapour equilibrium situation tends to arise between ambient air and the surfaces of fine-grained soil particles. Hence, only thin films of moisture, i.e. hygroscopic water, remain adsorbed on such surfaces. In the case of coarse-grained soil particles, and where extra moisture is available, water forms liquid rings surrounding contacts between grains. Such water is held by surface tension forces and may be referred to as capillary water. If, temporarily, water in excess of capillary water from rainfall or irrigation occurs in the soil-water zone, such gravitational water drains through the soil due to gravity.

The intermediate vadose zone extends from the lower edge of the soil-water zone to the upper limit of the capillary zone. Its thickness may vary from nothing (where the bounding zones merge with a high water table nearing the surface of the ground) to over 100 m under deep water table conditions. Where present, this zone links the zone near the surface of the ground with that near the water table and water moving downward passes through it. Some vadose water, of course, does not move, being held in place by hygroscopic and capillary forces. Temporary water excesses migrate down as gravitational water.

The capillary zone, sometimes termed the capillary fringe, extends down to the water table from the overlying limit of capillary rise of water. Figure 3.3 illustrates capillary rise of water in a glass tube. If this is taken as representing a void in the capillary part of the zone of aeration, water in it will rise as shown. The rise h_c depends upon an equilibrium between the weight of water raised and the surface tension; the appropriate expression is:

$$h_c = \frac{2v}{r\gamma} \cos \lambda$$

where v is the surface tension, γ the specific weight of the water, r the radius of the hypothetical tube and λ the angle of contact between the

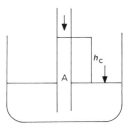

FIG. 3.3. Capillary rise of water in a glass tube. Arrow points indicate positions of equal (atmospheric) pressure; h_c is the height to which water rises. Point A is below atmospheric pressure by an amount equal to the weight of the water column above point B.

meniscus and the wall of the tube, this being different from 90°. In the case of pure water in a clean glass, the angle λ is 0 and, at 20°C, $v = 0.074$ g/cm and $\gamma = 1$ g/cm^3. Hence the capillary rise will be, roughly, given by:

$$h_c = \frac{0.15}{r}$$

The thickness of the capillary zone varies inversely with the pore size of a soil or rock, as demonstrated by the data shown in Table 3.5.[4]

TABLE 3.5
CAPILLARY RISE IN UNCONSOLIDATED MATERIALS[4]

Sample[a]	Grain size (mm)	Capillary rise (cm)
Fine gravel	2–5	2.5
Very coarse sand	1–2	0.65
Coarse sand	0.5–1	13.5
Medium sand	0.2–0.5	24.6
Fine sand	0.1–0.2	42.8
Silt	0.05–0.1	105.5

[a]All measurements were made after 72 days on samples of about the same porosity (40%).

In Chapter 1, reference was made to pollution in the unsaturated zone and a more detailed examination of this very important question must be made here. Soils may retain the water that enters through precipitation as well as absorbing it until plants draw on this resource during periods between rainfall or irrigation. This water-retaining capacity is expressed in terms of the available water, and runs through a range from field capacity to wilting point. The field capacity is defined as the quantity of water held in a soil after it has been wetted and subsequent to the decrease of drainage to a negligible level, a stage often assumed to have been reached after a couple of days. It is clear, however, that different soils possess varying rates of drainage; hence quantitative values may not always be comparable. The wilting point is defined as the water content of soils when plants which are growing in them reach permanent wilted states. Again, many factors influence this, including the age and type of plant involved. In the soil region and even deep into the zone of aeration, bacteria and viruses can survive for long periods until they are transported into the groundwater. Elimination is a combined result of physico-chemical and biological conditions and it has been found to accelerate at high temperature (37°C), at pH values of about 7, at low concentrations of oxygen and at high content of dissolved organic substances. Such water occurs in contaminated soils and groundwater, in which case the autochthonic bacteria become activated and can act adversely to pathogenic micro-organisms. The above-mentioned conditions determine the elimination constant which is specific for the various microbial species. To date, it is not feasible to predict elimination constants on the basis of controlling conditions and so they must be determined for each species and environment. Published values are contradictory and vary over a wide range. Actually, published data on removing bacteria and viruses from the subsurface are not consistent with the known long survival times, from which it is inferred that the survival time cannot alone be the criterion for the purifying effect of the subsurface passage of water. Physico-chemical transport processes must be considered; a contaminated plume will be subject to dispersion as it travels through the unsaturated zone and enters groundwater. The consequence is that the concentration in the contaminated plume decreases and there is an accompanying distribution of the contaminants spatially and in time as well as with the transport distance. J. Bear has given a general transport equation covering this:[5]

$$\frac{\partial C}{\partial t} = \text{div}\left(\frac{D}{R_d} \times \text{grad } C - \frac{v_w}{R_d} \times \text{grade } C\right) - \lambda C$$

where D is the coefficient of hydrodynamic dispersion ($=D' + D_d + D_e$), D' the coefficient of hydromechanic dispersion, D_d the diffusion coefficient, D_e the coefficient of active mobility of bacteria, grad C the concentration gradient, v_w the average velocity of the groundwater, R_d a retardation factor and λ an elimination constant.

The term referring to the active mobility of bacteria is temperature-related because this random tumbling decreases with lowering of temperature. An example of the process is that of *Escherichia coli* with velocities of 1·0 to 0·1 m/d at 20–10°C.

The lateral dispersion of bacteria has been found to be greater than that of a conservative tracer; for instance, the species mentioned above, *E. coli*, has a much broader lateral distribution than the chemical dye fluorescein or the radiotracer bromine-82. It appears that the bacteria are mostly transported in large pores. G. Matthess and A. Pekdeger in 1981 presented a set of concepts of a survival and transport model of pathogenic bacteria and viruses applicable to groundwater,[6] while G. Matthess *et al.* in 1979 noted that the velocity of seepage water in the unsaturated zone is usually extremely slow.[7] In humid conditions, the seepage velocity has been found to be only rarely more than 1 m annually and, in arid climates, lower annual groundwater recharge is associated with even lower rates. Of course, if heavy rainfalls occur or where sewage reclamation is carried out by means of land treatment, there may be accompanying acceleration of seepage velocity in coarse soils. In addition, desiccation or artificial cracks can permit a very rapid rate of infiltration and it has been recorded that, in 1927, a frost crack in soil in south-west Germany triggered a large-scale typhoid epidemic.[8]

If the transport time in the unsaturated zone is sufficiently long, i.e. more than 200 days, then there may occur a substantial elimination of the bacteria and viruses originally present. In the case of the groundwater, the velocities are usually much greater and, if the level of the water table changes, the bacteria may be immobilized in the capillary zone from which, if the water table rises, they may be remobilized subsequently. However, in the zone of aeration, when these micro-organisms travel they do so vertically, altering to an approximately lateral flow after entering groundwater.

Actually, the transport velocity of bacteria and viruses may be different from seepage and groundwater flow velocities and they may be subject to adsorption on underground particles, the equilibrium between the concentrations of suspended, C_s, and adsorbed, C_a, micro-organisms being described by the Freundlich isotherm, thus:

$$C_a = k \times C_s^n$$

The empiric constants k and n are assumed to be specific for the soil being investigated as well as the micro-organisms. Adsorption takes place rapidly, probably taking 24 h for bacteria and 2 h for viruses. No so much is known of the desorption velocity, but there is no doubt that continuous adsorption/desorption reactions are the cause of a retardation of the micro-organisms with respect to water transport. Such retardation can be described by a factor, R_d, the quotient of mean water velocity, v_w, to the mean transport velocity, v_m, of the micro-organisms:

$$R_d = \frac{v_w}{v_m}$$

Retardation factors are obtainable by means of laboratory and field tests and have been found to be between 1 and 2 for *E. coli* and the tracer *Serratia marcescens* where the scale was large enough, i.e. more than a few metres.

Model calculations which utilized the data of known adsorption coefficients and elimination constants have been made and demonstrated that the subterranean passage itself is capable of providing an extremely effective protection against virus contamination, although other observations indicated that, when the water chemistry changes, viruses can be desorbed again if the cation concentrations decrease, say as a result of very heavy rainfall, and hence can travel further. Perhaps this is the reason why land treatment of sewage is more effective in arid than in humid regions.

Turning now to the filtration processes, there is no doubt that removal of micro-organisms from subsurface water is complex and includes filtration as well as time-dependent processes already described. Filtration efficiency can be described by a factor λ_f and is indicated by a decrease in the initial concentration, C_0, to that observed at a distance x, C_x, thus:

$$C_x = C_0 \exp \lambda fx$$

Micro-organism transport can be restricted by the pore size of subsurface materials as well as by their own dimensions. However, mechanical filtration is not effective in sandy and gravelly situations precisely because the bacterial and viral diameters, 0·2–5 μm and 0·25–0·02 μm, respectively, are much smaller than the voids, which usually exceed 40 μm. In natural sediments with heterogeneity of grain size distribution, a percentage of the pore diameters, possibly more than 10%, will interfere with bacteria transport. An interesting comparison has been made between the filtration efficiency of clay minerals and that of bacteria; it showed that bacteria are eliminated more effectively than clay minerals, possibly being aided by their active mobility, ca 0·1 μm/d. The effective diameter of a bacterium is greater than 1 μm because of its irregular shape and the presence of filaments on its surface.

The degree of intensity with which bacteria and viruses are attached to subterranean solids depends upon the adsorption mechanism. Bacteria and viruses are usually negatively charged and hence strongly adsorbed by adsorbents of anions and only weakly by those of cations. Normally, underground solid substances are negatively charged, exceptions being the hydroxides of iron and manganese as well as humic substances at low values of pH. The negatively charged micro-organisms stay in suspension at high pH values, as the repulsive electrostatic forces are stronger than the Van der Waals force.

Dissolved cations in water decrease the repulsive forces of grain surfaces, monovalent ones being adsorbed by the solid substance with consequent decrease of their charge deficiency. Bivalent cations also may cause a positive charge deficiency so that the electrostatic forces can be more efficient for bacteria and virus adsorption. Under these conditions, the mass forces are more effective and an accumulation of particles may occur.

The duration of the contaminating process and the initial concentration of the contaminants are extremely important in regard to the filtration effectiveness. During a continuous process of contamination by micro-organisms and organic substances, the contaminated plume decreases in size with time because of the elimination and filtration mechanisms discussed earlier. If the initial concentrations are very high, flocculation and aggregation may take place at the source of the contamination, and then only a limited transport as far as an aquifer can occur. Aquifer water constitutes the zone of saturation, filling all the interstices, hence the (effective) porosity is a direct measure of the water contained per unit volume. While some water can be removed from below ground by pumping in a well, or by drainage, the bulk is held in place by the molecular and surface tension forces. Two terms need to be introduced, namely specific retention and specific yield.

The specific retention in a soil or rock, S_r, is the ratio of the volume of water which is retained after saturation against the force of gravity to its own volume:

$$S_r = \frac{w_r}{V}$$

where w_r is the volume occupied by retained water and V the bulk volume of the soil or rock in question.

The specific yield of a soil or rock, S_y, is the ratio of the volume of water which, after saturation, can be drained by gravity to its own volume:

$$S_y = \frac{w_y}{V}$$

where w_y is the volume of water drained. S_y and S_r can be expressed as percentages.

Specific yield values relate to grain size, shape and distribution of the pores, the degree of compaction of the relevant bed and also the time of drainage. They are low in the case of fine-grained materials, but coarse-grained ones release substantial quantities of water; in fact they can act as aquifers. Specific yields for thick, unconsolidated formations usually fall into the range 7–15% due to the mixture of grain sizes which occur in the beds. Values tend to decrease with depth, reflecting the effect of compaction. Specific yield can be measured in the laboratory, and also in the field by well-pumping tests which will be discussed later.

Some representative values given by A. I. Johnson in 1967 may be cited:[9] Gravels, 23–25%; sands, 23–27%; silt, 8%; clay, 3%; sandstone, 21–27%; limestone, 14%; loess, 18%; schist, 26%; siltstone, 12%; tuff, 21%; and till, 6–16%.

3.3. GROUNDWATER OCCURRENCE

Aquifers are geological formations which will yield significant amounts of water, either from wells or springs, and these are composed for the most part of unconsolidated rocks, i.e. materials which can store water in pores. Porosity may be due to intergranular spaces or fractures and the United Nations Organization, New York, issued in 1975 a set of data from which Table 3.6 is compiled, showing the important categories of rock involved, namely sedimentary, igneous/metamorphic and volcanic.[10]

It is estimated that probably 90% of all developed aquifers comprise unconsolidated rocks, mostly gravels and sands of alluvial deposits, which may be divided into several categories, namely water courses, abandoned or buried valleys, plains and intermontane valleys.[11]

TABLE 3.6
GEOLOGIC ORIGIN OF AQUIFERS FOLLOWING THEIR POROSITY AND ROCK TYPE

Porosity type	Sedimentary		Igneous/Metamorphic	Volcanic	
	Consolidated	Unconsolidated		Consolidated	Unconsolidated
Intergranular		Gravelly, clayey sandy and sandy clay	Weathered zone of granite–gneiss	Weathered zone of basalt	Volcanic ejecta, blocks and fragments
Intergranular and fracture	Breccia, conglomerate, sandstone, slate		Zoogenic limestone, oolitic limestone, calcareous grit	Volcanic tuff, cinder, volcanic breccia, pumice	
Fracture			Limestone, granite, dolomite, gneiss, dolomitic limestone, quartzite, diorite, schist, mica–schist	Basalt, andesite, rhyolite	

Water courses are made up of alluvium formed in and underlying stream channels as well as that formed in the neighbouring flood plains. The proximity to rivers means that wells excavated adjacent to them will be capable of producing large quantities of water, if the strata involved have a high permeability. Groundwater is augmented by infiltration from the streams.

Abandoned or buried valleys are old water courses with lower recharge and perennial yield than was formerly the case.

Large plains underlain by unconsolidated sediments occur in the USA, and in some localities beds of sand and gravel constitute important aquifers under them. However, elsewhere these beds are rather thin and show only a limited productivity.

Intermontane valleys may contain enormous masses of unconsolidated rock materials which originated through the erosion of the flanking mountains. There are many such semi-individual basins in the western part of the United States and their sand and gravel beds are sources of large amounts of water which is replenished by seepage from streams into alluvial fans at the mouths of mountain canyons.

Limestones are often important aquifers, although their density, porosity and permeability are very variable, according to the degree to which they are consolidated and to the extent to which permeable zones developed after deposition. It is suggested that those that are most significant as aquifers contain substantial proportions of the original rock which was dissolved and removed.[12] The openings to be found in limestones range from microscopic original pores to huge solution caverns which can form subterranean channels of sufficient size to transport the total flow of a stream, as R. W. Brucker *et al.* pointed out in 1972.[13] Streams which disappear underground in a limestone area are termed 'lost streams'. In the case of this rock, calcium carbonate solution by water produces hard water in its aquifers as well as increasing the available pore space and the permeability. Associated phenomena include the production of large springs and eventually a karst terrain; 'karst' is derived from the German form of a Slavic word, *kras*, a black, waterless place. Karst regions have been reviewed by H. E. LeGrand and V. T. Stringfield in 1973 and usually contain very large amounts of groundwater.[14]

Vast limestone aquifers are found in the south-eastern part of the USA and also in the Mediterranean region.

Volcanic rocks can become very permeable aquifers and basalt flows are often found to contain aquifers. This is because of the many interstitial voids in the uppermost parts of lava flows, the cavities between adjacent flows, shrinkage cracks, gas vesicles, post-cooling fissures and faults as well as cracking, together with the holes left after trees have burnt in them. Some of the largest American springs come from volcanic rocks. Apart from basalts, rhyolites also may be involved, although they are less permeable. However, shallow intrusive rocks are, to all intents and purposes, impermeable.

Sandstones and conglomerates represent the cemented forms of sand and gravel and, as such, they have a much reduced porosity and hence a much lower yield. In fact, the good sandstone aquifers release water through joint systems.

Igneous and metamorphic rocks are almost impermeable when solid, so they do not constitute good aquifers. However, when they are near the ground surface and have been weathered, they may be developed as domestic sources of water supply using small wells.

It is interesting that clays can have quite a high porosity and therefore contain a sizeable quantity of water; however, their pores are too small to permit exploitation and they are regarded as relatively impermeable. When mixed with other materials, as in clayey soils, they can provide small domestic water supplies from shallow wells with large diameters.

Various types of aquifer exist, most being of great extent; they constitute subterranean storage reservoirs of water which enters through the zone of aeration, if there is one, as a result of natural or artificial recharge. Water leaves under gravity or is extracted through wells. In normal circumstances, the annual volume of water which is removed or replaced represents only a tiny fraction of the total storage capacity. Classification of aquifers refers to their major properties and perhaps the most useful is their division into unconfined or confined, i.e. possessing or lacking a water table, a leaky aquifer combining both of these types.

Unconfined aquifers are those in which the water table exists as the upper surface of the zone of saturation and rises and falls in accordance with storage changes. The water table itself varies in form and also in slope, depending upon areas of recharge and discharge, pumping from wells and also permeability. Contour maps and profiles of the water table can be prepared from known water levels in wells which tap the aquifer and, in this way, it is possible to determine the amounts of water available as well as their distribution and movement. There is a special case of a confined aquifer, termed a 'perched aquifer'. Such an aquifer occurs wherever a groundwater body is separated from the main aquifer by a relatively impermeable bed of small areal extent, and from the surface of the ground by the unsaturated zone which lies between it and the main aquifer. Often, clay lenses in sedimentary deposits may have shallow, perched water bodies above them. They may be referred to as aquifuges, relatively impermeable formations which neither contain nor transmit water.

Confined aquifers, also termed artesian aquifers or pressure aquifers, are found in situations where groundwater is confined under a pressure exceeding atmospheric by overlying impermeable strata. If a well is drilled into such an aquifer, water rises above the base of the confining bed and may flow out at the ground surface. Of course, water can enter a confined aquifer where the confining bed reaches the surface and, if the confining bed terminates underground, then the aquifer becomes unconfined. Water entering at the ground surface does so through infiltration, exactly as noted earlier, and the region which supplied this water constitutes the recharge area; in addition, water may enter by leakage through a confining bed, and as juvenile water.

A confined aquifer possesses an imaginary surface, called the piezometric or potentiometric surface, which coincides with the hydrostatic pressure level of its water and is determined by the water level in a well penetrating the aquifer. If it is above the ground level, a flowing well is the result and, if it lies below the base of the upper confining bed, the aquifer ceases to be confined, becoming unconfined instead. Unconfined aquifers often overlie confined ones.

In nature, many leaky aquifers are found to occur. They may be regarded as semi-confined, constituting features common in alluvial valleys, plains or old lake basins where a permeable bed is either over- or under-lain by a semi-confining layer or semi-pervious aquitard. An aquitard may be defined as a relatively impermeable formation which does not yield any substantial amount of water to wells: a good example is clay.

In calculating factors such as storage in an aquifer, the assumptions are made that the aquifer is homogeneous and isotropic. Homogeneity implies identical hydrological properties in all parts, and isotropy that properties are independent of direction; of course, neither occurs in nature, where directional properties of hydraulic conductivity give rise to anisotropy. Nevertheless, the 'ideal' aquifer mentioned earlier can be utilized to provide reasonable quantitative approximations; that is why it is postulated.

It has been seen that bacterial and viral pollution of groundwater can take place from time to time. Groundwater pollution can be caused also by fluids which are immiscible with water as well as by dissolved gases. Considering pollution by fluids, it is useful to divide the various rock types into two categories, namely, porous and fissured (fractured) rocks. Similarly, it is also possible to divide the many fluids which are hazardous to groundwater into two categories, namely those which are miscible with water and those which are not (there are transitional forms also).

Miscible fluids migrate mainly through the process of advection and hydrodynamic dispersion. By contrast, the migration of fluids immiscible with water is much more complex. In aquifers, the flow initially occurs under the influence of gravity as a cohesive, two-phase flow, until the immiscible fluid in the pore spaces is not present as a cohesive phase, but rather dispersed into many isolated droplets and eventually attaining residual saturation. Thus, the fluid becomes practically immobile under normal conditions and, thereafter, it can only be further transported through solubility in groundwater, i.e. practically hardly at all. In order to analyze flow processes in an aquifer, the parameters density, kinematic viscosity, vapour pressure (volatility) and surface tension and physicochemical characteristics must be known.

Immiscible, pollutant fluids include organic substances, mostly mineral oil products and products of the chemical industry. Very important examples of the latter are the aliphatic chlorohydrocarbons, a group frequently included with organo-chlorine insecticides and polychlorobiphenyls, all of which are halogenated hydrocarbons.

In characterization of the fluid dynamic properties, the density is one of the most important parameters. The difference in density of the pollutant compared with that of groundwater determines the level at which migration in the aquifer occurs. Mineral oil products are usually much lighter than water, but some crude oils have almost the same density, 1·0 g/ml, and certain distillation products of coal (tar oils) attain values approaching 1·1. Aliphatic hydrocarbons are also dense, e.g. chloroform, $CHCl_3$, at 1·49. Viscosity determines the velocity of the flow process and the parameter considered is the kinematic viscosity, i.e. the quotient of the dynamic viscosity to the density. Results obtained by F. Schwille in 1981 demonstrate that, in a dry porous medium, chlorohydrocarbons and benzene would flow about 3 to 1·5 times faster and light heating oil or diesel fuel about 2 to 10 times more slowly than water.[15] Naturally, in water-saturated media, the degree of saturation of the immiscible fluids determines their relative flow velocities. Chlorohydrocarbons are highly volatile and hence, above ground, atmospheric-to-surface water transitions (and vice versa) take place rather rapidly. Transitions to the water of the unsaturated zone are similarly effected, and from it back to the surface. This also happens with benzene, but not with the mineral oil medium distillates.

The surface tension for the water/air boundary is 73 dyn/cm (mN/m), compared with 25–32 dyn/cm for oil and chlorohydrocarbons.

On the matter of physico-chemical properties, chlorohydrocarbons are either weakly polar or non-polar, oils also being non-polar for the most part. From this, Schwille concluded that in aquifers the sorption of dissolved substances must be small, but in organic soils a significant sorption is to be expected.

Whether groundwater is reached by infiltrating oil or chlorohydrocarbons depends to a great extent upon the rate of infiltration. There is a retention capacity for oil in the zone of aeration varying from 3–5 litre/m^3 in media of high permeability to 30–50 litre/m^3 in media of low permeability. Similar values were obtained for chlorohydrocarbons. As the unsaturated zone is heterogeneous, with layers of differing permeability, a lateral component of the percolation process is introduced and hence a broadening of the oil/chlorohydrocarbon body with increasing depth can be anticipated. This reduces the potential groundwater pollution since the penetration is lowered.

The behaviour of oil differs from that of the chlorohydrocarbons in the case where the infiltrated fluid volume exceeds the retention capacity of the unsaturated zone. As soon as oil reaches the upper boundary of the saturated zone within the capillary fringe, it spreads downwards as well as laterally; however, lateral spreading soon dominates. Oil tends to swim up to the surface of this zone even if injected below the groundwater table. Clearly, the determining surface for oil spreading is within the capillary fringe where the degree of saturation for water amounts to about 60–80%. The oil clings to this surface, which more or less parallels the groundwater table. Of course, the size and form of spreading depends upon the quantity of oil which infiltrates, the actual infiltration rate, the hydraulic gradient and the hydraulic conductivity. The same migration patterns apply to crude oils although when these cool some compounds may precipitate out, and thereafter they do not behave like Newtonian liquids.

When chlorohydrocarbons reach the surface of the saturated zone, their flow is halted because groundwater has to be displaced, but the laterally-effective component is not of much importance. The fluid sinks into the saturated zone; if the retention capacity of this zone is exceeded, the fluid will sink down to the aquifer bed, i.e. to its impermeable base, and then spread into a flattish mound which tends to parallel the base as oil does the groundwater table.

Recovery procedures may be utilized after pollution. Where local conditions permit and the water table is shallow it is feasible to pump out the oil collecting on it. If the water table is deeper, then both oil and chlorohydrocarbons may be recovered from deep wells. If these wells have penetrated to the base of the aquifer, however, they must fully reach and enter the impermeable layer, and they must be screened to the bottom in order to allow free passage of the fluid into the sumps. The fluid collecting there is recoverable using a special low-performance chlorohydrocarbon-proof pump or suitable collection devices. The type of chlorohydrocarbon is important: for instance, to flush out tetrachloroethylene (C_2Cl_4) takes roughly ten times as much water as is necessary to flush out 1,1,1-trichloroethane ($C_2H_3Cl_3$) under the same conditions.

Most groundwater contains dissolved gases derived from natural sources. Those arising from the geochemical cycle include the atmospheric gases, nitrogen, oxygen and carbon dioxide, but there are others, such as the inflammable gases hydrogen sulphide and methane arising from subterranean biochemical processes. Hydrogen sulphide is detectable in concentrations lower than 1 mg/litre and has a very distinctive bad odour. It may arise from the chemical reduction of sulphur-containing ions or from the action of certain bacteria on sulphides in groundwater. Methane is a more serious problem, partly because it is colourless, tasteless and odourless; occasionally it accumulates to a dangerous extent. It is a decomposition product of organic matter buried in geologically young, unconsolidated deposits. Water containing as little as 1 mg methane/litre can explode in an air space which is inadequately ventilated, so it is not surprising that explosions and fires occasionally occur in basements, water tanks, etc., as a result of its emission from groundwater. Safety measures include analyses to detect the gas, aeration of water before use and, of course, good ventilation.[16]

In sum, the number of possible pollutants in groundwater is enormous. D. K. Todd *et al.* in 1976 listed parameters and constituents which may be involved in the analysis of polluted groundwater.[17] That list was intended to be illustrative rather than representative so it is not quoted here, but it comprises the following categories: chemical–organic; chemical–inorganic; biological; physical; radiological. Instances of these have been given above, e.g. pesticides, oil, bacteria, odour, radium (for the use of radioisotopes in hydrology, see Section 1.2).

The same authors gave the principal causes and sources of pollution in groundwaters, and indicated disposal methods, in another publication in 1976.[18] They group the causes into four categories, namely municipal, industrial, agricultural and miscellaneous. Municipal sources include sewer leakage, liquid wastes and soil wastes; industrial also involve liquid wastes and leakage from tanks and pipelines as well as mining activities and oilfield brines. Agriculture produces pollution as a consequence of irrigation return flows, animal wastes, pesticides, etc. Under the heading of miscellaneous are listed spills and surface discharges, septic tanks and cesspools, roadway deicing, interchange through wells, and so on. Also, mention is made of saline intrusion, one of the biggest sources of pollution: saline water intrusions are discussed in more detail in Chapter 10. More recent work in several of the other fields may be mentioned here.

Nitrates are important pollutants of groundwater, and indeed of the environment in general. All over the world, according to J. L. Probst in 1985, an increasing input of fertilizers aimed at increasing agricultural output is occurring and concomitantly there is a general deterioration in the quality of both surface water and groundwater; today, in most rivers, there is an abnormal increase in nitrogen and phosphorus concentration.[19] Such an addition of nutritive elements can induce eutrophication and problems concerning the use of the water by human populations. Clearly, therefore, denitrification is desirable in many places. In the Chalk aquifer of east-central England, for instance, there has been serious pollution by nitrates leached from agricultural land so that its future as a potable source of water depends upon the nature and rate of denitrification. Some think that denitrification is occurring, hence the problem may only be a short-term one. However, more detailed work based on major-ion and environmental isotope data indicates that, on the contrary, denitrification is not significant; the apparent lowering of nitrate concentrations from >10 to <2 mg litre^{-1} in the apparent directions of flow is mainly due to mixing between waters of different origins. But there are reduced nitrogen species in some older waters (>4000 years) and these may show denitrification. According to K. W. F. Howard in 1985, denitrification cannot be relied on to reduce elevated concentrations of nitrate in modern recharge water.[20] It is clear that groundwater contamination by nitrates is a considerable health hazard because their presence at elevated concentrations in drinking water has been linked with the disease methaemoglobinemia, which is caused by bacterial reduction of nitrate to nitrite in the intestinal tract. Nitrate enters the blood stream and combines with haemoglobin to form methaemoglobin which reduce the capacity of the blood to carry oxygen. Reduction to nitrites occurs mainly in infants because the lower acidity of their gastric juices provides a better environment for nitrate-reducing bacteria.

It may be added that nitrate can also produce nitrosamines in the stomach and these are known carcinogens. There is evidence of a link between gastric cancer and high nitrate concentrations in ingested water. Thus, the leaching of nitrate from agricultural land is of great concern because it may lead to a contribution being made to the groundwater system. Some nitrate arises from natural soil nitrogen and an interesting demonstration may be cited. It is from C. W. Kreitler and L. A. Browning who, in 1983, published analyses of nitrate from two localities.[21] A set of 73 samples was analyzed from nitrate in the waters of the Cretaceous Edwards aquifer in Texas and they gave $\delta^{15}N$ values from -1.9 to $-10‰$, the average being $-6.2‰$. This indicated that they derived from naturally-occurring nitrogen compounds in recharge streams of which the waters had similar values. Measurements were made also on four samples from the Pleistocene-age Ironshore Formation in the Grand Cayman Islands, West Indies. These gave values for $\delta^{15}N$ of -18 to $-23‰$, indicating human waste (from cesspool and septic tank) as their origin.

Since nitrate pollution is so common, it is necessary to mention only one more case, this referring to the distribution, source and evolution of nitrate in a glacial till of southern Alberta, Canada, described by M. S. Hendry et al. in 1984.[22] The groundwaters of this till contain values up to two orders of magnitude greater than the recommended standard of the US Environmental Protection Agency (10 mg litre^{-1}), thus posing a serious health hazard. The nitrates result from the oxidation of ammonium present in the tills, this probably having taken place during the Holocene epoch when water tables were much lower than now. A potential for denitrification exists under the present water table in both the weathered and unweathered till as well as the bedrock. Isotope data showed that less denitrification may be occurring within the nitrate enclaves than in adjacent, down-gradient ones.

Finally, it should be added that J. C. Vogel et al. in 1981 cited gaseous nitrogen as evidence for denitrification in groundwater.[23]

Mention must be made of radioactive waste disposal, the object of which is to isolate radioactive wastes from the environment until there is no threat to public health and safety. The main option in the USA is to emplace them deep in geological media, using the concept of multibarriers. These would ensure that, to return to the biosphere, radionuclides must escape the waste form (glass or crystalline), the corrosion-resistant canister, and the overpack, and then pass through an engineered backfill of bentonite clay with additives such as quartz, zeolites, graphite, oxidizable metals and certain host rocks, thereafter somehow being transported back to the environment through the actual host rock. Among rock types considered as potential hosts are salt domes, bedded salt, granite, basalt and tuff. K. Cartwright et al. in 1981 mentioned that existing regulations assume that long-term isolation of toxic materials, including radionuclides, could be achieved in landfills with liners having very low hydraulic conductivity, adding that this is probably impossible in humid areas.[24] They advocated regulations which would provide performance standards on a site-to-site basis; among the factors to be taken into account would be site hydrogeology. This is significant because it governs the direction and rate of transport of contaminants. Another important factor is that release of unattenuated pollutants to both surface water and groundwater may take place and must be monitored. In fact, a watch-and-warn system would be necessary.

Turning to storage in aquifers, the storage coefficient is an important parameter and may be defined as the volume of water which an aquifer releases from or takes into storage per unit surface area per unit change in the component of head normal to that surface. This parameter is also termed the 'storativity'. Obviously, water which recharges an aquifer or water which discharges from it represents changes in the storage volume. In unconfined aquifers, the storage coefficient can be regarded as the product of the change in aquifer volume occurring between the water table at the commencement and at the end of a stated period of time and the average specific yield. However, in confined aquifers, if these stay saturated, pressure changes produce only very small changes in volume. Hence, hydrostatic pressure in the aquifer in part supports the superincumbent load, the rest of the support deriving from the solid structure of the aquifer itself. The associated compression actually forces water out and also the lowering of the pressure will cause a slight expansion and subsequent release of water. The water-yielding capacity of an aquifer is really expressed by its storage coefficient, S, and, in a confined aquifer, for a vertical column of unit area extending through it, this equals the volume of water released from it when the piezometric surface declines a unit distance. The storage coefficient is a dimensionless quantity involving a volume of water per volume of aquifer relationship and, in most confined aquifers, the range is $0.00005 < S < 0.005$. From this it is apparent that, in order to obtain substantial yields of water, very large pressure changes over very extensive areas are necessary. S can be determined from pumping tests in wells or from groundwater fluctuations responding to atmospheric pressure or ocean tide variations. It normally varies directly with the aquifer thickness. S. W. Lohman in 1972 gave an empirical expression for the correlation:[25] $S = 3 \times 10^{-8} b$, where b is the saturated aquifer thickness in metres applied for the purpose of estimation. For an unconfined aquifer, the storage coefficient corresponds to its specific yield (since it is the water table which declines a unit distance); see Section 3.2.

The concept of the groundwater basin is important because of hydraulic continuity in the groundwater resource contained therein; it refers to a hydrogeological unit comprising a large aquifer or several small, interconnected, interrelated ones which can supply a substantial amount of water. Such a basin may or may not coincide with a physiographic unit. Springs are surface manifestations of groundwater in many cases and are characterized by concentrated discharges which flow out. They are of various types, but K. Bryan in 1919 grouped them into two categories, namely those of non-gravitational and those of gravitational origin.[26] The first includes volcanic and fissure springs, usually thermal in nature. Volcanic springs are associated with volcanic rocks and fissure springs result from deep fissures in the terrestrial crust. Gravity springs arise from water flowing under hydrostatic pressure and include the following types. There are depression springs which arise when the ground surface intersects the water table, and contact springs resulting from a permeable, water-bearing formation overlying a less permeable one which intersects the ground surface. Artesian springs result when water under pressure is released from confined aquifers, and impervious rock springs pass through channels or fractures in impervious rock. Tubular or fracture springs issue from lava tubes or solution channels or impermeable rock connecting with groundwater.

O.E. Meinzer grouped all springs by discharge as shown in Table 3.7.[27] The discharge of any spring will depend upon factors such as the rate of recharge, which fluctuates in almost all springs; hence the rate of discharge also fluctuates. Relevant periods may vary from minute to

TABLE 3.7
SPRINGS CLASSIFIED BY DISCHARGE

Magnitude	Mean discharge
First	10 m^3/s
Second	1–10 m^3/s
Third	0·1–1 m^3/s
Fourth	10–100 litre/s
Fifth	1–10 litre/s
Sixth	0·1–1 litre/s
Seventh	10–100 ml/s
Eighth	10 ml/s

years, depending upon the hydrologic circumstances. However, areas of volcanic rock and sandhills are characterized by perennial springs with almost constant discharge.

Perennial springs derive from extensive permeable aquifers and discharge all the time. Intermittent springs discharge intermittently i.e. only during those parts of the year when enough groundwater is recharged to maintain flow. Springs exhibiting fairly regular discharge fluctuations not associated with rainfall or seasonal effects constitute periodic springs. Submarine springs also occur in coastal areas containing volcanic rock or limestone aquifers in regions such as the Mediterranean sea or Hawaii and, where the discharge is large enough, potable water can be taken directly from the surface of the sea.

Thermal springs are those which discharge water with a temperature exceeding that of the local groundwater; they comprise warm and hot springs with highly mineralized water which originated as meteoric water and was modified chemically during its subterranean passage.[28] Such springs are practically always in volcanic rock areas, concentrating in regions with large geothermal gradients. It may be inferred that the source water must have been able to percolate to considerable depths, perhaps as much as 3 km, such water subsequently creating a large convective current rising to supply hydrothermal areas. A geyser is a periodic thermal spring which arises from the expansive force of superheated steam in constricted subsurface channels. Water from the surface or shallow aquifers, or both, penetrates down into a deep, vertical tube where it heats to above boiling point. As the pressure mounts, steam pushes upwards and some water is released at the surface. This reduces the hydrostatic pressure, causing the deeper, superheated water to accelerate upwards and flash into steam. The geyser then surges into a full-scale eruption for a short time until dissipation of pressure is achieved. Afterwards, the cycle begins again.

Mudpots are another type of hot spring, and result where there is little water available. The limited quantity mixes with clay to form a muddy suspension at the ground surface through which water and steam bubble. A funarole is an opening emitting gases such as steam, carbon dioxide and hydrogen sulphide.

Thermal springs occur throughout the world, e.g. in Iceland, New Zealand, Japan, etc., and of course in the United States where, in Yellowstone National Park, Wyoming, there are thousands of hydrothermal features. The region is a remnant of a vast volcanic eruption which took place about 600 000 years ago and, even now, temperatures of 240°C exist at 300 m below the surface of the ground.

Internal heat from the Earth emerges at an average rate of 1.5×10^{-6} cal/cm^2 s and the resultant average geothermal gradient is 1°C/50 m, but in volcanic and active tectonic regions the heat flow is orders of magnitude higher. This matter will be discussed in greater detail in Chapter 8.

By contrast, in permafrost areas groundwater is restricted in movement because frozen ground creates an impermeable layer above it. Permafrost is perennially frozen ground defined as unconsolidated deposits or bedrock which have had a temperature below 0°C continuously for two to thousands of years.[29] Such frozen ground is impermeable and, in the continuous permafrost zones, is present everywhere to depths of 150–400 m, comprising a confining layer and practically eliminating shallow aquifers. This means that wells have to be drilled deeper than would be necessary in similar geological environments which do not have permafrost. Groundwater can be found above, below and sometimes within permafrost. In the continuous permafrost zone, water is obtained optimally from unfrozen alluvium beneath large lakes, in large valleys and adjacent to riverbeds. In the discontinuous permafrost zone, groundwater can be derived locally from shallow aquifers, but, because there may be pollution from the ground surface, sources beneath the frozen layer are preferable. J. R. Williams presented a map of the distribution of continuous and discontinuous permafrost in the Northern Hemisphere from which it is clear that the relevant zones are very widespread, occurring in Alaska, Canada, Greenland, northern Scandinavia, and, of course, the USSR.[29]

REFERENCES

1. MORRIS, D. A. and JOHNSON, A. I., 1967. *Summary of Hydrologic and Physical Properties of Rocks and Soil Materials as Analyzed by the Hydrologic Laboratory of the US Geological Survey, 1948–60.* US Geol. Surv. Water Supply Paper 1839–D, 42 pp.
2. MANGER, G. E., 1963. *Porosity and Bulk Density of Sedimentary Rocks.* US Geol. Surv., Bull. 1144-E, 55 pp.
3. SOIL SURVEY STAFF, 1951. *Soil Survey Manual.* US Dept of Agriculture Handbook No. 18, 503 pp.
4. LOHMAN, S. W., 1972. *Ground-water Hydraulics.* US Geol. Surv. Prof. Paper 708, 70 pp.
5. BEAR, J., 1972. *Dynamics of Fluids in Porous Media.* Elsevier, New York, p. 764.
6. MATTHESS, G. and PEKDEGER, A., 1981. Concepts of a survival and transport model of pathogenic bacteria and viruses in groundwater. *The Science of the Total Environment,* **21**, 149–59.
7. MATTHESS, G., PEKDEGER, A., RAST, N., RAUERT, W. and SCHULZ, H. D., 1979. Tritium tracing in hydrogeochemical studies using model lysimeters. *Isotope Hydrology 1978,* **2**, IAEA, Vienna, 769–85.
8. ALTHAUS, H., JUNG, K. D., MATTHESS, G. and PEKDEGER, A., 1982. Lebensdauer von Bakterien und Viren in Grundwasserleitern. *Umweltbundesamt Materialien,* 1/82, 190 pp.
9. JOHNSON, A. I., 1967. *Specific Yield — Compilation of Specific Yields for Various Materials.* US Geol. Surv. Water Supply Paper 1662-D. 74 pp.
10. DEPARTMENT OF ECONOMIC AND SOCIAL AFFAIRS, 1975. *Groundwater Storage and Artificial Recharge.* Nat. Resources, Water Series No.2, United Nations, New York, 270 pp.
11. TODD, D. W., 1980. *Groundwater Hydrology.* 2nd edn. Wiley, New York, 535 pp, p. 38.
12. WHITE, W. B., 1969. Conceptual models for carbonate aquifers. *Ground Water,* **7**, 3, 15–21.
13. BRUCKER, R. W. et al., 1972. Role of vertical shafts in the movement of ground water in carbonate aquifers. *Ground Water,* **10**, 6, 5–13.
14. LEGRAND, H. E. and STRINGFIELD, V. T., 1973. Karst hydrology — a review. *J. Hydrol.,* **20**, 97–120.
15. SCHWILLE, F., 1981. Groundwater pollution in porous media by fluids immiscible with water. *The Science of the Total Environment,* **21**, 173–85.
16. CARROLL, D., 1962. *Rainwater as a Chemical Agent of Geologic Processes. A Review.* US Geol. Surv. Water Supply Paper 1535-G, 18 pp.
17. TODD, D. K. et al., 1976. *Monitoring Groundwater Quality: Monitoring Methodology.* Rept EPA–600/4–76–026, US Env. Prot. Agency, Las Vegas, 154 pp.
18. TODD, D. K. et al., 1976. A groundwater quality monitoring methodology. *J. Am. Water Works Assn,* **68**, 586–93.
19. PROBST, J. L., 1985. Nitrogen and phosphorus exportation in the Garonne Basin (France). *J. Hydrol.,* **76**, 281–305.

20. HOWARD, K. W. F., 1985. Denitrification in a major limestone aquifer. *J. Hydrol.*, **76**, 265–80.
21. KREITLER, C. W. and BROWNING, L. A., 1983. Nitrogen-isotope analysis of groundwater nitrate in carbonate aquifers: natural sources versus human pollution. *J. Hydrol.*, **61**, 285–301.
22. HENDRY, M. S., MCCREADY, R. G. L. and GOULD, W. D., 1984. Distribution, source and evolution of nitrate in a glacial till of southern Alberta, Canada. *J. Hydrol.*, **70**, 177–98.
23. VOGEL, J. C., TALMA, A. S. and HEATON, T. H. E., 1981. Gaseous nitrogen as evidence for denitrification in groundwater. *J. Hydrol.*, **50**, 191–200.
24. CARTWRIGHT, K., GILKESON, R. H. and JOHNSON, T. M., 1981. Geological considerations in hazardous waste disposal. *J. Hydrol.*, **54**, 357–69.
25. LOHMAN, S. W., 1972. *Ground-water Hydraulics.* US Geol. Surv. Prof. Paper 708. 70 pp.
26. BRYAN, K., 1919. Classification of springs. *J. Geol.*, **27**, 522–61.
27. MEINZER, O. E., 1923. *Outline of Groundwater Hydrology with Definitions.* US Geol. Surv. Water Supply Paper 494, 71 pp.
28. WARING, G. A., 1965. *Thermal Springs of the United States and Other Countries of the World — A Summary.* US Geol. Surv. Prof. Paper 492, 383 pp.
29. WILLIAMS, J. R., 1970. *Groundwater in the Permafrost Regions of Alaska.* US Geol. Surv. Prof. Paper 696, 83 pp.

CHAPTER 4

The Movement of Groundwater

4.1. DARCY'S LAW

Possibly the most fundamental characteristic of aquifers, most of which are natural porous media, is that the groundwater is continuously moving in a manner governed by established hydraulic principles: this flow is expressible by Darcy's law, the application of which enables the rate and direction of flow to be assessed. A constant in the appropriate flow equation is hydraulic conductivity, which can be determined both in the laboratory and in the field and is a measure of the permeability of the medium. Dispersion, or mixing, which results from flow through porous media, causes flow irregularities which can be examined by using suitable tracers and the presence of air in the unsaturated zone also complicates water flow.

Henri Darcy, who was mentioned in Section 1.2, made the original statement of his law in his report in 1856.[1] Essentially, the law states that the flow rate through porous media is directly proportional to the head loss and inversely proportional to the length of the flow path. This is the foundation of existing knowledge about groundwater flow.

It can be verified experimentally by using water flowing at a rate, Q, through a cylinder of cross-sectional area A packed with sand and with piezometers at a distance L apart as shown in Fig. 4.1. A piezometer is an instrument for measuring pressure head. Total energy heads, or fluid potentials, above a datum plane can be expressed by the Bernoulli equation:

$$\frac{p_1}{\gamma} + \frac{v_1^2}{2g} + z_1 = \frac{p_2}{\gamma} + \frac{v_2^2}{2g} + z_2 + h_L \qquad (4.1)$$

where p is the pressure, γ is the specific weight of water, v is the velocity

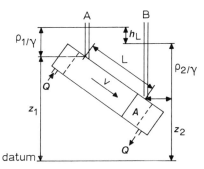

FIG. 4.1. The distribution of pressure and head loss in flow through a sand cylinder with piezometers A and B which are a distance L apart. The cross-sectional area is A, the rate of flow of water is Q, the velocity of flow is v, h_L is the head loss, ρ is pressure, γ is the specific weight of water, z is the elevation.

of flow, g is the acceleration due to gravity, z is the appropriate elevation and h_L is the head loss. The subscripts relate to the measurements shown in Fig. 4.1. Since velocities in porous media are usually low, velocity heads can be neglected without any appreciable error arising. Therefore, it is possible to re-write the equation so that the head loss becomes:

$$h_L = \frac{\rho_1}{\gamma} - z_1 - \frac{\rho_2}{\gamma} - z_2 \qquad (4.2)$$

and the consequent head loss is then defined as the potential loss within the sand cylinder, the energy lost through frictional resistance being dissipated as heat. The head loss, therefore, is independent of the inclination of the cylinder. Darcy's measurements indicated that the proportionalities $Q \propto h_L$ and $Q \propto 1/L$ exist; therefore a proportionality constant, K, may be introduced:

$$Q = -KA\frac{h_L}{L} \qquad (4.3)$$

which, generally, is

$$Q = -KA\frac{dh}{dl} \qquad (4.4)$$

and

$$v = \frac{Q}{A} = -K\frac{dh}{dl} \qquad (4.5)$$

where K is the hydraulic conductivity (a constant serving as a measure of the permeability of the porous medium), dh/dl is the hydraulic gradient and v is the Darcy velocity or specific discharge; this assumes that flow takes place through the whole cross-section of the medium without regard to the distinction between voids and solids. The negative sign shows that water flows in the direction of decreasing head. The equation states that the flow velocity equals the product of the constant K and the hydraulic gradient. Darcy velocity does not exist because the flow of water is restricted to voids; hence the average interstitial velocity is given by:

$$v_n = \frac{Q}{nA} \quad (4.6)$$

where n is the porosity. Obviously, if a sand has a porosity of 33%, the value of v_n will be $3v$.

The variability of void sizes causes the actual flow velocity to be non-uniform as well as causing continual changes in direction. In theory, it would be necessary to consider a precise point location within the medium when assessing the flow velocity. However, this cannot be done with geological materials, so actual velocities can only be quantified statistically.

The validity of Darcy's law must be examined. Velocity in laminar flow, such as occurs when water flows in a capillary tube, is proportional to the first power of the hydraulic gradient (Poiseuille's law). It is probable, therefore, that Darcy's law applies to laminar flow in porous media. A necessary concomitant of this is that there is no lower limit for Darcy's law because laminar flow can take place at extremely low velocities. The law is essentially one of percolation which is of laminar character; laminar flow is of course placid. There is an upper limit, however, and, in clarification, allusion must be made to flow in pipes, where the Reynolds number (which expresses the dimensionless ratio of inertial to viscous forces) constitutes a criterion for distinguishing between laminar and turbulent flow. This number has been utilized to establish the limit of flows described by Darcy's law corresponding to the value where the linear relationship is no longer valid. The Reynolds number is given by:

$$N_r = \frac{\rho v D}{\mu} \quad (4.7)$$

where ρ is the fluid density, v the velocity, D the diameter (of the pipe), and μ the viscosity of the fluid. In dealing with flow in porous media, the Darcy velocity is used for v and an effective grain size (d_{10}) is substituted for D. The diameter of a grain can do no more than represent an approximation of the critical flow dimension for which it is intended, and of course the measurement of pore size distribution is very difficult. It has been shown experimentally that Darcy's law is valid for $N_R < 1$ and does not diverge seriously as far as $N_R = 10^1$. Here, then, is the upper limit of Darcy's law mentioned above.

As inertial forces increase, there will be a gradual development of turbulence and so this upper limit spans a range of values. Clearly, turbulence will occur initially in larger voids, later spreading to small pores and, when it is completely developed, the head loss varies roughly with the second power of the velocity rather than in a linear manner.

It has been found that most natural subterranean water flow takes place with $N_R < 1$, so Darcy's law is applicable. Deviations may occur where there are steep hydraulic gradients, e. g. near pumped wells, and turbulent flow can occur in rocks such as basalt and limestone which possess large underground openings. With regard to rocks, there is evidence that Darcy's law may not be applicable when water flows through dense clay, an extremely slow process. The effect of electrically charged clay particles on water in minute pores produces non-linearities between flow rate and hydraulic gradient. It may be added that a defect in Darcy's original approach was alleged in 1977 by S. Schweitzer, who proposed a method to extend the law and obtain a first-order approximation to the equations governing the motion of flow through a porous medium under non-isothermal conditions.[2]

The permeability of a soil or rock dictates its capacity for transmitting a fluid and, as such, relates to the medium only, i.e. it has no connection with fluid properties. Intrinsic permeability, k, mentioned earlier (see Section 3.1), was proposed in order to make a distinction from hydraulic conductivity. Another important characteristic is transmissivity, T, which has been much utilized in groundwater hydraulics. It can be defined as the rate at which water of prevailing kinematic viscosity is transmitted through a unit width of aquifer under a unit hydraulic gradient. The following expression is relevant:

$$T = Kb \ (\text{m/day})(\text{m}) = \text{m}^2/\text{day} \tag{4.8}$$

where b is the saturated thickness of the aquifer.

Values of K, hydraulic conductivity, appear in Table 3.2. The parameter may depend upon a number of factors such as porosity, particle dimensions and distribution, their shape, etc. It can be determined in saturated zones by several methods which include calculation from appropriate equations, tracer tests and pumping tests in wells. The equation used takes the form:

$$k = f_s f_n d^2 \qquad (4.9)$$

where f_s is a pore or grain shape factor, f_n is a porosity factor and d is a characteristic grain diameter.[3,4] Hardly any equation provides very reliable results because there are so many imponderables to be considered. In the laboratory, a permeameter may be utilized. In this instrument, flow is maintained through a small sample of material while flow rate and head loss are measured. There are two types, the constant head and the falling head.

Figure 4.2 illustrates the constant head permeameter, which is able to measure hydraulic conductivities of consolidated and unconsolidated formations under low heads. Water penetrates the medium cylinder

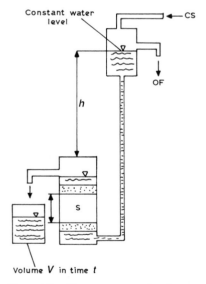 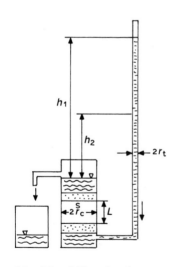

Fig. 4.2. The constant head permeameter: S = sample; CS, continuous supply, OF, overflow, L is the sample thickness and the porous plates are stippled.

Fig. 4.3. The falling head permeameter.

basally and is collected as overflow after passing upwards through the material. Using Darcy's law, it is possible to state hydraulic conductivity by the equation:

$$K = \frac{VL}{Ath} \tag{4.10}$$

where V is the flow volume in time t, A is the cross sectional area of the sample and the other dimensions are shown in the Figure. It is most important that the medium be completely saturated so that no air is entrapped in it. A number of differing heads in a set of tests gives a reliable result.

The falling head permeameter is shown in Fig. 4.3. In this case, water is added to a tall tube, flows upwards through a cylindrical sample and is collected as overflow. In the test, the rate of fall of the water level in the tube is measured and the hydraulic conductivity obtained by equating the flow rate, Q, in the tube:

$$Q = \pi r_t^2 \mathrm{d}h/\mathrm{d}t \tag{4.11}$$

with that in the sample which is, through Darcy's law:

$$Q = \pi r_c^2 K\, h/L \tag{4.12}$$

so that, equating and integrating:

$$K = \frac{r_t^2 L}{r_c^2 t} \ln \frac{h_1}{h_2} \tag{4.13}$$

where r_c, r_t and L are shown in the Figure and t is the time required for the water in the tube to fall from h_1 to h_2.

Unfortunately, permeameter data sometimes have only a remote relationship with hydraulic conductivities in the field. This is partly because undisturbed samples are extremely hard to obtain and disturbed ones have varied porosity, packing and grain orientation, all of which influence hydraulic conductivity. Also, a few samples can hardly represent the hydraulic conductivity of an aquifer.

Tracer tests can be utilized in the field. Dyes such as sodium fluorescein are cheap, safe and easy to detect. The technique necessitates measurement of the time taken for the tracer to travel between two test boreholes. If flow is taken to be average interstitial velocity, v_n, then:

$$v_n = \frac{K}{n} \frac{h}{L} \tag{4.14}$$

where K is the hydraulic conductivity, n is the porosity, h is the difference in water table level between the two boreholes after the experiment has been effected and L is the distance between them. Since v_n is also given by L/t, where t is the travel time of the tracer between the boreholes, equating the two equations produces the solution for K:

$$K = \frac{nL^2}{ht} \quad (4.15)$$

The procedure is simple, but there are many drawbacks. For instance, the boreholes must be near enough so that the time of travel of the tracer is not unduly long. Also, unless the direction of flow is well known, the tracer may miss the downstream borehole; this can be obviated only by using a number of sampling boreholes, which greatly increases costs. Finally, if (as is frequently the case) the aquifer is a stratified one, each layer will have a different hydraulic conductivity from its neighbours and the initial arrival of the tracer will give a conductivity much larger than the average for the aquifer.

Another tracer technique is the point dilution method described by E. Halevy in 1967.[5] The aim is to obtain an independent direct measurement of the filtration velocity in a water-bearing formation under natural or induced hydraulic gradient. The principle of this measurement lies in the relation between the observed decrease in the concentration of solution introduced into a filter tube and the filtration velocity of the groundwater flow in the aquifer. Assuming horizontal flow, the dilution of a tracer solution which is distributed homogeneously in a volume V in the borehole is given by:

$$v_a = \frac{-V}{Ft} \ln \frac{C}{C_0} \quad (4.16)$$

where v_a is the dilution rate (apparent velocity), V is the measuring volume, F is the cross-section of the measuring volume perpendicular to the direction of the groundwater flow, t is the time interval of the measurement, C_0 is the initial tracer concentration and C its concentration at the end of the measurement. The horizontal flow pattern in the aquifer is distorted by the borehole and the different flow within it. Thus, the measured velocity in the borehole, v_a, is related to the actual velocity, v_f, by a correction factor α defined as:

$$\alpha = \frac{Q_a}{Q_f} \quad (4.17)$$

where Q_a is the horizontal flow rate in the borehole and Q_f is the flow rate in the same cross-section in the aquifer formation. Hence:

$$v_f = \frac{-V}{\alpha F t} \ln \frac{C}{C_0} \qquad (4.18)$$

and, introducing to this equation the parameters of the borehole:

$$v_f = \frac{(r_1^2 - r_0^2)}{2\alpha r_1 t} \ln \frac{C}{C_0} \qquad (4.19)$$

where r_1 is the radius of the filter tube and r_0 is the radius of the measuring probe. The perforation of the filter tube may be around 30%. There are two operations, injection of the tracer into the borehole and measurement of its dilution. There are various ways of doing each.

Injection may be effected by pouring tracer solution into a thin pipe, by crushing a glass ampoule containing the tracer solution or by using a spatial injecting syringe. It should be done at one or, preferably, several depths in order to achieve the good mixing which is essential. Appropriate radiotracers should be used, such as bromine-82, which with others is described in Table 1.3. The temperature of the injected tracer should be the same as that of the water in the borehole so as to avoid any density current developing.

The tracer concentration can be measured directly in the borehole using a probe with a Geiger–Müller or scintillation detector and also by circulating the water between the isolated segment of a borehole and the ground surface using a small pump. In the latter case, measurement is made by a detector immersed in a container included in the circulation system, but a limitation on this method may be caused by vertical flow in the borehole. This can trigger a short circuit between aquifer layers of different permeability which results in vertical currents which can increase the filtration velocity. Even in the situation where a measured segment in a filter tube is isolated by inflatable packers, the vertical currents cannot be eliminated entirely. This is because there may be a gap between the filter tube and the wall of the borehole. Therefore, measurements of the vertical flow, if it is present, must precede determination of filtration velocity. This is feasible if the vertical velocity is only several times that of the horizontal velocity. The way in which this is done is to inject the radiotracer at a given depth. Four or five detectors can be placed at different distances from the point of injection and these will show the direction of the flowing cloud of activity. Groundwater flow direction will be discussed in detail in Section 4.3, but here it is appropriate to mention that radiotracing can be utilized in its determination also. The principle is that a suitable tracer is introduced into a segment of a borehole and subsequently it is carried away in the direction of flow as a result of horizontal flow. The radiotracer becomes adsorbed on the borehole wall in the direction of flow of the groundwater. If the activity is measured by a directionally-orientated collimated detector, then the direction of maximum activity corresponds to the direction of flow. Normally, a collimated probe is lowered into the borehole, employing a suspended rod which is rotated after each reading of the radiation intensity. $NH_4{}^{82}Br$, $Na^{131}I$, $^{198}AuCl_3$ and $^{51}CrCl_3$ radioisotopes have been utilized to depths of 50 m or more and the accuracy of the determination of the direction of flow of groundwater is better than $\pm 10°$.

Auger hole methodology entails the measurement of the change in water level after a volume of water is quickly taken from an unlined, cylindrical hole, which may need screening if the soil is loose. This approach is a simple one and it gives a value of K which relates to a horizontal direction near the hole. C. W. Boast and D. Kirkham in 1971 demonstrated that the appropriate equation is:

$$K = \frac{C}{864} \frac{dy}{dt} \qquad (4.20)$$

where dy/dt is the measured rate of rise in cm/s, the factor 864 gives K values in m/day, and C is a dimensionless constant referring to the ratios of depth to water against the radius of the hole, and depth of water below the water table against the depth to water, values related to these variables being presented for impermeable layers and infinitely permeable layers in an appropriate table by these authors.[6] Probably the best means of ascertaining the hydraulic conductivity in an aquifer is by pumping tests in wells which, on the basis of observations of water level near such wells, can yield an integrated value of K over a large aquifer section. The technique will be discussed in detail in Chapter 5.

In most hydrologic theory, the assumptions are made that the geological materials of an aquifer are homogeneous and isotropic, with the parallel inference that the hydraulic conductivity is the same in all directions. Of course, in nature this is practically never the case and aquifers are anisotropic so that K varies directionally. In alluvium, the causes are particle shapes and layering. Individual grains are not perfect spheres and so, after deposition, they tend to lie with flatter surfaces horizontal. Also, alluvium is usually layered; each layer contains different materials and therefore possesses a characteristic specific hydraulic conductivity.

4.2. ANISOTROPIC AQUIFERS

For simplicity, let us consider an anistropic aquifer model with only two horizontal layers, each of which is individually isotropic, having differing thicknesses and values of K as shown in Fig. 4.4.

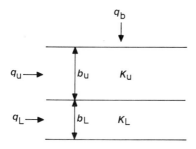

FIG. 4.4. Two isotropic and horizontal layers of an aquifer, each with different thicknesses and hydraulic conductivities.

In the case of horizontal flow parallel to these layers, the flow in the upper layer per unit width will be:

$$q_u = K_u i b_u \tag{4.21}$$

where q_u is the flow in the upper layer, K_u is the hydraulic conductivity, and b_u is the thickness as indicated in the figure; i, the hydraulic gradient, is the same in both layers for horizontal flow. Then the total horizontal flow, q_t, is:

$$q_t = q_u + q_L = i(K_u b_u + K_L b_L) \tag{4.22}$$

In the case of a homogeneous, isotropic aquifer, the expression is:

$$q_t = K_t i(b_u + b_L) \tag{4.23}$$

with K_t being the hydraulic conductivity in a horizontal direction for the aquifer as a whole. If these two equations for the two types of aquifer system are equated and solved for K, the resulting expression is:

$$K_t = \frac{K_u b_u + K_L b_L}{b_u + b_L} \qquad (4.24)$$

which is amenable to generalization for n layers in the form:

$$K_t = \frac{K_u b_u + K_L b_L + \cdots + K_n b_n}{b_u + b_L + \cdots + b_n} \qquad (4.25)$$

so defining the equivalent hydraulic conductivity in a horizontal direction for a layered material.

Turning to the matter of vertical flow through the two layers of the model anisotropic aquifer, the flow per unit horizontal area, q_v, in the upper layer is:

$$q_v = K_u \frac{dh_u}{b_u} \qquad (4.26)$$

where dh_u is the head loss incurred within the upper layer and:

$$dh_u = \frac{b_u}{K_u} q_v \qquad (4.27)$$

Clearly, q_v is the same in the underlying layer of the aquifer, so that the total head loss in the system must be:

$$dh_u + dh_L = \left[\frac{b_u}{K_u} + \frac{b_L}{K_L} \right] q_v \qquad (4.28)$$

In a homogeneous system, the expression for vertical flow is:

$$q_v = K_v \left(\frac{dh_u + dh_L}{b_u + b_L} \right) \qquad (4.29)$$

K_v being the hydraulic conductivity in a vertical direction for the aquifer as a whole. This may be re-written so that:

$$dh_u + dh_L = \left(\frac{b_u + b_L}{K_v} \right) q_v \qquad (4.30)$$

so that, after equating with eqn. (4.28) above, the following expression is obtained:

$$K_v = \frac{b_u + b_L}{\frac{b_u}{K_u} + \frac{b_L}{K_L}} \quad (4.31)$$

and can be generalized for n layers as:

$$K_v = \frac{b_u + b_L + \cdots + b_n}{\frac{b_u}{K_u} + \frac{b_L}{K_L} + \cdots + \frac{b_n}{K_n}} \quad (4.32)$$

so defining the equivalent vertical hydraulic conductivity for a layered material.

The horizontal hydraulic conductivity in alluvium is usually greater than that found to exist in the vertical direction, ratios of K_t/K_v normally falling into a range of about 2–10 according to D. A. Morris and A. I. Johnson in 1967.[7] If clay layers occur, values can reach or exceed 100. The situation becomes rather different when consolidated materials are considered. Here many variables occur; for instance, there may be fractures or solution openings. Also, the orientation of the beds is significant and structural conditions may exist which produce anisotropic results not possessing a horizontal component at all.

For two-dimensional flow in an anisotropic medium, in order to apply Darcy's law it is necessary to obtain the appropriate value of K for the direction of the flow. In the case of directions which are neither horizontal (K_t) nor vertical (K_v), the correct value of K may be derived from the following expression:

$$\frac{1}{K_\beta} = \frac{\cos^2 \beta}{K_t} - \frac{\sin^2 \beta}{K_v} \quad (4.33)$$

where K_β is the hydraulic conductivity in the particular direction which is at an angle β with the horizontal.

4.3. RATES AND DIRECTIONS OF FLOW IN GROUNDWATER

Darcy's law has been discussed above and its applicability depends upon validity restrictions. It indicates that the rate of groundwater flow will be subject to the hydraulic conductivity in an aquifer and also the hydraulic gradient. Natural flow rates are usually very slow, varying from metres per day to metres per year, so aquifers comprise vast, sluggishly moving subterranean water reservoirs. However, in certain circumstances, e.g. where groundwater flows through huge underground openings in limestones at turbulent velocities, rapid movement rates may be attained. In other situations the rate of groundwater movement may become so slow that it is practically negligible.

The point dilution method described by E. Halevy for obtaining a direct measurement of the horizontal filtration velocity (Darcy velocity), v_f, of water in an aquifer has been discussed in Section 4.2, and constitutes a valuable approach to the problem.[5]

Environmental radioisotope contents may also be utilized; thus, carbon-14 measurements at appropriate sampling points in wells, located optimally in an unconfined aquifer with negligible recharge in the sampling area or in a confined aquifer, provide an important means of assessing the velocity if the groundwater concerned is moving very slowly. Such an approach has been used in the Floridan aquifer, for instance, although where the water body is very large indeed it may not be applicable (see also Section 12.2). Thus, P. L. Airey et al. mentioned in 1983 that, in the Great Artesian Basin in Australia, chlorine-36 (with a half-life, $T_{1/2}$, of 301 000 years) had to be substituted for radiocarbon to date some groundwaters and a helium accumulation technique was employed with others.[8] However, G. E. Calf and M. A. Habermehl in 1983 investigated Division A of the Basin (Queensland), examining a number of non-flowing artesian water wells of which the ^{14}C-specific activity was interpreted as age. A flow rate was calculated for groundwater moving in a south-western direction as 2·2 m/y, which agreed well with 0·5 m/y for the south-westerly flow and 0·8 m/y^{-1} for southerly flow lines in Division B, west of A, in the Basin, both calculated from ^{36}Cl dating techniques.[9]

Of course, the conventional approach already mentioned (Section 4.1) may also be utilized. Its parameter v_n represents the pore velocity rather than the specific velocity, v, which is the time taken for water to pass between two reference points. The velocity v must be regarded as merely a value of Q/A, i.e. discharge divided by the cross-sectional area, and it is necessarily less than v_n. This is because water follows paths in porous media which almost invariably are longer than the straight line between the same two points. Hence, the parameter v_n can be described thus:

$$v_n = \frac{Q}{nA} = \frac{vA}{nA} = \frac{v}{n} \qquad (4.34)$$

Thus, the pore velocity (average) is the specific velocity divided by the effective porosity. The velocity distribution across a pore is probably parabolic (highest centrally and zero at the edges) and the maximum pore velocity is therefore about twice the pore velocity. This would lead to the conclusion that, where n is 33%, whilst v_n has a value of $3v$ as stated in Section 4.1, the maximum pore velocity v_m will have a value of $6v$. As noted, it is not possible to consider such microlocations within a porous medium such as an aquifer, so the discussion remains theoretical.

It may be salutary to recall the wise words of R. G. Kazmann, who wrote in 1980 that 'there is no virtue in going through elaborate mathematical reasoning if the results are either of no practical importance or, if of practical significance, the same results can be achieved more readily by other means.'[10]

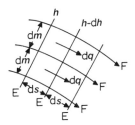

FIG. 4.5. An area of an orthogonal flow net comprising flow lines, F, and equipotential lines, E.

To the directions of flow of groundwater, flow nets are applicable. These represent graphically the flow characteristics and comprise flow lines and equipotential lines mapped in two dimensions. The two sets of lines make up an orthogonal pattern of small squares. They are usually derived by trial-and-error approximations or by model studies. H. R. Cedergren has offered suggestions for graphic construction of flow nets.[11] Figure 4.5 shows part of a flow net, in which the hydraulic gradient, i, is given by:

$$i = \frac{dh}{ds} \qquad (4.35)$$

and the constant flow, q, between two adjacent flow lines is given by:

$$q = K\frac{dh}{ds}dm \qquad (4.36)$$

for unit thickness. For the squares of the flow net, $ds \cong dm$ and the last equation will become:

$$q = K\, dh \qquad (4.37)$$

and, when this applied to a complete flow net, where the total head loss, h, is divided into n squares between any two adjacent flow lines, then:

$$dh = \frac{h}{n} \qquad (4.38)$$

and, if the flow is divided into m channels by flow lines, then the total flow

$$Q = m\, q = \frac{Kmh}{n} \qquad (4.39)$$

so that, in this way, the flow net geometry and the hydraulic conductivity plus head loss enable the total flow in the section to be computed directly. Naturally, in anisotropic media, the flow lines and equipotential lines are not orthogonal except in the case where the flow is parallel to one of the main directions, as J. Bear and G. Dagan indicated.[12] To calculate flows in this case, the boundaries of a flow section must be transformed so that an isotropic medium results. A typical case in alluvium is for $K_t > K_v$; all horizontal dimensions are reduced in a ratio K_v/K_t, creating a transformed section with an isotropic medium having an equivalent hydraulic conductivity

$$K' = \sqrt{K_t K_v} \qquad (4.40)$$

Using this transformed section, it is possible to draw the flow net and determine the flow rate. Once the flow net is defined, it can be converted back to the true anisotropic section by multiplying all horizontal dimensions by $\sqrt{K_t/K_v}$.

Some assertions are logically justifiable. Thus, since flow cannot cross impermeable boundaries, flow lines can be expected to parallel such a boundary. In addition, the water table in an unconfined aquifer will act as a boundary flow surface, if no flow transgresses it.

Under steady-state conditions, the elevation of any point on the water table equals the energy head, hence flow lines lie perpendicular to the water table contours. In the case of a confined aquifer, the flow lines are orthogonal to contours of the potentiometric surface.

Figure 4.6 shows how estimates of groundwater contours and direc-

THE MOVEMENT OF GROUNDWATER

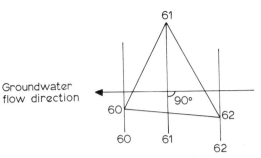

FIG. 4.6. Method of estimation of groundwater contours and the direction of groundwater flow from the depths-to-water-table data obtained at three wells.

tions of flow can be determined, if three groundwater elevations are known from wells in a particular area. Field measurements of static water level in these wells within a basin enable a water level contour map to be constructed a typical form of which is illustrated in Fig. 4.7.

Maps of this kind which show groundwater-level contour details and flow lines are very useful in locating new wells. Convex contours show

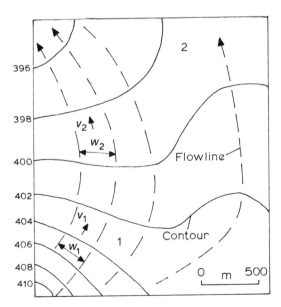

FIG. 4.7. Contour map of a groundwater surface with flow lines.

areas of groundwater recharge, while concave contours label areas of groundwater discharge; also, areas with favourable hydraulic conductivity can be inferred from the spacing of contours. Two adjacent flow lines can be regarded as impermeable boundaries in a sense, because there can be no flow across a flow line. With an aquifer of uniform thickness, then the flow at the sections 1 and 2 of Fig. 4.7 equals:

$$q = W_1 v_1 = W_2 v_2 \tag{4.41}$$

where v is the velocity and W the width of the flow section perpendicular to the flow. Taking Darcy's law into account:

$$W_1 K_1 i_1 = W_2 K_2 i_2 \tag{4.42}$$

which may be re-written:

$$\frac{K_1}{K_2} = \frac{W_2 i_2}{W_1 i_1} \tag{4.43}$$

where K is the hydraulic conductivity and i is the hydraulic gradient. Ratios of W_2 to W_1, also of i_2 to i_1, can be estimated from the water level contour map and, in the particular case of almost parallel flow, the above equation can be simplified to:

$$\frac{K_1}{K_2} = \frac{i_2}{i_1} \tag{4.44}$$

which it is possible to interpret as indicating, for an area of uniform groundwater flow, that areas with large spacing of contours and accompanying flat gradients possess higher hydraulic conductivities than those characterized by small spacing and accompanying steep gradients. Consequently, in the diagram Fig. 4.7, the chances of getting a productive water well are optimal near section 2.

As stated above, if flow does not cross it a water table functions as a groundwater boundary, but, in nature, flow may easily reach the water table. For instance, percolating water can do so and, in such cases, the flow lines will not parallel the surface as an impermeable boundary; in other words, a refraction effect arises.

Similarly, where flow transgresses a zone of hydraulic conductivity K_1 to another of K_2, there will be an accompanying change in flow direction. This can be derived from continuity considerations and expressed in terms of the two values of K. If a flow field such as that illustrated in Fig. 4.8 be considered, normal components of flow ap-

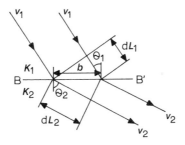

FIG. 4.8. Flow lines being refracted across a boundary, B–B′, separating two media of different hydraulic conductivities, K_1 and K_2.

proaching and quitting the boundary must be equal so that the normal velocities must be of a type permitting:

$$v_{n_1} = v_{n_2} \tag{4.45}$$

where v_n is the normal velocity concerned, or:

$$K_1 \frac{dh_1}{dL_1} \cos \theta_1 = K_2 \frac{dh_2}{dL_2} \cos \theta_2 \tag{4.46}$$

θ_1, θ_2 being the angles with the normal shown in the Figure. The distance, b, along the boundary between two adjacent flow lines is the same on both sides of this and:

$$b = \frac{dL_1}{\sin \theta_1} = \frac{dL_2}{\sin \theta_2} \tag{4.47}$$

which, re-writing, becomes;

$$dL_1 \sin \theta_2 = dL_2 \sin \theta_1 \tag{4.48}$$

If this is divided by eqn (4.46), since $dh_1 = dh_2$ between equipotential lines, the following may be obtained:

$$\frac{K_1}{K_2} = \frac{\tan \theta_1}{\tan \theta_2} \tag{4.49}$$

and, hence, for saturated flow passing from a medium of one hydraulic conductivity to another, a refraction effect again arises in the flow lines. It is of such a character that the ratio of the K values equals the ratio of the tangents of the angles which the flow lines make with the normal to the boundary. M. K. Hubbert in 1940 gave some excellent instances of refraction across layers of coarse and fine sand with a hydraulic conductivity ratio of 10.[13]

Whilst groundwater flow in shallow aquifers is sub-horizontal, more complex patterns of flow occur in the case of large-scale regional situations which necessarily involve very diverse field parameters, areas, magnitudes of recharge and discharge, topographical and stratigraphical considerations together with anisotropy. J. A. Toth showed that the variability of a water table can produce many flow patterns.[14,15] The approach has been extended to other subsurface boundary conditions.

Work on groundwater flow continually expands. Among some interesting recent papers are those of M. A. Sophocleous (on groundwater flow parameter estimation and quality modelling of the Equus beds aquifer in Kansas, USA),[16] and R. Singh et al. (on a groundwater model for simulating the rise of the water table under irrigated conditions).[17]

General flow equations entail examination of rectangular coordinates. Darcy's law may be written as:

$$v = -K\frac{\partial h}{\partial s} \qquad (4.50)$$

where s is the distance along the average flow direction. In the case of horizontal flow, generalization is feasible by considering the flow through a square element in an aquifer (Fig. 4.9). Inflow and outflow components in an x direction are given by:

$$q_{xi} = -T_x W \left(\frac{\partial h}{\partial x}\right)_i \qquad (4.51)$$

$$q_{x0} = -T_x W \left(\frac{\partial h}{\partial x}\right)_o \qquad (4.52)$$

where T_x is the transmissivity in the x direction, W is the length of a side of the square, and $(\partial h/\partial x)_i$ and $(\partial h/\partial x)_o$ define the hydraulic gradients at the entry and exit faces of the element, respectively. Of course, similar equations can be written for flow in the y direction. By continuity, the flow rate released from or stored in the element as a consequence of these flows equals:

$$(q_{xi} - q_{xo}) - (q_{yi} - q_{yo}) = -SW^2 \frac{\partial h}{\partial t} \qquad (4.53)$$

THE MOVEMENT OF GROUNDWATER

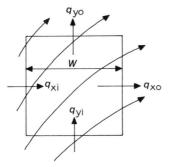

FIG. 4.9. Diagram illustrating horizontal flow through a square element of an aquifer.

where S is the storage coefficient. From this,

$$-T_x \frac{(\partial h/\partial x)_i - (\partial h/\partial x)_o}{W} - T_y \frac{(\partial h/\partial y)_1 - (\partial h/\partial y)_o}{W} = -S \frac{\partial h}{\partial t} \quad (4.54)$$

and, if the value of W is infinitesimal, the left-hand side derivatives become the second derivatives of h, hence:

$$T_x \frac{\partial^2 h}{\partial x^2} + T_y \frac{\partial^2 h}{\partial y^2} = S \frac{\partial h}{\partial t} \quad (4.55)$$

This is the general partial differential equation for unsteady flow of groundwater in a horizontal direction.

In the case involving three dimensions, utilizing an elemental cube instead of a square, then it can be demonstrated that:

$$K_x \frac{\partial^2 h}{\partial x^2} + K_y \frac{\partial^2 h}{\partial y^2} + K_z \frac{\partial^2 h}{\partial z^2} = S_s \frac{\partial h}{\partial t} \quad (4.56)$$

in which S_s is the specific storage (the volume of water which a unit volume of saturated aquifer releases from storage for a unit decline in hydraulic head).

In the case of steady flow, $\frac{\partial h}{\partial t} = 0$, so that:

$$K_x \frac{\partial^2 h}{\partial x^2} + K_y \frac{\partial^2 h}{\partial y^2} + K_z \frac{\partial^2 h}{\partial z^2} = 0 \quad (4.57)$$

and, in the case of isotropic homogeneous aquifers, this can be expressed as:

$$\frac{\partial^2 h}{\partial x^2} + \frac{\partial^2 h}{\partial y^2} + \frac{\partial^2 h}{\partial z^2} = 0 \qquad (4.58)$$

This is the Laplace equation for potential flow.[18-20]

The above argument can be applied to derive analytical solutions to particular groundwater flow problems, but in tackling these it is, unfortunately, always necessary to visualize an ideal aquifer and it is also a requisite of the operation that a similar attitude be taken to the boundary conditions of the flow system. Consequently, the results obtained can do no more than approximate the actual field conditions. To get a reasonably useful answer, therefore, it is practical to make some deviations from the ideal assumptions mentioned above. One of the most unreal of these is the isotropy and homogeneity of an aquifer; but, among those which may be modified, is infinite areal extent. Since some aquifers are vast in this respect, they can be taken as of infinite area for all practical purposes. If this cannot be done, the boundaries of an aquifer can be taken as either impermeable (e.g. aquicludes underlying or overlying the aquifer, faults, etc.) or permeable (e.g. the actual ground surface at which water may emerge from below). As well as the saturated flow, mostly sub-horizontal in nature, acting in the aquifer, unsaturated flow may also occur and involve downward movement of natural and artificial recharge together with the reverse, the upward, vertical movement of both evaporation and transpiration. In addition, there is the vertical downward movement of pollution from the surface of the ground and a degree of horizontal flow in the capillary zone lying above the water table. If the water table is lowered, then the capillary zone moves with it and water draining from above does so by a process of vertical percolation. Hence, the specific yield becomes an asymptotic function of time.

Such unsaturated flow in the unsaturated zone can be analyzed by the application of Darcy's law, but the unsaturated hydraulic conductivity, K_{us}, is a function of the water content and the tension. As a result of the presence of air in parts of the pore spaces, the cross-sections available for the flow of water diminish. Thus, K_{us} is invariably less than the saturated value of K.

There are hysteresis effects in the relations of K_{us} with water content and negative pressure, but empirical approximations show that the water

THE MOVEMENT OF GROUNDWATER 99

content fits the expression:

$$\frac{K_{us}}{K} = \left(\frac{S_s - S_0}{1 - S_0}\right)^3 \quad (4.59)$$

in which S_s is the degree of saturation and S_0 is the threshold saturation. K_{us} has a range of values. Thus, in the case where $S_s = S_0$, $K_{us} = 0$, and where $S_s = 1$ (100% saturation), $K_{us} = K$. Figure 4.10 illustrates eqn (4.59) graphically and shows the ratio of unsaturated to saturated hydraulic conductivity as a function of saturation; see also S. Irmay's paper published in 1954.[21]

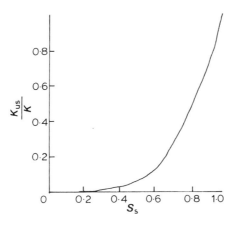

FIG. 4.10. The ratio of unsaturated to saturated hydraulic conductivity as a function of saturation.

The relationships between hydraulic conductivity and negative pressure are S-shaped, as illustrated in Fig. 4.11 which is after H. Bouwer (1964).[22] These relationships can be described approximately by a step function or by the expression:

$$\frac{K_{us}}{K} = \frac{a}{\frac{a}{b}(-h)^n + a} \quad (4.60)$$

a, b and n being constants varying with the sizes of particles of unconsolidated material and h the pressure head measured in cm. At atmospheric pressure, $h = 0$ and $K_{us} = K$. For different soils, the orders of magnitude of the constants are given in Table 4.1.[22]

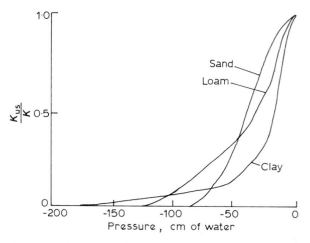

Fig. 4.11. Median relationships between soil-water pressure (tension) and hydraulic conductivity ratios. (Modified from ref. 22).

TABLE 4.1
CONSTANTS FOR EQUATION (4.60)

Materials	a	b	n
Medium sands	5×10^9	10^7	5
Fine sands, sandy loams	5×10^6	10^5	3
Loams and clay	5×10^3	5×10^3	2

The rate of lateral flow in the capillary zone above the water table decreases with the degree of saturation and hence the hydraulic conductivity. The fraction of flow effective above the water table can be calculated from an equivalent saturated thickness, as F. Mobasheri and M. Shahbazi indicated.[23]

In aquifers with substantial depths, the flow component above the water table may be regarded as negligible. However, this is not the case with shallow, unconfined aquifers.

Interestingly, R. D. Jackson and C. H. M. van Bavel in 1965 described as a survival technique the application of unsaturated flow under desert conditions.[24] Water is derived from the soil-water zone by a distillation method. A small, hemispherical hole is excavated, covered by a plastic film and this is held in position by placing a rock at its centre. When solar energy passes through the plastic it is absorbed by the soil, from which soil moisture evaporates and then condenses on the cooler plastic. Condensed water collects at the point of the inverted cone and can be drained into a centrally placed container (just under the stone). Yields of roughly one litre daily could be obtained and increased by 50% or so by adding cut plant material, especially cactus, peripherally under the plastic.

Some of the parameters discussed above, such as transmissivity and dispersivity, can be investigated using the isotopic hydrology approach. They will be discussed in Chapter 12.

4.4. VERTICAL MOVEMENTS OF GROUNDWATER

These involve rising and falling groundwater tables which can have serious and adverse effects.

The industrial decline in large cities in the western world may entail flooding threats caused by a drop in the quantities of groundwater used by factories. The phenomenon has been identified in the United Kingdom where, for two centuries or so, steel mills, breweries, etc., lowered the level of the water table through pumping in Liverpool, Manchester, Birmingham and London. However, reduced demand for their products has entailed diminution of water consumption and a corresponding rise in the water table to the point where foundations, tunnels and basements are endangered. British Rail has been forced to acquire additional pumping equipment for tunnels under the River Mersey and British Telecom is encountering difficulties with flooded exchanges and cable pipelines in Birmingham. Obviously, associated effects such as hydrostatic uplift, ground swelling and reduced bearing capacity are highly deleterious, particularly in London. Here, the existing groundwater level is about 55 m below the surface, but prior to the extensive pumping carried out in Victorian times it was probably only a couple of metres or so. If consumer demand continues falling, groundwater will rise. The Severn Trent Water Authority stated that such a rise has been established as taking place in Birmingham at an annual rate of approximately 25 cm. Similar effects have been noted in France and the Federal Republic of Germany as well as in New York City, where subway tunnels have been flooded.

An excellent instance of falling groundwater levels is provided by the historic Beacon Hill area of Boston, Massachusetts, where wood pilings supporting lower sections of this neighbourhood and the Back Bay section are rotting as a result. It was reported in July 1985 that the State Governor was asked to declare this part of the city a Federal disaster site. On one street alone, 17 homes have been condemned and a further 285 are under observation. Lower Beacon Hill and Back Bay were created between 1825 and 1870 by infilling part of Boston Harbour around an existing hill on which, fortunately, the State House was erected. Near the base of the Beacon Hill section, cracks 30 cm wide have split building foundation pilings, of which there are up to 200 under each residence. Although it was known as early as the 1920s that the groundwater level was dropping and the rate of decline is now some 0·6 m annually, the cause is unclear. However, if it persists, it is possible that the problem may spread beyond the presently affected region into other city neighbourhoods built on landfill such as the Fenway and the Boston University area.

REFERENCES

1. DARCY, H., 1856. *Les Fontaines Publiques de la Ville de Dijon*. V. Dalmont, Paris, 647 pp.
2. SCHWEITZER, S., 1977. On a possible extension of Darcy's law. *J. Hydrol.*, **22**, 29–34.
3. KRUMBEIN, W. C. and MONK, G. D., 1943. Permeability as a function of the size parameters of unconsolidated sand. *Trans. Amer. Inst. Min. Met. Engrs*, **151**, 153–63.
4. MASCH, F. D. and DENNY, K. J., 1966. Grain size distribution and its effects on the permeability of unconsolidated sands. *Water Res. Res., Bull.*, **2**, 665–77.
5. HALEVY, E. *et al.*, 1967. Borehole dilution techniques — a critical review. *Isotopes in Hydrology*, IAEA, Vienna, pp. 531–64.
6. BOAST, C. W. and KIRKHAM, D., 1971. Auger hole seepage theory. *Soil Sci. Soc. Amer. Proc.*, **35**, 365–73.
7. MORRIS, D. A. and JOHNSON, A. I., 1967. *Summary of Hydrologic and Physical Properties of Rock and Soil Materials, as Analyzed by the Hydrologic Laboratory of the US Geological Survey, 1948–60*. US Geol. Surv., Water Supply Paper 1839-D, 42 pp.
8. AIREY, P. L., BENTLEY, H., CALF, G. E., DAVIS, S. N., ELMORE, D., GOVE, H., HABERMEHL, M. A., PHILLIPS, F., SMITH, J. and TORGERSON, T., 1983. Isotope hydrology of the Great Artesian Basin, Australia. *Intl Conf. on Groundwater and Man*, Sydney, pp. 1–10.
9. CALF, G. E. and HABERMEHL, M. A., 1983. Isotope hydrology and hydrochemistry of the Great Artesian Basin, Australia. *IAEA/UNESCO Intl*

Symp. on Isotope Hydrology in Water Resources Development, IAEA, Vienna, SM-270/61.
10. KAZMANN, R. G., 1980. Review of Groundwater Hydraulics by V. Halek and J. Suce, 1980. J. Hydrol., **48**, 375–6.
11. CEDERGREN, H. R., 1977. Seepage, Drainage and Flow Nets. 2nd edn, John Wiley, New York, 534 pp.
12. BEAR, J. and DAGAN, G., 1965. The relationship between solutions of flow problems in isotropic and anisotropic soils. J. Hydrol., **3**, 88–96.
13. HUBBERT, M. K., 1940. The theory of ground-water motion. J. Geology, **48**, 785–944.
14. TOTH, J. A., 1962. A theory of groundwater motion in small drainage basins in Central Alberta, Canada. J. Geophys. Res., **67**, 4375–87.
15. TOTH, J. A., 1963. A theoretical analysis of groundwater flow in small drainage basins. J. Geophys. Res., **68**, 4795–812.
16. SOPHOCLEOUS, M. A., 1984. Groundwater flow parameter estimation and quality modeling of the Equus beds aquifer in Kansas, USA. J. Hydrol., **69**, 197–222.
17. SINGH, R., SONDHI, S. K., SINGH, J. and KUMAR, R., 1984. A groundwater model for simulating the rise of water table under irrigated conditions. J. Hydrol., **71**, 165–79.
18. HUBBERT, M. K., 1940. The theory of ground-water motion. J. Geol., **48**, 785–944.
19. HUBBERT, M. K., 1956. Darcy's law and the field equations of the flow of underground fluids. Trans. Amer. Inst. Min. Met. Engrs, **207**, 222–39.
20. TODD, D. K., 1980. Groundwater Hydrology, John Wiley, New York, 535 pp., p. 100.
21. IRMAY, S., 1954. On the hydraulic conductivity of unsaturated soils. Trans. Amer. Geophys. Union, **35**, 463–7.
22. BOUWER, H., 1964. Unsaturated flow in ground-water hydraulics. J. Hydraulics Divn, Amer. Soc. Civ. Engrs, **90**, HY5, 121–44.
23. MOBASHERI, F. and SHAHBAZI, M., 1969. Steady-state lateral movement of water through the unsaturated zone of an unconfined aquifer. Ground Water, **7**, 6, 28–34.
24. JACKSON, R. D. and VAN BAVEL, C. H. M., 1965. Solar distillation of water from soil and plant materials: a simple desert survival technique. Science, **149**, 1377–9.

CHAPTER 5

Groundwater Hydraulics

5.1. STEADY FLOW

This is a type of flow during which there is no change with time, although the actual conditions of flow are not the same in confined and unconfined aquifers. In the simplest case, the flow may be assumed to be unidirectional in character.

With confined aquifers, if the groundwater velocity is v in a direction x and the aquifer is of uniform thickness, then the equation for an elemental cube, eqn (4.56), if this flow is steady, can be reduced to:

$$\frac{\partial^2 h}{\partial x^2} = 0 \qquad (5.1)$$

the solution for which is:

$$h = C_1 x + C_2 \qquad (5.2)$$

where h is the head above a given datum, C_1 and C_2 being constants of integration. If it is assumed that $h=0$ when $x=0$ and $h/x = -(v/K)$ from Darcy's law, then:

$$h = -\frac{vx}{K} \qquad (5.3)$$

This is a statement of the fact that head decreases linearly.

With unconfined aquifers, no direct solution of the Laplace equation is possible. The problem is that the water table in the two-dimensional case represents a flow line, its shape determining the flow distribution. Simultaneously, the flow distribution governs the shape of the water table. In order to solve this, J. Dupuit in 1863 assumed that the flow velocity is proportional to the tangent of the hydraulic gradient, and that it is in a horizontal sense as well as being uniform everywhere in any vertical section.[1] The assumptions give a solution, but limit the applicability. In the case of unidirectional flow, the discharge per unit width, q, at any vertical section, is given by:

$$q = -Kh\frac{dh}{dx} \qquad (5.4)$$

where K is hydraulic conductivity, x the flow direction and h the height of the water table above an impervious base. On integration this becomes:

$$qx = -\frac{K}{2}h^2 + C \qquad (5.5)$$

and, if $h = h_0$ where $x = 0$, the Dupuit equation becomes:

$$q = \frac{K}{2x}(h_0^2 - h^2) \qquad (5.6)$$

indicating that the water table has a parabolic form.

In the case of flow between two fixed water bodies with constant heads h_0 and h_1, the slope of the water table at the upstream boundary of the aquifer is given by:

$$\frac{dh}{dx} = -\frac{q}{Kh_0} \qquad (5.7)$$

neglecting the capillary zone.

The boundary $h = h_0$ is an equipotential line since the fluid potential in a body of water is constant. Therefore, the water table is horizontal at this section, which is inconsistent with eqn (5.7). In the flow direction, the above-mentioned water table of parabolic form increases in slope. As this occurs, the Dupuit assumptions approximate more and more poorly to the actual flow, i.e. in the flow direction the water table in nature deviates more and more from the computed position, which lies below the real one. This can be explained through the assumed horizontality of the Dupuit flows; the true velocities of any given magnitude possess a downward vertical component, so there is a greater saturated thickness for the same discharge. At the downstream boundary, because there is no way that a consistent flow pattern can connect the water table directly to a downstream free water surface, there is a discontinuity in flow forms.

The water table approaches the boundary tangentially above the water body surface and a seepage face is produced. It is clear, therefore, that a parabolic form is not followed by the water table, but the equation is capable of accurately determining q or K for given boundary heads.

Base flow to a stream may occur, constituting groundwater recharge. It can be calculated by using the above analysis of unidirectional flow in an unconfined aquifer.

The subject of wells will be discussed in more detail later (Chapter 6), but here it is important to mention steady radial flow into a well. Pumping of a well will remove water from the aquifer in its immediate vicinity and will also lower either the water table or the piezometric surface. The distance to which these descend is referred to as the drawdown at a given point and an appropriate curve can be drawn to illustrate its variation with distance from the well (Fig. 5.1). Of course, in three dimensions the drawdown curve becomes an inverted conic shape, known as the cone of depression, and the outer limit of this cone delimits the area of influence of the well.

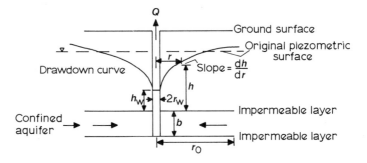

FIG. 5.1. Drawdown curve for steady radial flow to a well which penetrates a confined aquifer: b is the thickness of the confined aquifer.

In a confined aquifer containing a well which completely penetrates it, a radial flow equation can be constructed if the flow is assumed to be two-dimensional to the said well located on an island of circular shape. For this, the aquifer must be regarded as isotropic and homogeneous. The flow is taken as everywhere horizontal, so the assumptions of Dupuit will be applicable without any error. If plane polar coordinates are utilized with the well as the origin, discharge from it, Q, at any distance, r, equals:

$$Q = Av = -2\pi rbK \frac{dh}{dr} \qquad (5.8)$$

for steady radial flow to the well. If this is rearranged and integrated, taking into account the boundary conditions of the well, $h = h_w$ and $r = r_w$ and, at the edge of the island, $h = h_0$ and $r = r_0$, then:

$$h_0 - h_w = \frac{Q}{2\pi Kb} \ln \frac{r_0}{r_w} \qquad (5.9)$$

Alternatively,

$$Q = 2\pi Kb \frac{h_0 - h_w}{\ln(r/r_w)} \qquad (5.10)$$

with the negative sign neglected. In the more general case of a well which penetrates an extensive confined aquifer, there is no external limit for r and, hence, from eqn (5.10) at any given value of r:

$$Q = 2\pi Kb \frac{h - h_w}{\ln(r/r_w)} \qquad (5.11)$$

From this, it may be observed that h increases indefinitely with increase in r. It is possible, therefore, to state that theoretically there can be no steady radial flow in a very large aquifer, as a result of the fact that the cone of depression goes on enlarging with time. Nevertheless, in practice, h will approach h_0 as distance from the well increases and the drawdown varies with the logarithm of the distance from the well.

Equation (5.11) is termed the equilibrium equation or the Thiem equation, after G. Thiem who presented it in 1906.[2] It is valuable in that it permits the hydraulic conductivity or the transmissivity of a confined aquifer from a pumped well to be determined. Any two points will define the logarithmic drawdown curve; the approach entails measuring drawdowns in two observation wells at different distances from a well which is being pumped at a constant rate. It might be considered possible to utilize h_w at the pumped well itself, but here there may be losses incurred by flow through the well screen and errors may be induced inside the well. Thus, h_w must be avoided. The appropriate equation for transmissivity, T, is:

$$T = Kb = \frac{Q}{2\pi(h_2 - h_1)} \ln \frac{r_2}{r_1} \qquad (5.12)$$

where r_1 and r_2 are the distances and h_1 and h_2 are the heads of the respective observation wells. In practice, the drawdown s is measured rather than the head h; hence eqn (4.20) can be re-written thus:

$$T = \frac{Q}{2\pi(s_1 - s_2)} \ln \frac{r_2}{r_1} \qquad (5.13)$$

where s_1 and s_2 are as shown in Fig. 5.2. It is necessary for pumping to continue uniformly for long enough to approximate a steady-state condition, one in which the drawdown does not alter substantially with time, in order that this equation can be applied. The difference in drawdown becomes practically constant while both s_1 and s_2 are still increasing, so this can be done within a few days of pumping. It is important to locate the observation wells close enough to the pumping well so that drawdowns in them are measurable. The entire derivation, of course, assumes that the aquifer is isotropic and homogeneous, uniform in thickness and of infinite areal extent. It is taken that the well achieved complete penetration of the aquifer, also that the original piezometric surface is practically horizontal.

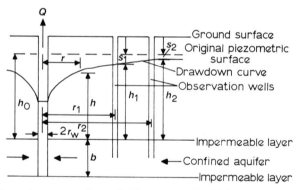

FIG. 5.2. Drawdown curve for radial flow to a well which penetrates a very large confined aquifer.

In the case of an unconfined aquifer, the Dupuit assumptions can be applied again. Figure 5.3 illustrates the well once more completely penetrating the aquifer right down to its horizontal base, and there is a concentric boundary of constant heads surrounding it. The discharge Q is given by:

$$Q = -2\pi Kh \frac{dh}{dr} \qquad (5.14)$$

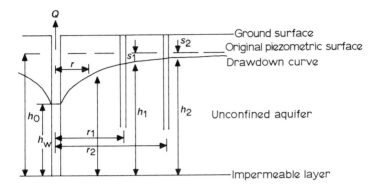

FIG. 5.3. Drawdown curve for radial flow to a well which penetrates an unconfined aquifer.

which, integrated between limits $h=h_w$ at $r=r_w$ (also $h=h_0$ at $r=r_0$) produces:

$$Q = \pi K \frac{h_0^2 - h_w^2}{\ln(r_0/r_w)} \qquad (5.15)$$

and, converting to heads and radii at two observation wells:

$$Q = \pi K \frac{h_2^2 - h_1^2}{\ln(r_2/r_1)} \qquad (5.16)$$

which, re-written to solve for hydraulic conductivity, produces:

$$K = \frac{Q}{\pi(h_2^2 - h_1^2)} \ln \frac{r_2}{r_1} \qquad (5.17)$$

Equation (5.17) does not describe the drawdown curve near the well in a precise manner because the large vertical flow components oppose the assumptions of Dupuit, but, in spite of this, it is found that the estimates of K obtained for any given head are good. But drawdowns must be small in relation to the saturated thickness of the unconfined aquifer. From eqn (5.17), an equation for transmissivity T is derivable:

$$T \cong K \frac{h_1 + h_2}{2} \qquad (5.18)$$

In those instances where the drawdowns are substantial, then the heads h_1 and h_2 in eqn (5.17) may be substituted by $(h_0 - s_1)$ and $(h_0 - s_2)$

respectively and thereafter the transmissivity for the total thickness becomes:

$$T = Kh_0 = \frac{Q}{2\pi\left[\left(s_1 - \frac{s_1^2}{2h_0}\right) - \left(s_2 - \frac{s_2^2}{2h_0}\right)\right]} \ln \frac{r_2}{r_1} \quad (5.19)$$

Finally, the case of an unconfined aquifer with a uniform recharge is illustrated in Fig. 5.4. The actual rate of the uniform recharge, which may be termed W, originates from water ingress derived from various sources such as rainfall or excess irrigation water. Flow, Q, towards the well increases as the well is approached and reaches a maximum which may

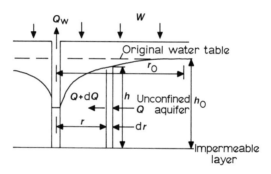

FIG. 5.4. Drawdown curve for steady flow to a well which penetrates an unconfined aquifer which is uniformly recharged.

be termed Q_w at the well. The increase in flow, dQ, through a cylinder with a thickness dr and a radius r originates from the recharged water which comes in from above, and so:

$$dQ = -2\pi r\, dr\, W \quad (5.20)$$

which, after integration, becomes:

$$Q = -\pi r^2 W + C$$

However, at the well, $r \to 0$ and $Q = Q_w$; hence:

$$Q = -\pi r^2 W + Q_w \quad (5.21)$$

and, if this equation be substituted into $Q = -2\pi Kh\, dh/dr$ (eqn (5.14)), the discharge in a well completely penetrating an unconfined aquifer as shown in Fig. 5.3, then:

$$-2\pi r K h \frac{dh}{dr} = -\pi r^2 W + Q_w \quad (5.22)$$

Integrating, since $h = h_0$ at $r = r_0$, the equation for the drawdown curve can be derived:

$$h_0^2 - h^2 = \frac{W}{2K}(r^2 - r_0^2) + \frac{Q_w}{\pi K} \ln \frac{r_0}{r} \quad (5.23)$$

When $r = r_0$, $Q = 0$; hence:

$$Q_w = \pi r_0^2 W \quad (5.24)$$

and it becomes apparent that the total flow of the well is equal to the recharge within the circle delimited by the radius of influence and, conversely, the radius of influence is a function of the pumping of the well and the rate of recharge. This causes a steady-state drawdown. It must be remembered, however, that the analysis is based on the assumption that there is an idealized circular outer boundary with a constant head and no flow: of course, such conditions are hardly ever found in nature.

In the instances cited above, drawdown curves depend upon the assumption that the groundwater surface is horizontal at the beginning, but this is not always so. Thus, wells may pump from an aquifer possessing a uniform flow field indicated by a uniformly inclined piezometric surface or water table. Here the circular area of influence which is characteristic of a radial pattern of flow is distorted. For most relatively flat slopes, however, it is permissible to apply the Dupuit radial flow equation without any significant error arising. Where wells pump from an area with a sloping hydraulic gradient, the hydraulic conductivity K can be obtained from:

$$K = \frac{2Q}{\pi r (h_u + h_d)(i_u + i_d)} \quad (5.25)$$

for an unconfined aquifer where Q is the pumping rate, h_u and h_d are the saturated thicknesses and i_u and i_d are the water table slopes at a distance of r upstream and downstream, respectively, from the well. In the case of a confined aquifer, the water table slopes become piezometric level slopes and $(h_u + h_d)$ is replaced by $2b$, where b is the thickness of the aquifer.

5.2. UNSTEADY FLOW

As well as the steady flow conditions discussed in Section 5.1, there is also the situation in which unsteady radial flow takes place in a confined aquifer. If an extensive body of this type is penetrated by a well which is pumped at a constant rate, the influence of the discharge which occurs extends outwards with time. The rate of decline of head multiplied by the storage coefficient summed over the whole area of influence gives the discharge. The relevant water derives from a decline in storage in the aquifer; therefore the head continues to decline as long as the aquifer is effectively infinite and, in consequence, unsteady (transient) flow operates. The rate of decline decreases continuously as the area of influence enlarges.

Examining the matter of radial coordinates, for a homogeneous and isotropic aquifer it is possible to show that the general partial differential equation for unsteady flow of groundwater in a horizontal direction (eqn (4.55)) is equivalent to:

$$\frac{\partial^2 h}{\partial r^2} + \frac{1}{r}\frac{h}{r} = \frac{S}{T}\frac{\partial h}{\partial t} \tag{5.26}$$

where r is the radial coordinate from the well. For steady flow, the appropriate expression is:

$$\frac{\partial h}{\partial r^2} + \frac{1}{r}\frac{h}{r} = 0 \tag{5.27}$$

There is an appropriate differential equation in plane polar coordinates:

$$\frac{\partial^2 h}{\partial r^2} + \frac{1}{r}\frac{h}{r} = \frac{S}{T}\frac{h}{t} \tag{5.28}$$

where h is the head, r is the radial distance from the pumped well, S is the storage coefficient, T is the transmissivity and t is the time elapsed since the commencement of pumping. C. V. Theis[3] in 1935 gave a solution for this equation which is based upon the analogy between the flow of groundwater and heat conduction. The well is assumed to be replaced by a 'mathematical sink' with constant strength. Imposing boundary conditions of $h = h_0$ for $t = 0$, and $h \to h_0$ as $r \to \infty$ for $t = 0$, the following expression may be derived:

$$s = \frac{Q}{4\pi T}\int_u^\infty \frac{1}{u}\exp(-u)\,du \tag{5.29}$$

GROUNDWATER HYDRAULICS   113

where s is the drawdown, Q is the constant well discharge and

$$u = \frac{r^2 S}{4Tt} \qquad (5.30)$$

Equation (5.29) is the non-equilibrium or Theis equation. The integral is a function of the lower limit u, and is termed an exponential integral; it can be expanded as a convergent series so that the eqn (5.29) becomes:

$$s = \frac{Q}{4\pi T}\left[-0.5772 - \ln u + u - \frac{u^2}{2\times 2!} - \frac{u^3}{3\times 3!} - \frac{u^4}{4\times 4!} + \cdots \right] \qquad (5.31)$$

The Theis equation is valuable because it allows determination of the formation constants S and T through pumping tests in wells. In practice, it is very widely employed and preferred over the equilibrium equations since it enables a value for S to be obtained, and it requires only one observation well and less pumping time. In addition, steady-state flow conditions do not have to be assumed.

There are several assumptions which have to be mentioned. Firstly, as usual, the aquifer is taken to be of infinite areal extent, isotropic, homogeneous and of uniform thickness. Secondly, the piezometric surface is horizontal prior to commencement of pumping. Thirdly, the well is pumped at a constant rate of discharge. Fourthly, the pumped well penetrates the whole aquifer, flow being everywhere horizontal within that body to the well. Fifthly, the diameter of the well is infinitesimal so that storage in it may be neglected. Sixthly, water removed from storage is discharged instantaneously with decline of head.

These assumptions are never wholly justified in nature and there are mathematical difficulties involved in applying either version of the Theis equation. Consequently, simpler solutions have been developed by Theis himself,[3] H. H. Cooper Jr and C. E. Jacob in 1946[4] and V. T. Chow in 1952.[5] These will be described briefly.

The Theis solution simplifies eqn (5.29) to:

$$s = \left(\frac{Q}{4\pi T}\right) W(u) \qquad (5.32)$$

where $W(u)$, the so-called well function, is a symbolic representation of the exponential integral. Equation (5.30) may be re-arranged to:

$$\frac{r^2}{t} = \left(\frac{4T}{S}\right) u \qquad (5.33)$$

and it is apparent that the relationship between $W(u)$ and u is like that between s and r^2/t because the terms in parentheses in the two equations are constants. In view of this, an approximate solution for S and T was proposed, based upon a graphical method of superposition.

A type curve is prepared, i.e. a plot on logarithmic paper of $W(u)$ against u. Values of drawdowns are plotted against values of r^2/t on logarithmic paper of the same size as that used for the type curve. Thereafter, the observed time-drawdown data are superimposed on the type curve, the coordinate axes of the two curves being kept parallel, adjustment then taking place empirically until most of the plotted points of the data observed fall on one segment of the type curve. A suitable point is selected and its coordinates are recorded. Thus, with values for $W(u), u, s$ and r^2/t available, S and T can be derived from the simplified equations.

L. K. Wenzel gave appropriate tables with values of $W(u)$ and u. The units of measurement in the USA and UK used to be as follows: drawdown was given in feet, discharge in gallons per minute, distance r in feet, time in minutes, hours or days and transmissivity in gallons per day per foot.[6] Metric measurements have largely superseded this practice, however.

In devising the Cooper–Jacob method, these investigators noted that for small values of r and large values of t, u is small. Hence the series terms in the expanded version of the Theis equation (eqn (5.31)) can be neglected beyond the first two. Therefore, the drawdown may be expressed as:

$$s = \frac{Q}{4\pi T}\left(-0.5772 - \ln\frac{r^2 S}{4Tt}\right) \qquad (5.34)$$

which can be re-written:

$$s = \frac{2.30 Q}{4\pi T} \log \frac{2.25 Tt}{r^2 S} \qquad (5.35)$$

and so a plot of drawdown s against the logarithm of t is a straight line. If this line is projected to $s=0$, where $t=t_0$, then:

$$0 = \frac{2.30 Q}{4\pi T} \log \frac{2.25 T t_0}{r^2 S} \qquad (5.36)$$

Thus:

$$\frac{2{\cdot}25Tt_0}{r^2 S} = 1 \qquad (5.37)$$

and, finally,

$$S = \frac{2{\cdot}25Tt_0}{r^2} \qquad (5.38)$$

A value for T can be obtained by noting that, where $t/t_0 = 10$, $\log(t/t_0) = 1$. Hence, if s is replaced by Δs, where Δs is the drawdown difference per log cycle of t, the following expression emerges:

$$T = \frac{2{\cdot}30Q}{4\pi\Delta s} \qquad (5.39)$$

Now it is possible to solve for both S and T, but the straight line approximation must be limited to small values of u (i.e. $u < 0{\cdot}01$) as in this way large errors can be evaded.

The Chow method is rather different and, in fact, has two advantages, namely that no curve fitting is necessary and that it is of unrestricted applicability. However, drawdown measurements have to be made in an observation well near to the well being pumped, the data then being plotted on semi-logarithmic paper as in the previous method. On the plotted curve, an arbitrary point is selected and the coordinates t and s are noted. Then, a tangent to the curve is drawn at the selected point and the drawdown difference Δs (in feet) determined per log cycle of time. $F(u)$ is calculated from:

$$F(u) = \frac{s}{\Delta s} \qquad (5.40)$$

and the corresponding values of $W(u)$ and u can be obtained from a table given by Chow.[5] Finally, it is possible to obtain T and S from the relevant preceding equations.

All these methods are extremely useful and they are widely utilized in practice.

Let us consider now what happens in a water well when the pumping test is ended. Clearly, the water level will rise (as it will do also in any adjacent observation wells), a phenomenon known as the recovery of groundwater levels. Drawdowns below the static water level which existed prior to pumping during the recovery period, i.e. residual drawdowns, are measured because analysis of the information so obtained gives a means of calculating transmissivity and an independent check upon the value derived from pumping tests. This procedure is much cheaper than a pumping test and in addition does not require attention being given to an observation well.

When a well is shut off after pumping for a known time interval, the subsequent drawdown will be the same as if the pumping were continued and a hypothetical recharge well with the same rate of flow were superimposed on the discharging well at the moment when the discharge is terminated. Theis showed that the residual drawdown s' can be obtained from the following expression:[3]

$$s' = \frac{Q}{4\pi T}\left[W(u) - W(u')\right] \quad (5.41)$$

where:

$$u = \frac{r^2 S}{4Tt} \text{ and } u' = \frac{r^2 S}{4Tt'}$$

Figure 5.5 defines t and t'. Where r is small and t' large, the well functions can be approximated from the first two terms of the expanded form of the Theis equation (eqn (5.31)), so eqn (5.41) becomes:

$$s' = \frac{2 \cdot 30 Q}{4\pi T} \log \frac{t}{t'} \quad (5.42)$$

so that a plot of residual drawdown s' against the logarithm of t/t' forms

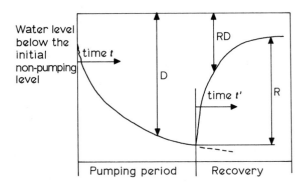

FIG. 5.5. The drawdown and recovery curves in an observation well near a pumping well: D = drawdown, RD = residual drawdown, R = recovery.

a straight line. The slope of this equals $2\cdot 30Q/4\pi T$; hence, for $\Delta s'$, the residual drawdown per log cycle of t/t', the transmissivity is given by:

$$T = \frac{2\cdot 30Q}{4\pi \Delta s'} \quad (5.43)$$

It is not possible to obtain a value for S using the recovery test method.

5.3. UNSTEADY RADIAL FLOW

This may be investigated in an unconfined aquifer and also in a leaky aquifer.

5.3.1. Unconfined Aquifers

The methods for the non-equilibrium equation which are applied in pump tests to confined aquifers can be extended to unconfined ones as long as the fundamental assumptions are satisfied and, generally, if the drawdown is small in relation to the saturated thickness, it is feasible to obtain reasonable approximations.

In unconfined aquifers, in the case of significant drawdowns, the assumption that water released from storage is discharged instantaneously with the decline of head is often untrue. Information from pump tests demonstrates that, as the water table declines, the drainage under gravity of water from the zone of aeration is a variable process termed delayed yield.[7] Time–drawdown curves for analyzing pumping test data from unconfined aquifers, taking into account delayed yield, were developed by N. S. Boulton,[8-10] and are shown in Fig. 5.6. Interpretation of any one curve may be considered in three time intervals. The first, occupying seconds to minutes, involves the release of water practically instantaneously from storage by aquifer compaction and the expansion of trapped air. This part of the curve can be fitted by a type curve with a storage coefficient equivalent to that of a confined aquifer. The second time interval shows a slope flattening which is attributed to drainage replenishment under gravity from the pore space above the cone of depression. In the third, lasting from several minutes to a number of days, equilibrium is approached between gravity drainage and the rate of decline of the water table. This state of affairs can be fitted by a type curve with a storage coefficient for an unconfined aquifer. The storage coefficient obtainable from the third time interval of the curve is

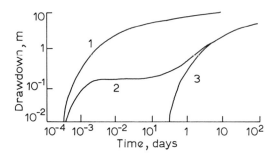

FIG. 5.6. Type curves of drawdown against time to show the effect of delayed yield for pumping tests in unconfined aquifers. 1, Theis type curve for the initial confined condition; 2, curve for delayed yield; 3, Theis type curve for the later unconfined condition.

the specific yield; pumping should go long enough to define this adequately, after which, by applying one of the methods of solution described earlier, a value for S can be derived.

T. A. Prickett[11] has given an empirical method for estimating the minimum length of a pumping test in an unconfined aquifer, relating the delay index t_d to the nature of the materials through which gravity drainage takes place and assessing t_d from these materials. Thereafter, if the distance, r, between pumping and observation wells is known and S and T are estimated, the approximate minimum pumping time, t_{min}, can be calculated.

There is an alternative approach, i.e. to make sure that the time taken by the pump test always exceeds the guidelines suggested by the US Bureau of Reclamation[12] and summarized in Table 5.1.

TABLE 5.1
GUIDELINES FOR MINIMUM PUMPING TIMES[12]

Main aquifer material	Minimum pumping time (h)
Silt, clay	170
Fine sand	30
Medium sand, coarser materials	4

5.3.2. Leaky Aquifers

When a leaky aquifer is pumped, water is taken from it and, in addition, from the saturated part of the overlying semi-pervious layer. This layer possesses a hydraulic gradient through lowering of the piezometric head in the aquifer below it. The result is that groundwater moves down vertically into the aquifer, the actual amount being proportional to the difference between the water table and the piezometric head, as R. L. Cooley and C. M. Case indicated in 1973.[13]

Steady-state flow to a well in a leaky aquifer occurs because of the recharge through such a semi-pervious aquitard and an equilibrium is reached when the discharge rate of the pump equals the recharge rate of vertical flow into the aquifer, assuming that the water table remains constant.[14,15]

When pumping commences in a well in a leaky aquifer, then the drawdown of the piezometric surface is given by:

$$s = \frac{Q}{4\pi T} W(u, r/B) \quad (5.44)$$

where s is the drawdown, Q is the well discharge and r is defined in Fig. 5.7, u being $r^2 S/4Tt$; r/B is given by:

$$\frac{r}{B} = \frac{r}{\sqrt{T/(K'/b')}} \quad (5.45)$$

T being the transmissivity and K' the vertical hydraulic conductivity of the semi-pervious layer. M. S. Hantush has provided values of the

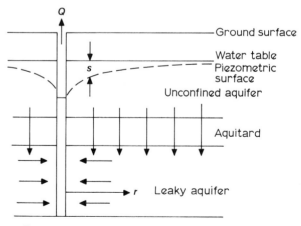

FIG. 5.7. A well pumping from a leaky aquifer.

function $W(u, r/B)$.[16] Equation (5.44) has the form of the Theis equation and, since in a confined aquifer K' is practically zero whereby B approaches infinity and r/B becomes zero, this eqn (5.44) reduces to the Theis equation.

5.4. THE METHOD OF IMAGES

If a well is being pumped near the edge of an aquifer, it is impossible to assume that the aquifer has an infinite areal extent. It is appropriate to postulate imaginary (image) wells, enabling the finite aquifer to be transformed into an infinite one to which the various methods of solution described earlier can be applied. Consider a well near a perennial stream: it is required to ascertain the head at any point influenced by pumping at a constant rate, Q, and to ascertain how much of the pumpage is derived from the stream. The real system is illustrated in Fig. 5.8 and Fig. 5.9 depicts an equivalent imaginary system in which an imaginary recharge well, i.e. one through which water enters an aquifer, is sited opposite to, and at the same distance as the real well from, the stream. The image well is taken as operating simultaneously and at the same rate as the real well; thus, the drawdown of head along the stream line is precisely equal and opposite to the increase of head around the recharge well. In this manner a constant head is established along the stream forming the boundary of the aquifer. Figure 5.10 shows a stylized plan view of the consequent flow net, a single equipotential line coinciding with the stream axis. The resulting asymmetrical drawdown of the actual well is given at any point by the algebraic sum of the drawdown of the real well and the buildup of the recharge well, just as if both were in an infinite aquifer. D. K. Todd in 1980 gave a very interesting set of

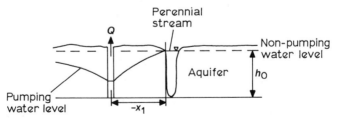

FIG. 5.8. Cross-section of a discharging well located adjacent to a perennial stream.

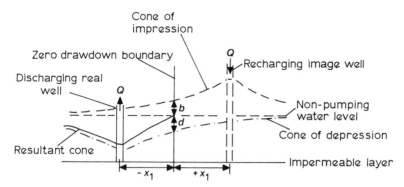

FIG. 5.9. Cross-section of a system equivalent to that of Fig. 5.8, but in an aquifer of infinite areal extent: d = drawdown component of the real well, b = buildup component of the image well.

hydraulically equivalent aquifer systems bounded by streams with various configurations including unidirectional, rectangular and parallel.[17] In the case of the single stream with a real well on one side and an image recharging well on the other, he noted that the steady-state drawdown at any point (x, y) is given by:

$$s = \frac{Q}{4\pi T} \ln \frac{(x+x_w)^2 + (y-y_w)^2}{(x-x_w)^2 + (y-y_w)^2} \tag{5.46}$$

where (x_w, y_w) are the coordinates of the pumped well. In his diagram, the image wells extend to infinity, but it is only necessary to include pairs of image wells closest to the real well because others will have a negligible effect on the drawdown. If Q_s is the flow from a stream, flow distribution analysis shows that Q_s/Q is the fraction of well discharge obtained from it. R. E. Glover and G. G. Balmer gave a graph which much simplified calculations, for determining the proportion of well discharge contributed by a nearby stream.[18]

In addition to the case of wells adjacent to streams, the method of images is applicable to well flow near other types of boundary. The approach is always the same, namely to replace the real situation by equivalent hydraulic systems, including imaginary wells, which allow solutions to be derived from those equations applicable solely to extensive aquifers. Some of the boundaries which may be encountered are single or double impermeable beds, i.e. impermeability on one or two sides of an aquifer, but there are many other possibilities in nature. J. E.

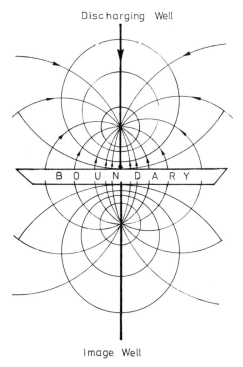

Fig. 5.10. Diagram of a flow net for a discharging real well and a recharging image well with the axis of an associated perennial stream shown as a stylized boundary layer. Along this boundary there is a constant head because, as the image well operates simultaneously with and at the same rate as the real well, the increase of head around the former exactly cancels out the decrease of head, the drawdown, along the line of the stream. The result is the constant head along the stream: this is equivalent to the constant elevation of the stream actually forming the boundary of the aquifer. The axis of the stream coincides with a single equipotential line bisecting the boundary in the diagram. The circles represent equipotential lines and the curves converging on the real and image wells represent flow lines.

Ferris *et al.* in 1962 described a number in a paper on the theory of aquifer tests.[19]

The T in the above equations refers to confined aquifers and, to adapt for unconfined aquifers, s is replaced by $s'' = s - s^2/2h_0$, where h_0 is the initial saturated aquifer thickness. Storage coefficients cannot be derived from steady-state boundary equations.

(a)

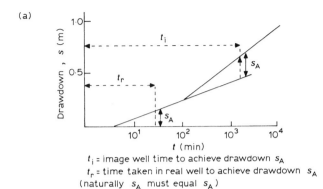

t_i = image well time to achieve drawdown s_A
t_r = time taken in real well to achieve drawdown s_A
(naturally s_A must equal s_A)

(b)

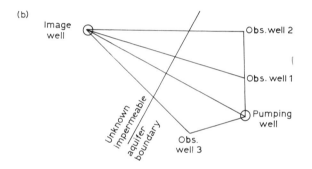

Fig. 5.11. Image well location.

It is worth mentioning that, whereas streams (permeable aquifer boundaries) are usually visible near a pumping well, some impermeable sub-surface boundaries such as faults, are not. Actually, these may be located by analyzing pumping data,[12,19] and their orientation may be assessed. In the Cooper–Jacob method (Section 5.2), the slope of the straight line on semi-logarithmic paper depends on the pumping rate and the transmissivity. In the case where an impermeable boundary is present, the rate of drawdown in an observation well doubles under the effect of an image pumping well. The image well location is determinable by fitting straight lines through the two legs of the data; an arbitrary drawdown, S_A, is chosen and a time, t_r, for this actually to take place under the influence of the real well is measured (Fig. 5.11a). In the same manner, a time, t_i, for the same drawdown to be produced by the image well is defined. As the distance r_r between the real well and the observation well is known, the distance r_i to the image well can be determined from:

$$\frac{r_i^2}{t_i} = \frac{r_r^2}{t_r} \tag{5.47}$$

This distance, r_i, marks the radius of a circle on which the image well is to be found. The image well may then be located by making measurements in a couple of additional observation wells so as to determine the intersection of three arcs. The boundary sought will be located at the mid-point of and perpendicular to a line which connects the real and the image wells (Fig. 5.11b).

5.5. INTERFERING WELLS AND THOSE ONLY PARTIALLY PENETRATING

If two pumping wells are too close, one will interfere with the other, i.e. their cones of depression will overlap because of the increased drawdown and pumping lift which is created. In a well field, as a consequence of the principle of superposition, at any point within the area of influence, the drawdown caused by the discharge of several of the wells can be found because it equals the sum of the drawdowns caused by each individual well. Hence,

$$s_r = s_a + s_b + \cdots s_n \tag{5.48}$$

where s_r is the total drawdown at any particular point and $s_a, s_b \ldots s_n$ are drawdowns at that point caused by the discharge of wells a, b, ... n, respectively. The number of wells and the total geometry of the well field are important factors in determining drawdowns and solutions may be based on either the equilibrium or the non-equilibrium equation. D. B. Rao et al. developed equations of well discharge for particular well patterns.[20] Wells in a field aimed at providing water supply should be so spaced that their areas of influence interfere with each other either not at all or only minimally, although sometimes, due to financial constraints, a non-conforming layout has to be adopted. On the other hand, if drainage wells installed to control water table elevations are to be drilled, it might be useful if their spacing actually increases the interference so as to increase the drainage effect.

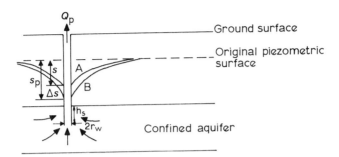

FIG. 5.12. Partially penetrating well in a confined aquifer: drawdown curve with full penetration, A, or with partial penetration, B.

Partially penetrating wells have a length of ingress for water which is less than the capability of the aquifer which they enter. Figure 5.12 shows the situation. Of course, the flow patterns to such wells are different from the radial horizontal flow which is assumed to exist around wells which penetrate fully. The average length of a flow line into a partially penetrating well exceeds that into a well which completely penetrates and so a greater resistance to flow is the result. This causes the following relationships between two similar wells, one partly and one completely penetrating the same aquifer: if $Q_p = Q$, then $s_p > s$ and, if $s_p = s$, then $Q_p < Q$, where Q is well discharge, s is drawdown at the well and the subscript p refers to the partly penetrating well. The effect of partial penetration on the flow pattern is unimportant; the drawdown beyond a radial distance exceeding 0·2–2 times the saturated thickness, b, depends upon the amount of penetration.

The actual drawdown, s_p, at the well face in the partly penetrating well in a confined aquifer is expressed as:

$$s_p = s + \Delta s \qquad (5.49)$$

where Δs refers to additional drawdown caused by partial penetration. It has been shown that, for steady-state conditions in a situation resembling that shown in Fig. 5.12,

$$\Delta s = \frac{Q_p}{2\pi T} \frac{1-p}{p} \ln \frac{(1-p)h_s}{r_w} \qquad (5.50)$$

where T is transmissivity, p is the penetration factor ($p = h_s/b$), h_s is the saturated thickness and r_w is the radius of the well. This equation can be modified to cover a well in an unconfined aquifer: thus,

$$2h_w \Delta s - \frac{Q_p}{\pi K} \frac{1-p}{p} \ln \frac{(1-p)h_s}{r_w} \tag{5.51}$$

where h_w is the saturated thickness of the well which fully penetrates, the hydraulic conductivity is $K = T/h_w$, and:

$$s_p^2 = s^2 + 2h_w \Delta s \tag{5.52}$$

There are elaborate ways of analyzing the effects of partial penetration on well flow for both steady and unsteady conditions in confined, unconfined, leaky and anisotropic aquifers: they have been described by various investigators.[21–23]

Solutions have been obtained for many of the special conditions that are sometimes involved, such as two-layered aquifers,[24] sloping aquifers[25] and large-diameter wells.[26]

5.6. WELL LOSSES AND SPECIFIC CAPACITY

A well loss results from flow through a well screen and flow inside the well to the pump intake; it adds to the drawdown and is associated with turbulent flow, so it may be indicated as proportional to the nth power of the discharge, Q^n, where n is a constant greater than unity. In 1947, C. E. Jacob proposed a value of $n = 2$, but M. I. Rorabaugh indicated that it can vary significantly from this.[27,28] There are probably too many differences between individual wells to give a precise value.

The total drawdown at a well, s_w, for a steady-state confined case can be expressed as:

$$s_w = \frac{Q}{2T} \ln \frac{r_0}{r_w} + CQ^n \tag{5.53}$$

where C is a constant governed by the geometry and state of the well and the well loss is taken into account. If

$$B = \frac{\ln(r_0/r_w)}{2T} \tag{5.54}$$

then

$$s_w = BQ + CQ^n \tag{5.55}$$

Hence, the total drawdown comprises the formation loss, BQ, and the well loss, CQ^n.

It must be noted that discharge has been found to vary only slightly with radius of the well; thus doubling the radius increases the discharge by a mere 10%. If pumping rates are large, then well loss may constitute a large part of the total drawdown but, if the well is properly designed and developed, it may be minimized. In older wells, the screens may have deteriorated or become clogged, which will increase losses. In this connection, W. C. Walton proposed limits for the well loss coefficient, C, in eqn (5.55),[29] and these are given in Table 5.2.

TABLE 5.2
DEPENDENCE OF WELL LOSS COEFFICIENT ON CONDITION OF WELL

Well loss coefficient, C (min^2/m^5)	Well condition
0·5	Correctly designed and developed
0·5–1·0	Mild deterioration or clogging
1·0–4·0	Severe deformation or clogging
4·0	Difficult to restore well to original capacity

In order to evaluate well loss, a step-drawdown pumping test is necessary. The procedure is to pump at a low rate to begin with, until stabilization of the drawdown is achieved.[30] Then the discharge is increased through a succession of steps and an incremental drawdown for each one is determined from roughly equal time intervals. Individual drawdown curves are extrapolated with a slope proportional to the discharge so as to measure the incremental drawdowns. From eqn (5.55), with $n=2$:

$$\frac{s_w}{Q} = B + CQ \quad (5.56)$$

By plotting s_w/Q against CQ and fitting a straight line through the points, the well loss coefficient, C, is given by the slope of the line and the formation loss coefficient, B, by the intercept at $Q=0$. M. I. Rorabaugh gave a modification of the graphic analysis described in order to determine n in cases where it deviates significantly from 2.[28]

The parameter specific capacity in a well refers to the ratio of discharge to drawdown; it is a measure of the productivity. Clearly, the greater the specific capacity, the better is the well. Recalling the simplified form of the Theis equation (eqn (5.35)) utilized for drawdown in the Cooper–Jacob method discussed earlier, see Section 5.2, if $n=2$, then:

$$s_w = \frac{2 \cdot 30 Q}{4\pi T} \log \frac{2 \cdot 25 T t}{r_w^2 S} + CQ^n \qquad (5.57)$$

and the expression for specific capacity is:

$$\frac{Q}{s_w} = \frac{1}{(2 \cdot 30/4\pi T) \log(2 \cdot 25 T t / r_w^2 S) + CQ^{n-1}} \qquad (5.58)$$

showing that the specific capacity decreases with decreasing Q and increasing t. However, for a given discharge, the well is frequently assumed to possess a constant specific capacity and, though this is not true, the change in time is small enough for the error to be disregarded.

A decline of any importance in the specific capacity in a well can be attributed to a reduction in transmissivity as a result of lowering of the groundwater level in an unconfined aquifer, or, alternatively, to an increase in well loss resulting from deterioration or clogging of the well screen. The actual efficiency of any well can be obtained by comparing the specific capacity measured on site (Q/s_w) against the theoretical value of Q/BQ calculated for known values of S and T in an aquifer.[31] For a specified time interval of pumping, the well efficiency, E_w, is given by:

$$E = 100 \frac{Q/s_w}{Q/BQ} = 100 \frac{BQ}{s_w} \qquad (5.59)$$

REFERENCES

1. DUPUIT, J., 1863. *Études Théoriques et Pratiques sur la Mouvement des Eaux dans le Canaux Découverts et à Travers les Terrains Perméables*, 2nd edn. Dunod, Paris, 304 pp.
2. THIEM, G., 1906. *Hydrologische Methoden*. Gebhardt, Leipzig, 56 pp.
3. THEIS, C. V., 1935. The relation between the lowering of the piezometric surface and the rate and duration of discharge of a well using groundwater storage. *Trans. Amer. Geophys. Union*, **16**, 519–24.
4. COOPER, H. H., Jr and JACOB, C. E., 1946. A generalized graphical method for evaluating formation constants and summarizing well-field history. *Trans. Amer. Geophys. Union*, **27**, 526–34.
5. CHOW, V. T., 1952. On the determination of transmissibility and storage

coefficients from pumping test data. *Trans. Amer. Geophys. Union*, **33**, 397–404.
6. WENZEL, L. K., 1942. *Methods for Determining Permeability of Water-bearing Materials with Special Reference to Discharging Well Methods*. US Geol. Surv. Water Supply Paper, 887, Washington, D.C.
7. EHLIG, C. and HALEPASKA, J. C., 1976. A numerical study of confined–unconfined aquifers including effects of delayed yield and leakage. *Water Res. Res., Bull.*, **12**, 1175–83.
8. BOULTON, N. S., 1954. The drawdown of the water table under non-steady conditions near a pumped well in an unconfined formation. *Proc. Inst. Civil Engrs*, **3**, III, 564–79.
9. BOULTON, N. S., 1963. Analysis of data from non-equilibrium pumping tests allowing for delayed yield from storage. *Proc. Inst. Civ. Engrs.*, **26**, 469–82.
10. BOULTON, N. S. and STRELTSOVA, T. D., 1975. New equations for determining the formation constants of an aquifer from pumping test data. *Water Res. Res., Bull.*, **11**, 148–53.
11. PRICKETT, T. A., 1965. Type-curve solution to aquifer tests under water-table conditions. *Ground Water*, **3**, 3, 5–14.
12. BUREAU OF RECLAMATION, 1977. *Ground Water Manual*. US Department of the Interior, 480 pp.
13. COOLEY, R. L. and CASE, C. M., 1973. Effect of a water table aquitard on drawdown in an underlying pumped aquifer. *Water Res. Res., Bull.*, **9**, 434–47.
14. HANTUSH, M. S. and JACOB, C. E., 1955. Non-steady radial flow in an infinite leaky aquifer. *Trans. Amer. Geophys. Union*, **36**, 95–112.
15. KRUSEMAN, G. P. and DE RIDDER, N. A., 1970. *Analysis and Evaluation of Pumping Test Data*. Intl Inst. for Land Reclamation and Improvement, Bull. No. 11, Wageningen. 200 pp.
16. HANTUSH, M. S., 1956. Analysis of data from pumping tests in leaky aquifers. *Trans. Amer. Geophys. Union*, **37**, 702–14.
17. TODD, D. K., 1980. *Groundwater Hydrology*, 2nd edn. John Wiley, New York, 535 pp., p. 143.
18. GLOVER, R. E. and BALMER, G. G., 1954. River depletion resulting from pumping a well near a river. *Trans. Amer. Geophys. Union*, **35**, 466–70.
19. FERRIS, J. E. et al., 1962. *Theory of Aquifer Tests*. US Geol. Surv. Water Supply Paper 1536-E, 69–174.
20. RAO, D. B. et al., 1971. Drawdown in a well group along a straight line. *Ground Water*, **9**, 4, 12–18.
21. HANTUSH, M. S., 1966. Wells in homogeneous anisotropic aquifers. *Water Res. Res., Bull.*, **2**, 273–9.
22. HANTUSH, M. S. and THOMAS, R. G., 1966. A method for analyzing a drawdown test in anisotropic aquifers. *Water Res. Res., Bull.*, **2**, 281–5.
23. LAKSHMINARAYANA, V. and RAJAGOPALAN, S. P., 1978. Type-curve analysis of time–drawdown data for partially penetrating wells in unconfined anisotropic aquifers. *Ground Water*, **16**, 328–33.
24. NEUMAN, S. P. and WITHERSPOON, P. A., 1972. Field determinations of the hydraulic properties of leaky multiple aquifer systems. *Water Res. Res., Bull.*, **8**, 1284–96.

25. HANTUSH, M. S., 1964. Drawdown around wells of variable discharge. In: *Advances in Hydroscience*, ed. V. T. Chow, Vol. 1. Academic Press, New York, pp. 281–432.
26. HANIUSH, M. S. and PAPADOPOULOS, I. S., 1962. Flow of ground water to collector wells. *J. Hydraulics Division, Amer. Soc. Civ. Engrs*, **88**, HY5, 221–44.
27. JACOB, C. E., 1947. Drawdown test to determine effective radius of artesian well. *Trans. Amer. Soc. Civ. Engrs.*, **122**, 1047–70.
28. RORABAUGH, M. I., 1953. Graphical and theoretical analysis of step-drawdown test of artesian well. *Proc. Amer. Soc. Civ. Engrs*, **79**, Leaflet no. 362, 23 pp.
29. WALTON, W. C., 1978. Comprehensive analysis of water-table aquifer test data. *Ground Water*, **16**, 311–17.
30. SHEAHAN, N. T., 1971. Type-curve solution of step-drawdown test. *Ground Water*, **9**, 1, 25–9.
31. BIERSCHENK, W. H., 1964. Determining well efficiency by multiple step-drawdown tests. *Intl. Assn. Sci. Hydrol., Publ.*, **64**, 493–507.

CHAPTER 6

Water Wells

6.1. INTRODUCTION

Water wells may be defined as vertical holes excavated in order to obtain groundwater, artificially to recharge an aquifer by introducing water into it, to disperse sewage, to dispose of industrial waste products or to effect subterranean exploration. Shallow varieties, i.e. those under 15 m deep, may be hand-dug, bored, driven or jetted.

Before any permanent well is constructed, test boreholes are usually put down, their purpose being to ascertain the depth to groundwater, the quality of the water and the type and thickness of the relevant aquifer. A test borehole is cheaper to install and therefore valuable because if the data provided by the temporary borehole are unfavourable and the conditions are unsuitable a permanent well need not be drilled. The diameters of such test boreholes are normally less than 20 cm and, although they can be constructed by any of the methods utilized for wells, it is customary to employ cable tool, rotary and jetting techniques. In the event that such a test hole gives satisfactory information and a permanent well is decided upon, it is feasible to use the already established hole by reaming it with hydraulic rotary equipment, thus converting it into the permanent well.

In constructing a well or a test hole, a record (the log) is maintained showing the various geological formations encountered, and the relevant depths. Subsequent grain size distribution analyses and also details of the structure of the various beds which have been penetrated may be compiled. If the hydraulic rotary technique is utilized, the analysis must be very carefully effected because drilling mud is mixed with each sample. Here, a drilling-time log may be valuable. This comprises an accurate record of the time, in minutes and seconds, necessary to drill each unit depth of the borehole.[1]

6.2. CONSTRUCTION TECHNOLOGY

In the simplest cases, wells may be hand-dug: such wells are mentioned in the Bible. They have yielded water successfully ever since, sometimes in large quantities, and their diameters can be anything between 1 and 10 m. Clearly, they derive the water from shallow depths; S. B. Watt and W. E. Wood have indicated their value in areas of unconsolidated glacial and alluvial deposits.[2] The large diameters are useful because they allow a great amount of water to be stored, provided that the well extends below the water table. An advance in hand digging was the introduction of pick and shovel and, for large wells, portable excavating equipment may be employed, including clamshell and orange-peel buckets. In order to prevent any cave-in from taking place, it is possible to brace the hole by insertion of wood or sheet piling. Modern wells are lined permanently with a curb or casing of wooden staves, brick or rock, metal or concrete. Such lining should be perforated so that water may enter and, also, they should be sealed basally. Dug wells should extend some metres below the water table. Gravel should be back-filled around the casing and at the bottom of the well. This will obviate cave-in and control and entry of sand. A dug well associated with a permeable aquifer can produce as much as 7500 m^3 of water daily.

Bored wells are installed by using either hand-operated or powered earth augers. The former come in various shapes and sizes, but they all use cutting blades at the bottom which bore into the material with a rotary motion. As soon as the augers are filled with loose earth, they are pulled out and cleared. Hand augering usually penetrates no more than 15 m depth and gives wells usually less than 20 cm diameter. As might be expected, power augering is on a larger scale, producing wells with depths in excess of 30 m and diameters exceeding those of the manually augered variety and sometimes attaining 1 m. The auger itself comprises a cylindrical steel bucket with a cutting edge which projects from a basal slot. This bucket is filled by in-hole rotation by a drive shaft of which the length is adjustable. As soon as it is filled, the auger is brought up to the surface of the ground and emptied through hinged openings on the side or bottom of the bucket. Attached to the top of this are reamers which can enlarge the hole beyond the actual size of the auger. Also, there is a continuous-flight power auger which has a spiral extending from the base of the hole to the surface, cuttings being conveyed upwards and extra sections being available for addition as the depth of the hole increases. Such equipment is simple, may be mounted on a lorry and can be handled by a single man. It functions down to at least 50 m in unconsolidated materials devoid of big boulders.

All augers are optimally utilized where formations which do not easily cave in are concerned, and occasionally they can be utilized as a supplement to other types of well drilling.

Driven wells are produced by insertion of a set of connected pipe lengths into the ground through repetition of impacts, water obtaining ingress through a sand point at the lower end of the resultant well. Such a point comprises a screened cylindrical section which is protected during the impacting process by a bottom-placed steel cone. Driven wells have small diameters (3–10 cm) and are usually less than 15 m deep. Their casing consists of standard weight water pipe with threaded couplings. In consequence of the fact that suction-type pumps extract water from driven wells, the water table must be near the ground surface if a continuous water supply is to be made available. This means that it should not be lower than 5 m depth. The yields which are derived from such wells are small, normally about 100–250 m^3 daily, and hence they are mostly used for domestic water supply or for temporary sources of water. During the lowering of the water table, a necessary preliminary to many foundation-installation activities in civil engineering, well-point systems have to be used to de-water the excavation zone and, for such work, batteries of driven wells connected by a suction header to one pump are efficient, as D. G. Noble has indicated.[3] As regards the environment of operation, driven wells are applicable only to unconsolidated formations free of any large boulders which could damage the drive point. The actual impact is achieved by using a sledge hammer, drop hammer or air hammer and an outer protective casing is normally installed at least down to 3 m depth. The technique can be considered a very valuable one because it is quick, relatively economical and can be utilized by one man alone, if necessary.

Jetted wells are obtained as a result of the cutting action of a downwardly directed water jet at high velocity which washes out earth at the same time as the casing which is lowered into the hold is conducting both cuttings and water out of the deepening well. Although quite large diameters up to 30 cm can be obtained, usually small-diameter holes are drilled, from 3 to 10 cm, and depths exceeding 15 m are attainable. Such wells have low yields and again are only to be installed in unconsolidated beds, but they have a great advantage in that they can be produced rapidly by easily transportable equipment; therefore they often act as exploratory test holes or observation wells and, occasionally, even as members of well-point systems.[4] There are various types of bit available according to the different materials which may be encountered. In dealing with clays and hardpan, the drill pipe effects sharp percussive blows, thus proceeding by shattering. During jetting, the pipe is turned slowly in order to maintain the straightness of the hole. After the casing extends below the water table, the well pipe, with a screen attached, is lowered to the bottom of the hole inside the casing. The outer casing is then retracted, gravel is inserted in the outer space and the well is ready to be pumped.

Only shallow wells have so far been considered. In the case of deep wells, drilling is necessary to install them and either the cable tool or one of several rotary methods can be utilized; since each has specific advantages, it is sometimes the practice to make equipment available for various techniques.[5] From the undated well drilling manual of the Speedstar Division, Koehring Co., Enid, Oklahoma, USA, Table 6.1 provides details regarding the performance of the various drilling methods in the types of geological formation cited.[6]

The cable tool approach is sometimes termed the percussion or standard method; the equipment comprises a standard well-drilling rig, percussion tools and a bailer.[7] Holes up to 60 cm diameter can be drilled in consolidated rocks down to 600 m. The approach is much less efficient in unconsolidated materials such as sand and gravel; this is because loose material falls in around the bit. The drilling is carried out by regularly lifting and dropping a string of tools with a sharp chisel-edge bit at the lower end which breaks up rock through impact.

The tools comprise a swivel socket at the top followed by a set of jars, a drill stem and the bit with a total weight which may reach thousands of kilograms. They are made of steel and are joined together using tapered box-and-pin screw joints. Obviously, the most important tool is the bit: these are available in lengths of 1–3 m and can weigh up to 1500 kg. They vary in shape according to the type of formation in which they are to be used. The drill stem is a long steel bar adding both length and weight to the drill in order to facilitate rapid vertical cutting. The set of jars comprises a pair of narrow, connecting links designed to loosen the tools if they get stuck in the hold; with normal tension, they remain fully extended. However, if sticking occurs, they open to their full length as the line slackens and thereafter an upstroke of the line makes the upper section of the jars impart an upward impact upon the tools. Finally, the swivel socket functions in attaching the drilling cable to the string of tools. The bailer is the means of removing drill cuttings from the well and

TABLE 6.1
PERFORMANCE OF VARIOUS DRILLING METHODS

Formation type	Drilling methodology		
	Cable tool	Rotary	Rotary percussion
Dune sand	Difficult	Rapid	Not suitable
Loose sand, gravel	Difficult	Rapid	Not suitable
Quicksand	Difficult	Rapid	Not suitable
Loose boulders in alluvial fans or glacial drift	Slow	Difficult	Not recommended
Clay and silt	Slow	Rapid	Not recommended
Firm shale	Rapid	Rapid	Not recommended
Sticky shale	Slow	Rapid	Not recommended
Brittle shale	Rapid	Rapid	Not recommended
Poorly cemented sandstone	Slow	Slow	Not recommended
Well cemented sandstone	Slow	Slow	Not recommended
Chert nodules	Rapid	Slow	Not recommended
Limestone	Rapid	Rapid	Very rapid
Fractured or cracked limestone and limestone with chert nodules	Rapid	Slow	Very rapid
Cavernous limestone	Rapid	Slow to impossible	Difficult
Dolomite	Rapid	Rapid	Very rapid
Basalts, thin layers in sedimentary rocks	Rapid	Slow	Very rapid
Basalts, thick layers	Slow	Slow	Rapid
Metamorphic rocks	Slow	Slow	Rapid
Granite	Slow	Slow	Rapid

it consists of a pipe section with a bottom valve and a ring at the top for attachment to the bailer line. If lowered into the well, the valve allows the cuttings to enter the bailer, but prevents them from escaping. When it is full, the bailer is brought up to the surface and emptied. There is a range of diameters available for bailers and lengths may range from 3 to 8 m, capacities going up to 0.25 m^3.

The drilling rig is composed of a mast, a multiline hoist, a walking beam and an engine, the whole assembly usually being lorry-mounted so that it can be transported easily.

In the operation of drilling, 20–40 strokes per minute are made by the tools through 40–100 cm length, and the drilling line is rotated in order that the bit creates a round hole. If there is insufficient in the hole, water must be added to mix with the cuttings and reduce friction on the bit. After the initial 1–2 m penetration of a formation, the string of tools is raised to the surface and the hole is bailed out. Casing may have to be inserted near the base of the hole to prevent caving in unconsolidated materials. Where casing is driven deep into such materials, vibration may trigger collapse of the sides of the hole against it and the frictional forces may increase so as to prevent further driving. The solution adopted in that case is to insert a smaller-diameter casing inside the one already in the hole and the drilling is then continued, using a smaller bit. In fact, such action may have to be repeated several times.

All in all, this method of drilling is very satisfactory as regards results, but there are some disadvantages. For instance, the depth of penetration is restricted and the rate of drilling is relatively slow; also, casing must be driven at the same time as the drilling is proceeding in unconsolidated beds, and it is difficult to remove from deep holes.

The rotary method is more rapid in unconsolidated strata, and deep wells with diameters up to 45 cm are possible. A mixture of clay and water, drilling mud, is forced through a hollow, rotating bit and material loosened in this way is carried upwards by the ascending mud. There is usually no need for casing since the mud forms a mud cake lining on the well wall by filtration. As a consequence, the wall is sealed and this obviates cave-in, groundwater entry and mud loss.

Again, a variety of bits is available including the fishtail type, the conetype and the carbide button type. Typically, the string of tools comprises the bit, a drill collar which contributes extra weight to the bit as well as assisting in maintaining the alignment of the hole, and a drill pipe extending to the ground surface. At the top, the drill pipe is attached to a section of drill rod called the kelly around which is rotating a table which fits it closely, this permitting the drill rod to move downwards as the hole gets deeper. The associated rig is composed of a mast (derrick), the pump for the drilling mud, a hoist and an engine.

The drilling mud is made of a suspension of water, bentonite, clay and some organic additives. Its properties of viscosity, weight, low proportion of suspended solids, etc., must be preserved if the drilling is to proceed smoothly.[8] A new development is the introduction of special organic additives which degrade with passing time, thus promoting the breakdown of mud cake within a few days. Drilling mud emerges from the drill pipe through the bit, thereafter cooling and lubricating the cutting surface, and later enmeshes cuttings which are carried up to the surface within the annular space which separates the drill pipe from the wall of the hole. When it gets out at the surface, the mud is conducted into a settling ditch in which cuttings separate. Later, the mud is taken up by the pump and re-circulated. This technique is utilized in oil wells, but its application to water wells is increasing. The advantages are rapidity, lack of necessity for a casing and the convenience for electrical logging, but the drawbacks include high cost, greater operational complexity, loss of circulation in cavernous or unusually highly permeable formations and the need to take out mud cake in the development of the well.

As well as mud, air can be used. This is the compressed air rotary technique and it is extremely rapid for constructing small diameter holes in consolidated formations where there is no need for a clay lining to support the walls against cave-in. M. D. Campbell and J. H. Lehr in 1973 stated that larger-diameter holes can be drilled by using foams and other additives.[9] Extension to depths as great as 150 m or more may be possible.

A variant technique involves a rotary-percussion approach which utilizes air as the drilling fluid: this is the fastest way of entering hard rocks. A rotating bit acts like a pneumatic hammer and delivers 10–15 impacts per second to the bottom of the hole. Penetration rates as high as 0·3 m/min have been achieved.

Finally, the reverse-circulation rotary technique must be mentioned, as it has proved to be a useful means for drilling large-diameter holes in unconsolidated formations. The method is to pump water upwards through the drill pipe, using a centrifugal pump with a large capacity or a jet pump; discharge from the hole enters a pit where the cuttings settle. The water then flows through a ditch and re-enters the hole so that the water level in the hole is kept at ground surface.

If the velocity of the water is too high, there may be adverse effects on the side of the hole; therefore it has to be kept low, and so the minimum hole diameter is around 40 cm. Drilling bits cover a range of 0·4–1·8 m in diameter and the velocity of the water up the drill pipe is usually greater than 2 m/s. To obtain an effective differential of head between the well and the aquifer, the water table must be 3–4 m below the surface of the ground, in which case fine particles suspended in the water act to stabilize the walls. In the case where the water table is closer to the surface, casing may be extended above the ground so as to increase the head. In the opposite case, where the water table is deep, it may be necessary to emplace surface casing so as to minimize water loss.

This reverse-circulation rig is the fastest drilling equipment available for unconsolidated beds. A large volume of water must be available. Depths of 125 m can be attained and modifications with air lift enable greater depths to be reached. The completed wells are usually packed with gravel because, with this technique, they have large diameters.

No matter how a water well is made, its completion is an important matter. This may involve inserting casing, emplacing screens and gravel packing. In hard rocks, the well can be left just as an open hole, so such measures are not required.

Casing acts as a liner which keeps the hole open from the aquifer back to the surface of the ground. Also, it keeps out surface and undesirable groundwater, additionally supplying structural support against caving. This entails the material of the casing which may be wrought iron, alloy and non-alloy steel and ingot iron.[10] Joints are often welded in order that watertightness may be achieved. With the cable tool method, casing is driven in, but, in rotary methods, because it is smaller than the actual hole, it can be lowered into place. In the case of shallow observation wells with small diameters, polyvinyl chloride pipe is utilized for casing.

Surface casing is installed from the surface of the ground, passing through upper beds of fractured material into impermeable material. It can fulfill various requirements; thus it may serve as a reservoir for a gravel pack or reduce the loss of drilling fluids.

Pump chamber casing refers to all that casing which occurs above the screen in wells of uniform diameter. In telescoping wells, i.e. those mentioned on p. 136, in which smaller-diameter casing is placed inside that already in the hole, sometimes more than once, the pump chamber casing is that in which the pump bowls are set. It should normally possess a nominal diameter at least 5 cm greater than the nominal diameter of the pump bowls. To avoid corrosion non-metallic pipes, e.g. concrete or plastic, may be used but the strength is then reduced.

Cementing is effected round the casing in the annular space so as to preclude unsatisfactory water ingress; it also protects against corrosion and, where necessary, stabilizes the beds against cave-in. It is possible to emplace cement grout using a tremie pipe, by pumping or utilizing a dump bailer.[10] The grout is introduced at the base of the appropriate space so that the sealing is properly effected.

Screens are required in wells which penetrate unconsolidated formations so that stabilization can be achieved, sand entry prevented and the greatest quantity of water allowed to enter with as little hydraulic resistance as possible. In the cable tool drilling technique, screens may be placed by the pullback method in which, after the casing is inserted, the screen is lowered inside and the casing thereafter pulled up to the top of the screen. On top of the screen there is a lead packer ring which is flared out so as to create a seal between the inside of the casing and the screen. There is also a rotary method of drilling without casing: the screen is lowered into place as the drilling mud is diluted; again, a lead packer effects a seal between it and the casing. Another installation method for screens is the baildown technique, which requires bailing out sediment below the screen until the latter is lowered to the desired depth in the aquifer.

Casing is usually perforated prior to installation. The slots, made by machining or sawing, are of uniform size in the range 1·6 mm. It may be noted, however, that manufactured screens are preferable because the openings may be made specially for the particular circumstances of the aquifer concerned. They can be louvred, punched, stamped and so on. Various screen diameters are available, ranging from 5 to 40 cm nominal.

It has been determined that the entrance velocity of water must be kept within specified limits in order to minimize well losses and clogging of the screen. Those aquifers which comprise fine-grained material have a tendency to clog screens more easily than is the case with those made of coarser materials and it has been shown empirically that there is a definite relationship between the screen entrance velocity and the hydraulic conductivity of the aquifer. W. C. Walton in 1962 gave relevant data which are cited in Table 6.2.[11]

TABLE 6.2
OPTIMUM SCREEN ENTRANCE VELOCITY FOR VARIOUS HYDRAULIC CONDUCTIVITIES[11]

Hydraulic conductivity of aquifer (m/day)	*Optimum screen entrance velocity (m/min)*
> 250	3·7
250	3·4
200	3·0
160	2·7
120	2·4
100	2·1
80	1·8
60	1·5
40	1·2
20	0·9
< 20	0·6

The velocities in Table 6.2 can be expressed thus:

$$v_s = \frac{Q}{c \pi d_s L_s P}$$

where v_s is the optimum screen entrance velocity, Q is the well discharge, c is a clogging coefficient (for which an estimate is 0·5, since about half of the open area of a screen will be blocked by material of the aquifer), d_s is the diameter of the screen, L_s is the length of the screen and P is the percentage of open area in the screen (which can be obtained from the manufacturer's specifications). With this information, it is possible to select a well screen with the correct dimensions for any given aquifer material, aquifer thickness, well yield and so on. The screens themselves are made of various metals, metal alloys, plastics, concrete, wood, fibreglass-reinforced epoxy, etc.[8] They are subject to corrosion and therefore non-ferrous metals and plastics are preferred if the screen is to last. The US Bureau of Reclamation has provided some pertinent data listed in Table 6.3.[8] The last three alloys in the Table are recommended only in areas where there is unusually high corrosion potential. The first four are not recommended for installations where there are sulphate-reducing or similar bacteria; also, they are not advised for places where water contains more than 60 mg/litre SO_4^{2-}.

Perhaps the most significant property of the well screen is its slot size and, if a grain size analysis for the particular aquifer is supplied, the manufacturer may recommend the optimum. In the case where the uniformity coefficient of an aquifer is 5 or less for a naturally developed well without a gravel pack, the slot size chosen should retain 40–50% of

TABLE 6.3
ACID AND CORROSION PROPERTIES OF SCREEN MATERIAL[8]

Material in order of increasing cost	Acid resistance	Corrosion resistance in normal groundwater
Low-carbon steel	Poor	Poor
Toncan and Armco iron	Poor	Fair
Admiralty red brass	Good	Good
Silicon red brass	Good	Good
304 Stainless steel	Good	Very good
Everdue bronze	Very good	Very good
Monel metal	Very good	Very good
Super nickel	Very good	Very good

TABLE 6.4
CRITERIA FOR SELECTING GRAVEL PACK MATERIAL[a8]

Uniformity coefficient of aquifer, U_c	Criteria for the gravel pack
2·5	(i) U_c between 1 and 2·5 with 50% size not exceeding six times the 50% size of the aquifer. (ii) Where (i) is not available, U_c between 2·5 and 5 with 50% size not exceeding nine times the 50% size of the aquifer.
2·5–5	(i) U_c between 1 and 2·5 with 50% size not exceeding nine times the 50% size of the aquifer. (ii) Where (i) is not available, U_c between 2·5 and 5 with 50% size not exceeding 12 times the 50% size of the aquifer.
5	(i) Multiply the 30% passing size of the aquifer by 6 and 9 and locate the points on the grain-size distribution graph on the same horizontal line. (ii) Through these points draw two parallel lines representing materials with $U_c = 2·5$. (iii) Select gravel pack material that falls between the two lines.

[a] Screen slot size: 10% passing size of the gravel pack.

the aquifer material. The screen allows fine material to enter the well and this is later removed by bailing, the coarser material remaining outside and forming a permeable envelope embracing the well. This constitutes a natural gravel pack. An artificial version may be introduced also. It aims at stabilizing the aquifer, minimize pumping of sand, and facilitating the employment of a large screen slot with a maximum open area; also it gives an annular zone of high permeability, thus increasing the effective radius and yield of the well. The criteria for selection of gravel pack material have been given by the US Bureau of Reclamation; they are listed in Table 6.4.[8]

The maximum size of grain of a pack should be approximately 1 cm and the thickness should be within the range 8–15 cm. The gravel chosen has to be washed and should comprise siliceous, rounded, abrasive-resistant dense material. Emplacement should be carried out in such a manner that all the annular space around the well is filled; usually, this may be done by extending two tremie pipes to the bottom of the well on opposing sides of the screen and withdrawing them step-by-step as gravel is introduced, either by washing or by pumping it in.

6.3. DEVELOPMENT OF WELLS

This is a process that follows their completion and involves increasing of specific capacity, obtaining maximum life and obviating sanding. There are various approaches, including pumping, surging, the utilization of compressed air, introduction of chemicals and even the employment of explosives.

Pumping means proceeding in a set of steps from low discharge to discharge which exceeds the design capacity and, for maximum effect, the intake area of the pump must extend to near the centre of the screened section. At each one of the steps, pumping goes on until the water clears; then the power is switched off, whereupon water in the pump column re-enters the well. The step is repeated so as to obtain only clear water at the wellhead and then the rate of discharge is increased and the entire procedure is carried out again until the final rate is the maximum capacity of the pump or the well. Such an irregular pumping will stir up fine material around the well so that it enters the well and is pumped out. Coarser material accrues basally and may be removed by a bailer.

Surging involves up-and-down motion of a surge block which is attached to the bottom of a drill stem. A cable tool rig is appropriate. Solid or vented surge blocks or a flap-valve bailer are used and can be obtained commercially. A typical cylindrical block is anything up to 5 cm smaller than the well screen. It is fitted with rubber or leather belting which should not damage the screen. As it moves up and down in the latter, it imparts a surging action to the water and the downstroke makes the backwash break up and possibly causes bridging. The upstroke pulls loose sand grains into the well. At the beginning, the surging consists of a slow stroke at the bottom of the screen, but gradually the strokes become progressively more rapid. The operation is accomplished when the quantity of material accumulating at the well bottom becomes insignificant. Surging may also be carried out with air; for this purpose, an air compressor is connected to an air pipe in the well. This air pipe, and a discharge pipe around it, can both be moved vertically. At first, they extend almost to the base of the screened section. Initially the air pipe is closed and the air pressure permitted to build up to about 10^6 Pa. Then it is released suddenly into the well through a valve, triggering the surge, which first increases and then decreases the pressure as the water is forced up the discharge pipe. In the process, fine material surrounding the perforations is loosened and brought into the well. The same is done at intervals along the screened section by moving the pipes accordingly, and goes on until accretion becomes negligible.

Among the chemicals which may be introduced, hydrochloric acid may be added to well water so as to develop open-hole wells in limestone or dolomite formations, the solvent action removing fine particles and also tending to widen fractures which lead into the well bore. In the case of rocks containing silicates, hydrofluoric acid may be substituted. Another technique is to add polyphosphates which act as deflocculants and dispersants of clays and other fine-grained materials, thus promoting development of the well. Occasionally, blocks of dry ice (solid carbon dioxide) are put into a well after acidifying and surging with compressed air in order to complete development. Gaseous carbon dioxide will accumulate and build up pressure which, being released, produces an eruption of mud and water from the well. Detonation of explosives in rock wells may increase the yields by enlarging the hole or by increasing the size and number of fractures as well as by removing fine-grained material on the face of the well bore.

When a well is developed, it must be tested so as to determine its yield and drawdown, data which provide a basis for ascertaining the water supply available and also for choosing the type of pump required. The test consists of first measuring the static water level, then pumping the well at maximum rate until the water level stabilizes, when the depth-to-water is observed. The difference in these two depths is the drawdown and the discharge-to-drawdown ratio, Q/s_w (see Section 5.6), permits an estimate of the specific capacity of the well to be made. Discharge is easily determined through any measuring device connected to the discharge pipe.

6.4. THE EQUIPMENT FOR PUMPING

The flow in a well is produced by a pump which transforms mechanical into hydraulic energy. Many types of pump exist and the choice of an

TABLE 6.5
PUMP DATA[12]

Pump	Suction lift[a](m)	Usual well-plumbing depth(m)	Usual pressure head(m)	Advantages	Disadvantages
1. Reciprocating					
(i) Shallow well	6–7	6–7	30–60	Positive action; discharge against variable heads; pumps water containing sand and silt; especially adapted to low capacity and high lifts.	Pulsating discharge; subject to vibration and noise; maintenance cost may be high; may cause destructive pressure if operated against closed valve.
(ii) Deep well	6–7	Up to 180	Up to 180 above cylinder		
2. Centrifugal					
(i) Shallow well Straight centrifugal (single-stage)	6 max.	3–6	30–45	Smooth, even flow; pumps water containing sand and silt; pressure on system is even and free from shock; low-starting torque; usually reliable with good service life.	Loses prime easily; efficiency depends on operating under design heads and speeds.
Regenerative vane turbine type (single impeller)	8 max.	8	30–60	Same as straight centrifugal except not suitable for pumping water containing sand or	Same as straight centrifugal except maintains priming easily.

(ii) Deep well				
Vertical line shaft turbine (multistage)	Impellers submerged	15–90	30–250	Same as shallow well turbine.	Efficiency depends on operating under the design head and speed.
Submersible turbine (multistage)	Pump and motor submerged	15–120	15–120	Same as shallow well turbine; easy to frost-proof installation; short pump shaft to motor.	Repair to motor or pump requires pulling from well; sealing of electrical equipment from water vapour critical; abrasion from sand.
(iii) Jet					
Shallow	4–6 below ejector	Up to 4–6 below ejector	25–45	High capacity at low heads; simple to operate; does not have to be installed over well; no moving parts in well.	Capacity reduces as lift increases; air in suction or return line will stop pumping.
Deep	4–6 below ejector	7–35, 60 max.	25–45	Same as shallow well jet.	Same as shallow well jet.
(iv) Rotary					
Shallow well (gear type)	7	7	15–75	Positive action; discharge constant under variable heads; efficient operation.	Subject to rapid wear if water contains sand or silt; wear of gears reduces efficiency.
Deep well (helical rotary type)	Usually submerged	15–150	30–150	Same as shallow well rotary; only one moving pump device in well.	Same as shallow well rotary except no gear wear.

[a] This is at sea level and should be reduced 0·3 m per 300 m elevation.

appropriate one depends upon its capacity, the diameter and depth of the relevant well, the depth and variability of the pumping level, the straightness of the well, the pumping of sand, the total pumping head, the available power and, of course, the cost.

The total pumping head of a pump represents the total vertical lift of the water from the well. There are three components, namely drawdown in the well together with aquifer and well losses, static head (i.e. the difference between the static groundwater level and the static discharge elevation), and friction losses caused by flow through the intake and discharge pipes.

There are two main categories of environment to be considered, namely shallow and deep.

From shallow wells, only small discharges are required and so hand-operated pitcher pumps, turbine pumps or gear pumps can be used. For larger discharges, a centrifugal pump is utilized, the assembly being mounted with a horizontal or a vertical shaft. The horizontal shafts are efficient, easy to install and maintain, and are normally connected with an electric motor. As such pumps have a low suction head, they are placed near the water level in large-diameter wells.

Deep wells need high lift, so large-capacity pumps are mandatory. There are several suitable types, for instance plunger, displacement, airlift, jet, deep-well turbine and submersible.

The most interesting is the submersible, which is a deep-well turbine close-coupled to a small-diameter, submersible electric motor. It has the advantage that it is effectively cooled through being submerged. The deep-well turbine itself is widely used and has an impeller which is suspended vertically on a long drive shaft inside a discharge pipe. The impeller is within the bowl of the pump and this also contains the guide vanes. A multiple-stage variety has several bowls which are connected in series so as to provide higher heads. The electric motor is on the ground in this case, and the pump is connected by a long vertical shaft which is positioned by bearings within the discharge pipe. No priming is necessary and such pumps can operate over a great range of water levels without having to be re-set.

The US Public Health Service has given important data on pumps, as noted in Table 6.5.[12]

6.5. THE PROTECTION AND RESTORATION OF WELLS

If the groundwater produced by a water well is to be used by human beings, it must be of suitable sanitary quality. Some of the potential pollutants have been discussed earlier in Sections 1.2, 3.2 and 3.3; they arise mostly from waste disposal entering on or below the ground.

Surface pollution can enter into wells through the annular space outside the casing, or directly through the well top. Measures can be taken against entry by these routes. Thus, the annular space may be infilled with cement grout and the top of the well can be sealed with a watertight cover.[13] In fact, some pumps are available with closed metal bases which act to provide the required sealing and, if they do not have such a feature, a metal or possibly asphalt seal has to be provided. Any reinforced concrete cover slab surrounding a surface seal must be elevated above the adjacent level of the ground and it must also slope away from the well.

Disinfection of a newly completed or restored well is carried out by adding a chlorine compound to the water and agitating it, after which the well must be pumped until all trace of the disinfectant is removed. Actually, bacteriological analysis of a sample by an appropriate laboratory can give a final test of potability. In areas where frost occurs in winter, the pump and water lines must be frost-proofed.

Abandonment of a well must be followed by sealing it up with clay or concrete. This obviates surface contamination getting into it and, in addition, accidents are avoided.

Well recovery is a consequence of deterioration, which sometimes entails a lowering of water output with time; perhaps this is due to the pump, which can be checked, but often it is attributable to factors in the well itself, for instance a depletion in the supply of the groundwater. This particular fault can be rectified by decreasing the pump draft or by deepening the well. However, some difficulties arise when the casing connections are defective, when something is wrong with the screen or if the gravel pack is not properly emplaced. In order to assess the situation inside the well, it is possible to lower a television camera into it. Appropriate instruments with wide-angle facilities, usually under 7 cm diameter, can give continuous visual inspection facilities which may be videotaped.[14] As a result of the data obtained, it may be feasible to repair the well, but it is sometimes necessary to replace it. One of the most prevalent causes of such a failure is corrosion, which may arise either from the chemical action of the groundwater itself or from an

TABLE 6.6
METHODS OF RECOVERING WELLS[16]

Recovery method	Unconsolidated aquifers	Consolidated sandstone	Consolidated limestone
Muriatic acid (not usable with concrete screens)	Removes iron, sulphur, carbonate deposits.	Not usually effective.	Occasionally beneficial; best results by pressure acidification.
Polyphosphate followed by chlorine	Removes fine silt, clay, colloids, disseminated shale, soft iron deposits.	Not usually effective.	Not usually effective.
Dynamiting	Not recommended.	Effective for all types of well-screen deposits.	Effective when very large charges are used.
Compressed air	Removes plugging deposits of silt and fine sand in areas adjacent to screens.	Not used.	Not used.
Dry ice	Same as compressed air.	Used rarely to remove cuttings from the face of a new production well.	Not usually effective.
Surging Chlorine (usually in a concentration of 500 mg/litre	Same as compressed air. Removes iron and slime-producing bacteria.	Rarely used. Removes iron and slime-forming bacteria.	Rarely used. Removes iron and slime-forming bacteria.
Caustic soda	Removes oil scum left by oil-lubricating pumps.	Removes oil scum left by oil-lubricating pumps.	Removes oil scum left by oil-lubricating pumps.

electrolytic action triggered by the occurrence of two different metals in the well. Corrosion can be reduced if non-metallic screens or screens made of corrosion-resistant metal, such as copper or nickel, are installed. There is another problem — incrustation, which is produced by precipitation on or near screens and which is due to materials in solution in the groundwater.[15] Thus, water entering a well as a result of heavy pumping encounters a rapid drop in pressure and thereby liberates carbon dioxide, causing precipitation of calcium carbonate. Also, oxygen which is present can alter soluble ferrous ions to ferric hydroxide which is insoluble. Screens so affected can be cleaned by adding hydrochloric acid (also employed to develop open-hole wells in limestone; see Section 6.3) or sulphamic acid, H_2NSO_3H, to the well, succeeded by agitation and surging. If organisms enter which deposit slime and block screens, chlorine gas or solutions of hypochlorite can be employed to clear them.[8] Rock wells may also be treated by using explosives.

C. R. Erickson in 1961 described appropriate well recovery methods, which are listed in Table 6.6.[16]

6.6. NON-VERTICAL WELLS

Some wells are horizontal because of adverse subsurface circumstances such as permafrost or saline water which preclude installing vertical wells. Horizontal wells constitute infiltration galleries of which qanats (see Section 1.2, p. 9), are an example. Such galleries intercept and collect water by gravity flow.[17] Normally, they are constructed at the elevation of the water table and discharge into a sump from which the water is pumped to the surface of the ground to be used. In both the USA and Europe, such galleries are installed parallel to riverbeds and often provide a good perennial supply of water. The approach has been utilized in Alaska (where permafrost underground does not make a contribution of groundwater), south-eastern England (the chalk aquifers, where the galleries are termed adits, are about 2 m in diameter and extend as far as 2 km, intersecting fissures from which most of the water is derived), Hawaii (unlined installations in basalts) and Barbados (unlined in coral limestone).[18,19]

As well as galleries, small-diameter horizontal and perforated pipes can be emplaced in specially (rotary) drilled holes in order to obtain groundwater which would otherwise be discharged by seepage or small springs.[20]

Finally, there are collector wells. If a collector well is located next to a surface water source, it lowers the water table and so induces infiltration of surface water through the bed of the actual water body to the well. A central concrete caisson cylinder of diameter about 5 m is sunk into the aquifer by excavation of the inside earth. After attaining the requisite depth, a thick concrete plug is poured in to seal the bottom. Then, perforated pipes of 15–20 cm diameter are jacked hydraulically into the water-bearing formation through pre-cast portholes in the caisson so as to form a radial pattern of horizontal pipes. While construction is proceeding, fine-grained material is washed into the caisson, and thus natural gravel packs are formed around the perforations. Of course, the number, length and the radial pattern of the pipes can be varied in order to attain optimum results.

Since the collector well has a very large area of exposed perforations, the inflow velocities are low; therefore, there is a much-reduced risk of incrustation or clogging. Also, river water which is polluted is filtered as a result of its passage through the unconsolidated aquifer to the well. Yield is variable, the average being about $27\,000\,m^3$ daily. Collector wells can also function in permeable aquifers.

REFERENCES

1. KIRBY, M. E., 1954. Improve your work with drilling-time logs. *Johnson Nat. Drillers J.*, **26**, 6, 6, 7, 14.
2. WATT, S. B. and WOOD, W. E., 1976. *Hand Dug Wells and Their Construction.* Intermediate Technology, London, 234 pp.
3. NOBLE, D. G., 1963. Well points for de-watering. *Ground Water*, **1**, 3, 21–6.
4. MATLOCK, W. G., 1970. Small diameter wells drilled by jet percussion method. *Ground Water*, **8**, 1, 6–9.
5. HUISMAN, L., 1972. *Groundwater Recovery.* Winchester Press, New York, 336 pp.
6. SPEEDSTAR DIVISION. *Well Drilling Manual.* Koehring Co., Enid, Okla., USA, 72 pp.
7. US ENVIRONMENTAL PROTECTION AGENCY, 1976. *Manual of Water Well Construction Practices.* Rept, EPA 570/9-75-001, Washington, DC, 156 pp.
8. US BUREAU OF RECLAMATION, 1977. *Ground Water Manual.* US Department of the Interior, 480 pp.
9. CAMPBELL, M. D. and LEHR, J. H., 1973. *Water Well Technology.* McGraw-Hill, New York, 681 pp.
10. AMERICAN WATER WORKS ASSOCIATION, 1973. *Ground Water.* AWWA Manual M21, 130 pp.
11. WALTON, W. C., 1962. *Selected Analytical Methods for Well and Aquifer*

Evaluation. Bull. 49, Illinois State Water Survey, Urbana, Ill., 47 pp.
12. US PUBLIC HEALTH SERVICE, 1962. *Manual of Individual Water Supply Systems.* Publn No. 24, 121 pp.
13. BERNHARD, A. P., 1973. Protection of water supply wells from contamination by wastewater. *Ground Water,* **11**, 3, 9–15.
14. GORDER, Z. A., 1963. Television inspection of a gravel pack well. *Amer. Water Works Assn, J.,* **55**, 31–4.
15. MOGG, S. L., 1972. Practical construction and incrustation guide lines for water wells. *Ground Water,* **10**, 2, 6–11.
16. ERICKSON, C. R., 1961. Cleaning methods for deep wells and pumps. *Amer. Water Works Assn, J.,* **53**, 155–62.
17. BENNETT, T. W., 1970. On the design and construction of infiltration galleries. *Ground Water,* **8**, 3, 16–24.
18. FEULNER, A. J., 1964. *Galleries and Their Use for Development of Shallow Ground-water Supplies with Special Reference to Alaska.* US Geol. Surv. Water Supply Paper 1809-E, 16 pp.
19. PETERSON, F. L., 1972. Water development on tropical volcanic islands — type example: Hawaii. *Ground Water,* **10**, 5, 18–23.
20. WELCHERT, W. T. and FREEMAN, B. N., 1973. Horizontal wells. *J. Range Management,* **26**, 253–6.

CHAPTER 7

Environmental Effects on Groundwater

7.1. GROUNDWATER LEVELS

These indicate the elevation of atmospheric pressure of an aquifer and any agency which causes a change in pressure on groundwater will alter its level. Hence, the difference between input of water and its withdrawal from such a water-bearing body will do so; variations in streamflow are also related to groundwater levels. Various meteorological and tidal events, seismic tremors and the activities of Man have parallel effects, as does land subsidence.

Variations in groundwater level may be secular, i.e. they encompass a number of years, and they often result from alternations between wetter and drier years. However, rainfall does not correlate simply with these changes in level; rather, they are governed by recharge, if annual withdrawals remain constant. Marked long-term trends may be noted in basins which are over-exploited, when decline continues for years, especially in urbanized areas with ever-mounting demands for water.

In addition, the levels of groundwater can demonstrate a seasonal fluctuation pattern resulting from factors such as precipitation and irrigation pumping. Even smaller-scale fluctuations may also be observed, sometimes daily or weekly where municipal or industrial users, respectively, are involved.

Groundwater may discharge into a stream with a channel directly connecting to an unconfined aquifer or this may be recharged from the stream. Gaining streams are those receiving groundwater, and losing streams recharge the aquifer. Of course, changes from one type to the other may take place at any time when the stream stage alters.[1]

Rising water is a term applied to sizeable increases in streamflow in reaches in which a subsurface block of some sort drives groundwater to the surface.

If flooding takes place, then inflow from a stream causes a rise in groundwater levels near the channel, a temporary phenomenon but one providing storage water that is later released, after the flood. It is hard to evaluate this bank storage, but it has to be recognized.

In 1963, H. H. Cooper, Jr, and M. I. Rorabaugh derived solutions for changes in groundwater head near the stream as well as for groundwater input to it and bank storage.[2] They also produced a set of asymmetrical flood-wave stage hydrographs to assist in studies on the results which may occur in groundwater as a consequence of a number of shapes of floods.

A very important parameter is base flow, i.e. streamflow arising from the discharge of groundwater, a minor component during precipitation, but one which may become significant when it is dry — even to the extent of comprising the total streamflow. Usually, base flow is not subject to big fluctuations and indicates the characteristics of an aquifer in a basin.[3] At any one moment, streamflow may include groundwater which has entered at various times and at different points within a basin; in order to separate it from surface runoff in floods, measurements of chemical concentrations may be undertaken.[4] The concentration of total dissolved solids or of any major ion can be utilized for the equation:

$$C_{TR}Q_{TR} = C_{GW}Q_{GW} + C_{SR}Q_{SR} \qquad (7.1)$$

where C is the ionic concentration and Q is the streamflow; the subscript TR indicates the total runoff, GW the base flow (contribution from groundwater) and SR the surface runoff. If solved for the base flow, this gives:

$$Q_{GW} = (C_{TR} - C_{SR})/(C_{GW} - C_{SR})Q_{TR} \qquad (7.2)$$

where

$$Q_{TR} = Q_{GW} + Q_{SR} \qquad (7.3)$$

The values of C_{GW} are measured when a dry period occurs and C_{SR} is measured in a tributary stream during a storm. C_{TR} is measured during the peak flow period in the main stream.

In 1969, G. F. Pinder and J. F. Jones determined the groundwater component of peak discharge in three small basins (6–13 km^2 in area) from the chemistry of the total runoff and showed that groundwater contributed from 32 to 42% of the total flow at peak discharge.[5]

Recession curves can be constructed for base flow in order to show its variation with time during periods of low or no rainfall over a drainage basin, and the drainage rate of the groundwater storage is measured from this.[6] The continuation of base flow in prolonged drought conditions is possible if the supply arises from extensive and very permeable aquifers, but this is not the case where small, low-permeability aquifers are involved.

Streamflow hydrograph analysis demonstrates that the recession curve may often be expressed by:

$$Q = Q_0 K^t \tag{7.4}$$

where Q is the streamflow at time t in days after a given discharge Q_0 and K is a recession constant which is consequent upon the hydrological features of the basin and determinable empirically from the slope of a straight line fitted to a series of consecutive discharges plotted on semi-logarithmic paper (typical values lie in the range 0·89–0·95). It has been shown that recession curves of base flow are dependent upon the extent of entrenchment of the stream channel in the aquifer. For one which penetrates fully, the recession curves do not plot as straight lines on semi-logarithmic paper, but continuously decrease with time and produce a concave curve. On the other hand, if penetration is only partial, then a straight-line approximation is applicable. The value of K in eqn (7.4) in fact varies directly with the extent of stream entrenchment.

It has been assumed above that groundwater drains only towards the stream channel, but it can also flow downwards to an underlying leaky aquifer, and be lost as well by evapotranspiration into the atmosphere, as K. P. Singh indicated in 1968.[7] If these diversions become significant, the recession curve is deflected downwards. In semi-arid areas, where evapotranspiration is marked, streamflow being intermittent, losses through this route are sufficient to steepen the recession curve until the streamflow finally ends.

7.2. EVAPOTRANSPIRATIVE FLUCTUATIONS

It was noted previously that evaporation can affect the yield from subterranean water (see Section 1.2 p. 11), and it may cause daily fluctuations in unconfined aquifers having water tables near to the surface of the ground. Transpiration can have a similar effect and has a parallel diurnal fluctuation because it is also correlated with temperature. Evaporation increases as the water table nears the ground surface, but its rate depends upon the structure of the soil, which controls the capillary tension above the water table and hence its hydraulic conductivity. In isothermal conditions the upwards movement is in the liquid phase, but, in a dry soil, the water may respond to a vapour pressure gradient and ascend in the vapour phase. In experiments related to this matter lysimeters are used to measure evaporation.

The rate of transpiration is practically equalled by the uptake of water by plant roots in the saturated bed (if they reach it). The fluctuations in transpiration rate relate to season, weather and the kind of vegetation involved.

Generally, it is not possible to separate evaporation and transpiration and the combined loss is studied, i.e. the evapotranspiration or consumptive use. Investigations demonstrate that the deeper roots go, the greater is the depth at which water losses take place. It seems that, even in the case of very deep water tables, evapotranspiration still occurs to some extent in the vapour phase.

The amount of groundwater which is withdrawn by evapotranspiration daily can also be computed. It is assumed that the process is negligible from midnight to 0400 hours, also that the water table level during this time interval approximates to the daily mean so that the hourly recharge from midnight to 0400 represents the average rate for the day. If h is the hourly rate of rise of the water table from midnight to 0400 and s is its net fall or rise during the entire 24-hour period, to a

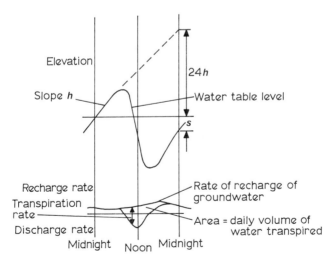

FIG. 7.1. The interrelationships between the water table level, the recharge and evapotranspiration fluctuations.

reasonable approximation the daily volume of groundwater discharge per unit area is:

$$V_{ET} = S_y(24h \pm s) \qquad (7.5)$$

where S_y is the specific yield near the water table. In fact, the rate of groundwater recharge to the vegetation varies inversely with the level of the water table. Consequently, the difference between this and the slope of the groundwater level curve provides the rate of evapotranspiration and this is illustrated in Fig. 7.1. The area between the curves is a measure of the quantity of water released daily to the atmosphere.

7.3. METEOROLOGICAL FLUCTUATIONS

Barometric changes cause substantial fluctuations in wells which penetrate confined aquifers.[8] Atmospheric pressure increases produce decreases in water levels and vice versa, i.e. the relationship is an inverse one. If atmospheric pressure changes are expressed in terms of a column of water, the ratio of change in water level to pressure change is a statement of the barometric efficiency of an aquifer. Hence:

$$B = \frac{\gamma \Delta h}{\Delta p_a} \qquad (7.6)$$

where B is the barometric efficiency, γ is the specific weight of water, Δh is the change in the piezometric level and Δp_a is the change in the atmospheric pressure.

The reason for the phenomenon of variable barometric efficiency lies in the fact that aquifers are elastic bodies, as O. E. Meinzer indicated as long ago as 1928.[9] He referred to studies made on part of the Dakota sandstone: up to that time, it had been thought that these water bodies were rigid and incompressible. If Δp_a is the change in the atmospheric pressure and Δp_w is the resultant change in the hydrostatic pressure at the top of a confined aquifer, then:

$$\Delta p_a = \Delta p_w + \Delta s_c \qquad (7.7)$$

where Δs_c is the increased compressive stress on the aquifer. With a well penetrating the confined aquifer, there is the following relation:

$$p_w = p_a + \gamma h \qquad (7.8)$$

ENVIRONMENTAL EFFECTS ON GROUNDWATER   157

If the atmospheric pressure increases by Δp_a, then:

$$p_w + \Delta p_w = p_a + \Delta p_a + \gamma h' \qquad (7.9)$$

and, substituting for p_w from eqn (7.8):

$$\Delta p_w = \Delta p_a + \gamma(h' - h) \qquad (7.10)$$

However, from eqn (7.7), it is clear that $\Delta p_w < \Delta p_a$; thus $h' < h$. Normally, therefore, the water level in a well falls with an increase in barometric pressure and vice versa. It is interesting to note that nuclear explosions in the USSR originated atmospheric pressure waves which triggered piezometric surface fluctuations in limestone aquifers in England, according to J. Ineson.[10] There are expressions which relate the barometric efficiency of a confined aquifer to its properties[11] and it has been demonstrated that soil moisture changes in response to infiltration are able to affect the magnitude of barometric efficiency.[12]

In unconfined aquifers, the changes in atmospheric pressure are transferred at once to the water table so that no pressure difference arises. However, trapped air can be affected in pores underneath the water table, where there may be fluctuations, but they are much smaller than those taking place in confined aquifers. Where air is so trapped, temperature fluctuations in the capillary zone can induce fluctuations in the water table.

On small and permeable oceanic islands, the water tables are much influenced by fluctuations in atmospheric pressure. Sea-level response is isostatic, adjusting to a constant mass of the column of the ocean and atmosphere. The ocean then acts as an inverted barometer with the sea level rising roughly 1 cm to compensate for a drop in the atmospheric pressure of 100 Pa. Such fluctuations total approximately 20 cm in the open ocean and are transmitted as long-term tides to the water table.

Rainfall is not an accurate index of the recharge of groundwater because of the surface and sub-surface losses it incurs, and because of the transit time for vertical percolation. This can vary from minutes (in shallow water table in permeable aquifer situations) to years (in deep water tables which underlie sediments having low vertical permeability). Of course, in arid regions, there may be practically no recharge from rainfall. Droughts of long duration cause considerable decline in water levels.

Wind can be an agent of fluctuation of water levels, by blowing over well tops. This causes the air pressure in the well to drop suddenly and hence the water level rises rapidly. Later, when the wind has passed, the air pressure in the well rises and the water level declines.

Frost layering above the water table in very cold areas may result in a gradual decline of the table through the winter; it is succeeded by a sharp rise in the early spring, i.e. before any surface recharge could affect it. The process is as follows. In the winter, the water moves upwards from the water table by capillary movement and vapour transfer to the frost layer, and there it freezes. In the spring, at that time when the air temperature reaches 0°C, i.e. rather early, thawing commences at the bottom of the frost layer and this meltwater percolates downwards and re-joins the water table.

7.4. TIDAL FLUCTUATIONS

Two tides are involved, namely oceanic and earth tides.

In the first case, coastal aquifers which are in contact with the ocean respond to tides with sinusoidal fluctuations in the levels of groundwater. Sinusoidal waves are propagated inland from the submarine outcrop of the aquifer when the sea level varies with a simple harmonic motion and, as distance increases, the inland amplitude of the waves decrease with an accompanying increase in the time lag of a given maximum. The effect may be considered as analogous to heat conduction in a semi-infinite solid which is subject to periodic variations of temperature normal to the infinite dimension.[13]

In the same manner as atmospheric pressure changes cause variations in piezometric levels, tidal fluctuations vary the load on confined aquifers which extend under the floor of the ocean.

However, the tidal effects are direct and, as the sea level increases, so does the level of the groundwater. The term tidal efficiency is applied to describe the radio of the piezometric level amplitude to the tidal amplitude, and C. E. Jacob demonstrated that it relates to the barometric efficiency thus:

$$C = 1 - B \quad (7.11)$$

where C is the tidal efficiency and B is the barometric efficiency.[14] Tidal efficiency, therefore, is a measure of the incompetence of overlying, confining strata to resist pressure changes.

Earth tides are manifested by regular, approximately semi-daily fluctuations of small magnitude which take place in piezometric surfaces of confined aquifers which may be located very far away from oceans.[15]

Allowing for changes in atmospheric pressure, such fluctuations appear to be quite distinct in some wells. The earth tides that cause them are in turn produced by the attraction of the moon on the crust of the Earth; to a lesser extent, the attraction of the sun is also involved. Observations suggest that there are two daily cycles of fluctuation which occur some 50 min later each day (as the moon does also). In addition, the average daily retardation of cycles is in close agreement with the transit of our satellite and the daily troughs of the level of water coincide with those of the transits of the moon at upper and lower culminations. Finally, the periods of large and regular fluctuations coincide with periods of new and full moon, periods of small and irregular fluctuations coinciding with periods of first and third quarters of the moon. Hence, wells are seen to be sensitive indicators of this dilatation of the crust of the Earth.[16] At new and full moons ocean tides are greater than usual because moon and sun act together. When the moon is in the first and third quarters, the tide-producing forces of the sun and moon act perpendicularly to each other with the result that the ocean tides are smaller than average. This coincidence of the moon's transit with low water can be explained by the fact that horizontal tidal attraction of groundwater is then maximum, the overburden load on the aquifer is lessened and hence the aquifer is permitted to expand slightly.

7.5. POPULATION CENTRES

In towns and cities there arises a greater water demand coupled with a reduced component of recharge due to the much greater interception and runoff, compared with country communities; there, a hydrologic balance is based upon supply from shallow wells more or less being restored to the ground as a result of disposal of waste water through cesspools or septic tanks.

As the numbers of people grow in the process of urbanization, shallower wells are abandoned and fewer, but deeper, sources are exploited. At a later stage, sewer systems come into being and the storm waters plus waste waters discharge into surface water bodies. Consequently, there is no longer a hydrologic balance and a decline in the groundwater level may occur. Three relevant factors are adverse: firstly, the reduced recharge due to interception and runoff; secondly, the increased exploitation of the groundwater by pumping deeper wells; and thirdly, a decrease in groundwater recharge resulting from the conduction of waste water through sanitary sewers. The effects have been demonstrated on Long Island, New York,[17] where decline in water table levels has been accompanied by increase in pollution.

7.6. SEISMICITY

As might be expected on *a priori* grounds, earthquakes exert a variety of effects on groundwater. Of these, the most visible are rapid rise and fall of water in wells or in the output of springs, together with mud and water eruptions from the ground. More frequent are slight fluctuations in well water levels, however, these hydroseisms operating in wells penetrating confined aquifers. The mechanism is dilatation, i.e. compression and expansion of elastic, confined aquifers through Rayleigh (earthquake) waves travelling at about 200 km/min, fluctuations then appearing roughly an hour after the event, even in the case of very distant earthquakes.

It is interesting to note that Man's activities can trigger earthquakes also — and not only by nuclear testing. Thus, injection of waste water into a deep well can produce shock waves, caused by reduced frictional resistance to faulting and increasing pore pressure. In fact, these observations led to investigations into the feasibility of injecting water deliberately into potentially dangerous fault zones, in which stress could be released thereafter in a set of minor earthquakes. This is preferable to the sudden liberation of accumulated stress which is characteristic of the disastrous variety of earthquake.

One of the best-known instances of microseisms was the earthquake at Anchorage, Alaska, USA, which occurred on 27 March, 1964 with a value of 8·4–8·8 on the Richter scale. Levels of groundwater throughout the USA were affected by this disturbance and the largest fluctuation exceeded 7 m at a well in South Dakota.[18] Hydroseisms were recorded also as far away as Egypt, South Africa and Australia.

Another example of a cause of a fluctuating piezometric surface is railway train traffic. As mentioned above, aquifers are elastic and hence they are subject to hydrostatic pressure changes when the load values change. A superincumbent load imposed upon an aquifer compresses it and increases the hydrostatic pressure temporarily. Later, interior water flows away radially from the point of application of the load and the hydrostatic pressure declines almost to its original value. It may be inferred from this that the load is first shared by the material of the aquifer and its confined water, but in this later stage, the geological material takes over most of the load.

7.7. SUBSIDENCE

Subsidence may result from changes in the level of groundwater or the subsurface conditions of moisture; it can be very deleterious to wells and interfere adversely with drainage, flood protection and water-conveying structures. Often, it accompanies a sizeable lowering of the piezometric surface in areas where heavy pumping is being effected from confined aquifers. The process is explicable in terms of soil mechanics. The geostatic pressure at any one depth is given by:

$$p_t = p_h + p_i \qquad (7.12)$$

where p_h is the hydraulic pressure, p_i is the intergranular pressure and p_t is the total pressure (geostatic pressure) at that depth. If pumping is conducted in a confined aquifer and the piezometric surface is lowered, the water table may remain unaltered; if there is an impermeable clay layer separating two aquifers, an upper, unconfined and a lower, confined then eqn (7.12) becomes:

$$p_t = p'_h + p'_i \qquad (7.13)$$

where p'_h and p'_i are the hydraulic and intergranular pressure, respectively, in a confined aquifer. For both the confined aquifer and the clay layer, $p'_h < p_h$ and $p'_i > p_i$. Instantaneous adjustment to the new distributions of pressure occurs in the permeable, coarse-grained aquifer. However, such an adjustment may take months or years to take place in a rather impermeable, fine-grained clay layer. Clay is highly compressible and the increase in inter-granular pressure $(p'_i - p_i)$ compacts the clay which reduces its porosity, water from the relevant pores being squeezed down into the confined aquifer. Obviously, the total volume of displaced water will be equal to the sum of the reduction in the volume of the clay and the volumetric change due to vertical subsidence of the land. It is clear that the degree of compaction relates to the thickness of the clay and its vertical permeability, also to the time and extent of decline in the piezometric surface as well as to the microstructure of the clay itself. Incidentally, increased intergranular pressure has almost no effect on an aquifer where sand and gravel are involved because they are relatively incompressible.

There are many examples of land subsidence through compaction of fine-grained materials; of these perhaps the most fascinating is the case of Venice, of which a mathematical simulation was given by G. Gambolati and R. A. Freeze in 1974.[19] Many such studies have been made and J. F.

Poland in 1972 gave a list of areas of major land subsidence due to overdrafting (too much removal) of groundwater; this is given in Table 7.1.[20]

Since the compaction of clays is both inelastic and permanent, the sole possible control is to increase the piezometric levels by reducing the pumping, and by recharging water through special injection wells. The removal of oil and gas produces the same subsidence difficulties and, of course, there are major affected areas in the USA as well as in Venezuela, Japan and elsewhere.

Ground surface collapse may result also from hydrocompaction, a phenomenon taking place when water is applied to soils. The soils most liable to this are loose ones such as alluvium and loess which may be deficient in moisture; some may never have been saturated since their formation, desiccation being accompanied by high voids content and low density (1·1 to 1·4 g/cm^3). Problems arise from irrigation: when water is applied, the interior void structure of the material is broken up and an irregular subsidence of the surface of the ground may result.

The contrary activity, the de-watering of organic soils such as peat, is also a cause of subsidence. De-watering results from desiccation, compaction, erosion by wind, burning, biochemical oxidation, etc., and it has been shown empirically that the rate of subsidence is proportional to the depth of the water table. Subsidence of this type has been observed in many countries, including the USA, the USSR and the Netherlands.

Sinkholes are manifest in catastrophic land subsidence with which may be associated also decline in the levels of groundwater. The process of formation is in itself slow; solube limestones and dolomites dissolve in groundwater and the land surface sinks into a cup-shaped depression. Covering large areas, there may form a huge karstic sinkhole plain, of which most of the drainage is subsurface. The lowering of water tables by pumping can promote the formation of new sinkholes and this has happened in South Africa where, in the Far West Rand, pumping of groundwater from a dolomite aquifer for mine de-watering caused eight sinkholes to form within three and a quarter years. None is less than 50 m in diameter and all exceed 30 m in depth.

An opposite effect is the uplift of the crust which has been noted as taking place over large land areas which have been heavily pumped for groundwater. This phenomenon is a tectonic one which entails an elastic expansion of the lithosphere rebounding after huge volumes of groundwater have been withdrawn. It has been recorded in parts of Arizona, California and Texas where the groundwater obtained from aquifers is

ENVIRONMENTAL EFFECTS ON GROUNDWATER 163

TABLE 7.1
AREAS OF MAJOR LAND SUBSIDENCE DUE TO OVERDRAFTING OF GROUNDWATER[20]

Location	Depositional type and age	Depth range of compacting beds (m)	Max. subsidence (m)	Area of subsidence (km^2)	Main dates
Japan					
Osaka	Alluvial, shallow marine: Quaternary	10–400	3	190	1928–68
Tokyo	Alluvial, shallow marine: Quaternary	10–400	4	190	1920–70
Mexico					
Mexico City	Alluvial, lacustrine: late Cenozoic	10–50	9	130	1938–70
Taiwan					
Taipei basin	Alluvial, lacustrine: Quaternary	10–240	1.3	130	1961–69
USA					
Arizona, central	Alluvial, lacustrine: late Cenozoic	100–550	2.3	650	1948–67
California, Santa Clara valley	Alluvial, shallow marine: late Cenozoic	55–300	4	650	1920–70
San Joaquin valley (three sub-areas)	Alluvial, lacustrine: late Cenozoic	60–1000	2.9–9	11 000	1935–70
Lancaster area	Alluvial, lacustrine: late Cenozoic	60–300(?)	1	400	1955–67
Nevada Las Vegas	Alluvial: late Cenozoic	60–300	1	500	1935–63
Texas Houston–Galveston area	Fluvial, shallow marine: late Cenozoic	60–600(?)	1–1.5	8 860	1943–64
Louisiana Baton Rouge	Fluvial, shallow marine: Miocene to Holocene	50–600(?)	0.3	650	1934–65

removed by the evapotranspiration of crops.[21] In the Santa Cruz River Basin in Arizona, USA, a crustal uplift was noted of 6 cm between 1948 and 1967. During this period, 43.5×10^9 tons of groundwater were pumped from the area of just over 8000 km^2.

REFERENCES

1. NORRIS, S. E. and EAGON, J. B., Jr, 1971. Recharge characteristics of a water-course aquifer system at Springfield, Ohio. *Ground Water*, **9**, 1, 30–41.
2. COOPER, H. H., Jr, and RORABAUGH, M. I., 1963. Groundwater Movements and Bank Storage due to Flood Stages in Surface Streams. US Geol. Surv. Water Supply Paper 1536-J, pp. 343–66.
3. JOHNSTON, R. H., 1971. Base Flow as an Indicator of Aquifer Characteristics in the Coastal Plain of Delaware. U.S. Geol. Surv. Prof. Paper 750-D, 212–15.
4. VISOCKY, A. P., 1970. Estimating the ground-water contribution to storm runoff by the electrical conductance method. *Ground Water*, **8**, 2, 5–10.
5. PINDER, G. F. and JONES, J. F., 1969. Determination of the ground-water component of peak discharge from the chemistry of total runoff. *Water Res. Res.*, **5**, 438–45.
6. FREEZE, R. A., 1972. Role of subsurface flow in generating surface runoff *Water Res. Res.*, **4**, 985–99.
7. SINGH, K. P., 1968. Some factors affecting baseflow. *Water Res. Res.*, **4**, 985–99.
8. CLARKE, W. E., 1967. Computing the barometric efficiency of a well. *J. Hydr. Divn, Amer. Soc. Civ. Engrs*, **93**, HY4, 93–8.
9. MEINZER, O. E., 1928. Compressibility and elasticity of artesian aquifers *Econ. Geol.*, **23**, 263–91.
10. INESON, J., 1963. Form of groundwater fluctuations due to nuclear explosions. *Nature*, **198**, 22–3.
11. JACOB, C. E., 1939. Fluctuations in artesian pressure produced by passing railroad trains as shown in a well on Long Island, New York. *Trans. Amer. Geophys. Union*, **20**, 666–74.
12. GILLILAND, J. A., 1969. A rigid plate model of the barometric effect. *J. Hydrol.*, **7**, 233–45.
13. GREGG, D. O., 1966. An analysis of ground-water fluctuations caused by ocean tides in Glynn County, Georgia. *Ground Water*, **4**, 3, 24–32.
14. JACOB, C. E., 1940. On the flow of water in an elastic artesian aquifer. *Trans. Amer. Geophys. Union*, **21**, 574–86.
15. ROBINSON, E. S. and BELL, R. T., 1971. Tides in confined well aquifer systems. *J. Geophys. Res.*, **76**, 1857–69.
16. BREDEHOEFT, J. D., 1967. Response of well–aquifer systems to earth tides. *J. Geophys. Res.*, **72**, 3075–87.
17. FRANKE, O. L., 1968. *Double-mass-curve Analysis of the Effects of Sewering on Ground-water Levels on Long Island, New York.* US Geol. Surv. Prof. Paper 600-B, pp. 205–9.

18. VORHIS, R. C., 1967. *Hydrologic Effects of the Earthquake of March 27 1964 Outside Alaska*. U.S. Geol. Surv. Prof. Paper 544-C, 54 pp.
19. GAMBOLATI, G. and FREEZE, R. A., 1973. Mathematical simulation of the subsidence of Venice. *Water Res. Res.*, **9**, 721–33; **10**, 563–77.
20. POLAND, J. F., 1972. Subsidence and its control. In: *Underground Waste Management and Environmental Implications*. Amer. Assn Pet. Geologists, Memoir 18, pp. 50–71.
21. HOLZER, T. J., 1979. Elastic expansion of the lithosphere caused by groundwater depletion. *J. Geophys. Res.*, **84**, 4689–98.

CHAPTER 8

Geothermal Waters

8.1. OCCURRENCE

The normal geothermal gradient is approximately 1°C per 30 m depth, but this is far exceeded in certain regions often characterized by surface manifestations such as volcanoes, geysers, fumaroles, hot springs and boiling mud pools which indicate additional, subterranean, terrestrial heat energy. These tend to follow planetary paths also involving seismic activity such as the so-called Pacific Belt or Girdle of Fire, the ribbon-like border of that ocean. The plate tectonics theory explains the association by ascribing it to the relative motions of plates which may originate by sea-floor spreading, by-pass each other, or be destroyed by subduction. Thus, it is possible to divide the surface of the Earth into non-thermal and thermal parts, the former with normal geothermal gradients and the latter with the very high ones mentioned above. H. Christopher H. Armstead has termed the thermal regions 'hyperthemal' and it is possible to consider them as comprising vapour-dominated and water-dominated systems.[1] A great deal of investigation has taken place, mostly in fumarolic and hot-spring areas; where exploration is involved, the problem is to evaluate the cause of the surface temperature anomalies, i.e. to ascertain whether the temperature in springs results from deep circulation in a locality where the geothermal gradient is only a little above normal or whether the water is derived from a shallow aquifer possessing relatively high temperatures and cools while ascending to the ground surface.

8.2. CHEMICAL AND ISOTOPIC THERMOMETRY

To use geochemical indicators to evaluate subsurface temperatures, several assumptions are made which may preclude wider application. They are as follows. Firstly, the chemical reactions which affect concentrations of dissolved chemical components in hot waters must depend upon temperature and all relevant components must be present in adequate amounts. Secondly, the water–rock and gas reactions must attain equilibrium which must occur at the temperature of the aquifer. Thirdly, re-equilibration at temperatures less than that in the aquifer must not take place during the movement of the fluid from the water reservoir to the surface. Finally, the fluid from depth must not enter into mixing processes with shallower, colder fluids. However, R. O. Fournier and A. H. Truesdell in 1974 believed that wider applicability was justified, although C. Panichi in 1976 did not.[2,3]

8.3. QUANTITATIVE CHEMICAL GEOTHERMOMETERS

It has been stated that high-temperature waters in a hydrothermal area are saturated in silica (assuming equilibrium with quartz), and a chemical method for determination of the temperatures of the waters supplying hot springs and geothermal wells was devised.[4]

In laboratory experiments, it was shown that the solubility values for silica in water are governed by temperatures and an appropriate graph provides the most probable value of the temperature as the average of two water–quartz equilibration curves, relative to cooling by conduction or by isoenthalpic expansion of the water flowing surfacewards. It has been observed that, when equilibrium is attained in high-temperature aquifers, the silica may precipitate quickly when cooling of the water to 180°C takes place; hence the approach cannot be utilized in estimation of actual temperatures in the aquifer when these exceed 180°C. However, the rate of precipitation of silica decreases rapidly when the temperature is less than 180°C. The method is probably not applicable where the waters have an acidic pH and low concentrations of chloride ions, because these waters, at temperatures around 100°C or lower, react readily with silicates and acquire a large amount of dissolved amorphous silica (which is more soluble than quartz).

Hot waters flow up to the surface with accompanying hydrostatic pressure decrease, so that boiling may occur when the vapour and gas pressure is greater than hydrostatic. As some of the water is removed as steam and silica is retained in solution if the transition is fast, solutions result which are supersaturated in quartz and, at a later stage, in amorphous silica as well. If the upwards movement of the water is slow, silica concentration alters towards the solubility values for quartz at the lowest temperatures of the shallower zones, the excess silica precipitating.

The kinetics of the equilibrium reaction of water with quartz is rapid, taking hours or days, not years. It must be noted that in New Zealand apparently only rarely does silica, as quartz and in the amorphous form, appear in quantity in wells that are almost continuously used. Where wells underwent repeated shutting down, however, quartz occurred in deeper, hotter layers and amorphous silica in the shallower layers just above the zone in which boiling initially begins.

It is important to remember that thermal waters may mix with non-thermal and cause erroneous estimates of temperature. In New Zealand, curiously, the reverse of the normal is the case in that the cold waters have lower silica concentrations than the thermal ones; thus the composition of the mixture would lead to temperature estimates for the aquifer that are lower than the true values. In the thermal springs of Yellowstone National Park, where the composition implies mixing of deeper, hotter waters and surface, colder ones, the original temperature of the deeper constituent and also the proportion of colder water in the mixture have been assessed from the measured temperatures and the silica content of the thermal springs together with the silica content in the non-thermal springs of the region.[1]

Considering the sodium–potassium and sodium–potassium–calcium ratios, appropriate equilibrium constants for exchange and hydrothermal alteration are dependent upon temperature, i.e. the component ratios in solution vary with the equilibrium temperature.

It has been determined empirically that the hot-water reactions with volcanic rocks produce sodium–potassium ratios in solution which correlate with ratios noted in thermal spring waters of which the temperatures are known.[5] Where a coexistent sodium and potassium feldspars system occurs, there is a sodium–potassium ratio in solution which is a function of the temperature of the system. However, D. E. White in 1970 stated that the sodium–potassium molal ratios are important only when the variations range between 8:1 and 20:1 and suggested that temperatures so evaluated are invalid for spring waters with acidic pH.[6]

With deep waters which cool when flowing to the surface through initial boiling, exchange reactions which govern the sodium–potassium ratios take longer to attain new equilibria than is the case with silica solubility. As a result, it has been proposed that they represent a better source for deep aquifer temperatures than the latter.[3] The sodium–potassium ratios do appear optimal for evaluating deep temperatures, and silica geothermometry may be more suitable for well waters in the absence of evaporation. In a water–silicate system, it was shown that calcium content may be significant, but the quantity actually in solution depends upon the presence of carbon dioxide.[7] Calcium concentrations derived from carbonates decrease with increasing temperatures and may also vary when the partial pressure of carbon dioxide alters through pH variations. Consequently, it was asserted that the sodium–potassium ratios cannot be employed for evaluation of deep temperatures in cases where there are high partial pressures for the carbon dioxide, as these cause greater retention of calcium in solution.[3] Sodium–potassium–calcium molal concentrations for a number of geothermal waters lie on a straight line when a graph is constructed of the function log K against the reciprocal of absolute temperature:

$$\log K = \log \frac{Na^+}{K^+} + \beta \log \frac{\sqrt{Ca^{2+}}}{Na^+} = -2\cdot 24 + \frac{1647}{T} \qquad (8.1)$$

where β is 1/3 for equilibrated waters at a temperature above the boiling point and 4/3 for equilibrated waters at lower temperatures, which are given in K.

Sodium–potassium–calcium ratios are said to provide better results than the sodium–potassium ratios, especially in the case of geothermal waters which have circulated in calcium-rich environments, never having undergone calcium carbonate precipitations after leaving the main aquifer.[3] In the opposite case, of course, the geothermometer will provide anomalously high aquifer temperatures.

In 1975, T. Paces showed that, with waters which circulate in an acidic environment (e.g. one rich in carbon dioxide), the sodium–potassium–calcium thermometer is unusable unless the solution has attained equilibrium with minerals which contain these elements.[8] It is found that waters in which the carbon dioxide partial pressures exceed 5×10^{-3} atm give temperatures higher than the real ones: this probably results from their failure to reach equilibrium with rocks due to the dissociation of carbonic acid.

In applying quantitative geothermometers in areas of geothermal waters where little is known of the hydrology, the following procedure can be followed. With springs at boiling point, there are two cases to be considered, namely low and high rates of flow. With low flow rates, cooling probably occurs primarily as a result of conduction, and the chemical indicators can be employed by assuming very little or no water loss by evaporation. With high flow rates, an adiabatic cooling is assumed and the indicators can be used if a maximum vapour loss is assumed.

With springs below boiling point, low and high flow rates may again be examined. With low flow rates, interpretation is hard and the chemical indicators are applicable only where it is known that a hot water has undergone cooling by conduction. With high flow rates, a nonconductive cooling is assumed; hence, if the sodium–potassium–calcium thermometer is used and the data show an equilibrium at higher temperatures than were noted in the spring, the sample may be regarded as a mixed water and the evaluation must be made on this basis.

8.4. QUALITATIVE CHEMICAL GEOTHERMOMETERS

High or low temperatures are indicated by assessing the contents of calcium and bicarbonate, and the ratios of magnesium, calcium and sodium, and calcium together with Cl^- ($HCO_3^- + CO_3^{2-}$) and Cl^-/F^-. The results are a control on those derived by applying the silica and the sodium–potassium–calcium thermometers. The drawback is that it is necessary to understand the chemical behaviour of calcium and carbon dioxide in the geothermal waters and this is highly complex. In the geysers, there are waters cooled from over 170°C and here the total calcium and carbon dioxide concentrations are lower than in non-thermal waters with equal salinity. However, in water from springs which deposit travertine, the calcium and carbon dioxide contents are higher, in agreement with actual subsurface temperatures obtained from boreholes or estimated from silica concentrations.

Thermal spring analyses both in Japan and in New Zealand have shown that low magnesium contents and low values of the magnesium–calcium ratio characterize high-temperature systems.

The $Cl^-/(HCO_3^- + CO_3^{2-})$ ratio is higher in high-temperature waters, but the actual temperture dependence is not fully understood. The control factor may be carbonic acid which is undissociated at high temperatures. High values of the Cl^-/F^- ratio represent high temperatures at depth; fluorite solubility decreases with depth.

8.5. ISOTOPE THERMOMETERS

Equilibrium constants for isotopic exchange reactions are temperature-dependent; the fractionation between two molecules in a chemical reaction usually decreases with increasing temperature. This provides a means of evaluating temperatures in steam-producing geothermal systems where the chemical indicators cannot be employed because of their low volatility, which leads to their practical absence from the vapour which is produced. A consequence is that the water discharged on the surface has acquired its chemical composition after condensation of the vapour in the upper subterranean layers, and thus it does not indicate actual aquifer conditions. The gas content varies locally, but carbon dioxide is probably always predominant, perhaps to as much as 90% or more, with subsidiary amounts of hydrogen sulphide, methane, hydrogen and nitrogen. It may be that there is an isotopic equilibrium situation in the aquifer of the type:

$$CO_2 + 4H_2 = CH_4 + 2H_2O$$

involving isotopic fractionations between the stable carbon, hydrogen and oxygen isotopes as well.

CO_2 and CH_4 isotopic fractionation has been widely used in evaluating deep temperatures for geothermal fluids. The appropriate equilibrium constants were calculated by Y. Bottinga in 1969 for the temperature range of 0–700°C.[9] In the geothermal range, say from 100 to 400°C, values are expressible by the equation:

$$-1000 \ln = -9{\cdot}01 + 15\,301\,\frac{10^3}{T} + 2361\,\frac{10^6}{T^2} \qquad (8.2)$$

where T is the temperature in K, the enrichment factor α being defined by $1 + \alpha \doteq R_A/R_B$ where α is the fractionation factor and R_A and R_B are the $^{13}C/^{12}C$, $^{18}O/^{16}O$ and D/H ratios of the components A and B in the given system, values of α being expressed per mille (‰).

The basic assumption in applying the thermometer is that there is an isotopic equilibrium between CO_2 and CH_4. Such an isotopic equilibrium is attained where the reaction kinetics are sufficiently fast at the relevant temperatures, and the time of contact between the reacting isotopes is also relevant. Carbon dioxide and methane in natural gas mixtures obtained from the same geochemical process are probably in isotopic equilibrium, but in the case of geothermal fluids there is evidence that temperature values derived are higher than those measured on the surface, implying that there may not be such an equilibrium. With hydrogen fractionation geothermometers have not been so widely utilized, and the reaction rate of oxygen fractionation in the CO_2–H_2O system is very rapid, which is a disadvantage in using it. Geothermal waters undergo alterations in their oxygen isotope composition because of exchange reactions with oxygen in host rocks during deep circulation and the rate of these exchanges is temperature-related.

Attempts have been made to evaluate subsurface temperatures using the oxygen isotope fractionation between dissolved sulphate and geothermal waters. G. Cortecci in 1974 employed appropriate equations for the HSO_4–H_2O and $CaSO_4$–H_2O systems and derived isotopic temperatures for water samples from wells at Larderello in Italy which agreed rather well with those actually observed.[10] This particular fractionation process was said by C. Panichi to be not only temperature-dependent, but also pH-dependent.[3] He quoted times for the isotope exchange reaction between sulphate and water to attain equilibrium: at pH 7, it attains equilibrium after nine years at 200°C, and at pH 3·8, it does so at the same temperature after 1·5 years; at pH 7, at 330°C, the time taken is estimated as 0·6 years, and at pH 3·8 at 330°C, as 0·08 years.

8.6. HYDROTHERMAL REACTIONS AND RESERVOIR TEMPERATURE ESTIMATES

Clearly, the composition of a geothermal fluid is controlled by temperature-dependent reactions between minerals and the fluid. P. R. L. Browne in 1978 summarized those factors which affect the formation of hydrothermal minerals as temperature, pressure, rock type, permeability, fluid composition and duration of activity.[11] The effects of rock type *per se* appear most pronounced at low temperatures and are disregarded as insignificant above 280°C. In fact, above this temperature and to at least 350°C, the typical stable mineral assemblage in active geothermal systems does not depend upon the original rock type and includes albite, potassium feldspar, chlorite, iron epidote, calcite, quartz, illite and pyrite, epidote not forming below 240°C.

The equilibrium required in using geochemical indicators as subsurface temperature evaluation criteria (see Section 8.2) is probably only attainable in conditions of high permeability. As seen earlier, the chemical isotopic reactions most widely used as geothermometers include silica, sodium–potassium, sodium–potassium–calcium and sulphate–oxygen isotope, $\Delta^{18}O$ (SO_4^{2-}–H_2O). Some pertinent data are summarized in Table 8.1.

R. O. Fournier in 1981 stated that the quartz geothermometer is best suited to use for well waters where the subsurface temperatures exceed 150°C, but that it can give erroneous results if employed indiscriminately.[12] Quartz is the most stable and least soluble polymorphic variety of silica in the temperature and pressure range of geothermal systems and, in natural waters above 150°C, appears to control the dissolved silica concentration. However, in special circumstances, other silica species may do so. Thus, S. Arnórsson in 1975 showed that, in the basaltic terrain of Iceland, chalcedony usually controls aqueous silica at temperatures under 110°C and occasionally at temperatures up to 180°C as well.[13] Most groundwaters which have not reached temperatures in excess of 80 or 90°C possess concentrations of silica greater than those predicted by the solubility of quartz. As stated above, some of these have equilibrated with chalcedony, but the silica concentrations in many groundwaters are a result of non-equilibrium reactions whereby silica is released to solution during the acid alteration of silicate minerals. At low temperatures, the rates of quartz and chalcedony precipitation are extremely slow and so aqueous silica values may become high in the case where acid is supplied by an outside source, e.g. the decay of organic matter, the oxidation of sulphides or the influx of carbon dioxide.

Fournier believed that the sodium–potassium geothermometer gives good results when the waters originate in high-temperature environments (180–200°C).

The sodium–potassium–calcium geothermometer handles calcium-rich waters which give anomalously high calculated temperatures using the sodium–potassium method. The effect on it of dilution is usually negligible if the high-temperature geothermal water is considerably more saline than the water of dilution. As seen earlier, if a particular water comprises a mixture of hot and cold with the former making up less than 20–30%, the effects of mixing on the sodium–potassium–calcium geothermometer are observable. If one of the two hypothetical waters commences at 210°C and the other at 10°C, then the mixture of equal proportions would have an actual temperature of approximately 110°C and give a $\beta = 4/3$ temperature of 143°C and a $\beta = 1/3$ temperature of 198°C, the latter being selected as the more likely.[7] Hence, the sodium–potassium–calcium geothermometer would yield a temperature which would be a

TABLE 8.1
EQUATIONS SHOWING TEMPERATURE DEPENDENCE OF SELECTED GEOTHERMOMETERS[12]
C = the concentration of dissolved silica. All concentrations are given in mg/kg.

Geothermometer	Equation	Restrictions
Quartz — no steam loss	$t°C = \dfrac{1309}{5.19 - \log C} - 273.15$	$t = 0\text{--}250°C$.
Quartz — maximum steam loss	$t°C = \dfrac{1522}{5.75 - \log C} - 273.15$	$t = 0\text{--}250°C$.
Chalcedony	$t°C = \dfrac{1032}{4.69 - \log C} - 273.15$	$t = 0\text{--}250°C$.
α-Cristobalite	$t°C = \dfrac{1000}{4.78 - \log C} - 273.15$	$t = 0\text{--}250°C$.
α-Cristobalite	$t°C = \dfrac{781}{4.51 - \log C} - 273.15$	$t = 0\text{--}250°C$.
Amorphous silica	$t°C = \dfrac{731}{4.52 - \log C} - 273.15$	$t = 0\text{--}250°C$.
Na/K (Fournier)	$t°C = \dfrac{1217}{\log(\text{Na}^+/\text{K}^+) + 1.483} - 273.15$	$t = 150°C$.
Na/K (Truesdell)	$t°C = \dfrac{855.6}{\log(\text{Na}^+/\text{K}^+) + 0.8573} - 273.15$	$t = 150°C$.
Na–K–Ca	$t°C = \dfrac{1647}{\log(\text{Na}^+/\text{K}^+) + \beta\{\log\sqrt{\text{Ca}^{2+}/\text{Na}^+} + 2.06\} + 2.47} - 273.15$	$t < 100°C, \beta = 4/3$; $t > 100°C, \beta = 1/3$.
$\Delta^{18}\text{O}(\text{SO}_4^{2-} - \text{H}_2\text{O})$	$1000 \ln \alpha = 2.88(10^6 T^{-3}) - 4.1$ $\alpha = \dfrac{1.000 + \delta^{18}\text{O}(\text{HSO}_4^-)}{1.000 + \delta^{18}\text{O}(\text{H}_2\text{O})}$ and T is in K	

mere 12°C below the real one. However, if the quantity of cold water in the mixture exceeds 75%, the $\beta = 4/3$ geothermometer temperature would drop to under 100°C and a hot spring would emerge at less than 60°C. As the $\beta = 4/3$ temperature is less than 100°C, it would probably be selected as the more likely temperature of water–rock equilibrium instead of the $\beta = 1/3$ temperature.

R. O. Fournier and R. W. Potter II in 1979 indicated that the sodium–potassium–calcium geothermometer gives anomalously high results when it is applied to waters which are rich in magnesium, temperature corrections to subtract from Na–K–Ca-calculated ones being available.[14] In addition, temperature corrections can be calculated by employing their equations, as follow:

For R between 5 and 50,

$$\Delta t_{Mg} = 10\cdot 66 - 4\cdot 7415 R + 325\cdot 87 (\log R)^2 - 1\cdot 032 \times 10^5 (\log R)^2 / T$$
$$- 1\cdot 968 \times 10^7 (\log R)^2 / T^2 + 1\cdot 605 \times 10^7 (\log R)^3 / T^2 \qquad (8.3)$$

and, for $R > 5$,

$$\Delta t_{Mg} = -1\cdot 03 + 59\cdot 971 \log R + 145\cdot 05 (\log R)^2$$
$$- 36711 (\log R)^2 / T - 1\cdot 67 \times 10^7 \log R / T^2 \qquad (8.4)$$

where $R = [Mg^{2+}/(Mg^{2+} + Ca^{2+} + K^+)] \times 100$, with concentrations expressed in equivalents, Δt_{Mg} is the temperature correction (°C) that must be subtracted from the Na–K–Ca-calculated temperature and T is the Na–K–Ca-calculated temperature (K). In certain conditions these two equations give negative values for Δt_{Mg} and, in this event, a Mg^{2+} correction is not applied to the sodium–potassium–calcium geothermometer. As with other geothermometers, the Mg-corrected sodium–potassium–calcium geothermometer is liable to error because of continued water–rock reaction as an ascending water cools down. If the Mg^{2+} concentration rises during the upward flow, applying the magnesium correction leads to an anomalously low calculated reservoir temperature. Therefore, the decision whether to apply such a correction is not always easy.

The $\Delta^{18}O(SO_4^{2-}-H_2O)$ geothermometer is based upon experimental work by R. M. Lloyd in 1968; he measured the exchange of ^{16}O and ^{18}O between H_2O and SO_4^{2-} at 350°C.[15] He also derived the rate of equilibration. In the pH range of most of the deep geothermal waters, rates of the sulphate–oxygen isotope exchange reactions are very slow compared with silica solubility and cation exchange reactions. This is an advantage because, once equilibrium is reached after a prolonged residence time in a reservoir at high temperature, there is little re-equilibration of the oxygen isotopes of sulphate as the water cools during its progress to the surface (unless this movement is extremely slow). However, if steam separation takes place during cooling, the oxygen isotopic composition of the water alters. The fractionation of ^{16}O and ^{18}O between liquid water and steam is temperature-dependent and comes to equilibrium practically at once at temperatures even as low as 100°C; hence, if a water cools adiabatically from a high temperature to 100°C, that liquid which remains at the end of boiling will have a different isotopic composition according to whether steam escaped continuously over a range of temperatures or whether all the steam stayed in contact with the liquid and separated at the last temperature (100°C). Clearly, boiling makes the interpretation difficult, but it does not preclude the use of this geothermometer and, in fact, it has been shown that sulphate–oxygen isotopic geothermometer temperatures can be calculated for three end-member models, namely, conductive cooling, one-step steam loss at any specified temperature and continuous steam loss.[16] The correct model to employ is quite clear where water is produced from a well and steam is separated at a known temperature or pressure. Conductive cooling is assumed for springs which emerge at temperatures well under the boiling point. Of course, the validity of calculated temperatures is affected in an adverse manner if different waters mix, e.g. hot and cold waters. Even if the cold part of the mixture contains no sulphate, calculated sulphate–oxygen temperatures will be erroneous unless the isotopic composition of the water is corrected back to the composition of the water in the hot part before the mixing occurred. This has been done for boiling and mixing effects in waters from the Yellowstone National Park and Long Valley, California. Another difficult problem is the formation of sulphate by oxidation of hydrogen sulphide since a small quantity of low-temperature sulphate causes a large error in the result obtained by the geothermometer. In fact, if only a couple of springs are involved, the addition of sulphate by the oxidation of hydrogen sulphide is frequently hard to observe; if all the springs have the same Cl^-/SO_4^{2-} ratio, however, it is likely that this H_2S is not important. If variations in Cl^-/SO_4^{2-} occur in waters from hot springs within an area, the water with the highest Cl^-/SO_4^{2-} ratio has the greatest possibility of being unaffected by the oxidation of hydrogen sulphide.

8.7. SUBTERRANEAN MIXING AND BOILING PHENOMENA

As might be expected, mixing is probably a very common occurrence when hot water rises into the shallower parts of hydrothermal systems and encounters cold groundwater. However, it is believed that mixing also takes place at depth. Some effects of this have been discussed earlier and the major problem is to recognize such mixed waters at the surface; this becomes more difficult where water–rock re-equilibration has happened after the mixing. Total or partial chemical re-equilibration becomes more likely if the post-mixing temperature exceeds the range 110–150°C or if the mixing takes place in an aquifer with a long fluid residence time. Indications of mixing are numerous, fortunately. Thus, in boiling springs, variations in chloride contents which are too large to be explained by steam loss is one such indicator. Another is variations in the proportions of isotopes of oxygen and hydrogen, particularly tritium. A third is systematic variations of the compositions of springs and their measured temperatures. Usually, the cold water is more dilute than the hotter water.

As mentioned earlier (Section 8.3; ref. 1), the dissolved silica concentration of a mixed water can be utilized in order to determine the temperature of the hot-water component, and the simplest method of calculation uses a plot of dissolved silica against enthalpy of liquid water. Although temperature can be measured whereas enthalpy is a derived property (obtained from steam tables if temperature, pressure and salinity are known), the latter is favoured as a coordinate rather than temperature. The reason is that the combined heat contents of two waters at different temperatures are conserved when those waters are mixed, but the combined temperatures are not.

Upward flowing and hot solutions may boil as they ascend because of the decreasing hydrostatic head. If the rate of upflow is sufficiently rapid, the cooling of the fluid may be roughly adiabatic. In the event of boiling, there will be a partitioning of dissolved elements between the steam and the residual liquid, with dissolved gases and other relatively volatile components concentrating in the steam and the non-volatile components concentrating in the liquid in proportion to the quantity of steam which separates. It is possible to calculate change in concentration resulting from boiling: for non-volatile components which remain with residual liquid as steam separates, the final concentration, C_f, after single-stage steam separation at a given temperature, t_f, is given by:

$$C_f = \frac{(H_s - H_f)}{(H_s - H_i)} C_i \qquad (8.5)$$

where C_i is the initial concentration before boiling, H_i is the enthalpy of the initial liquid before boiling and H_f and H_s are the enthalpies of the final liquid and steam at t_f. For solutions possessing salinities under about 10 000 mg/kg, enthalpies of pure water were given by J. H. Keenan et al. in 1969 in steam tables which can be used to solve eqn (8.5).[17]

For higher salinities, enthalpies of sodium chloride solutions have been tabulated by J. L. Haas, Jr, in 1976.[20] Another approach is to solve the equation graphically using a plot of enthalpy against a non-reactive dissolved constituent such as a chloride. Where a variety of chloride concentrations in hot springs seems to result mainly from different amounts of boiling, this can provide information regarding the minimum temperature of the reservoir which supplies the springs, as was the case with two chemically distinct types of hot spring water from Upper Basin in Yellowstone National Park plotted at the enthalpies corresponding to liquid water at the measured temperature of each spring. Figure 8.1 shows these enthalpy–chloride relationships. In the two waters considered, there are different data. The Geyser Hill type has ratios of $Cl^-/(HCO_3^- + CO_3^{2-})$, expressed in equivalents, exceeding four and chloride concentrations ranging from 352 to 465 mg/kg. The Black Sand

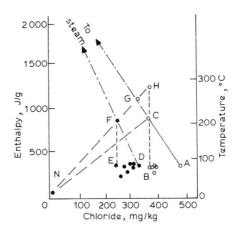

FIG. 8.1. Enthalpy–chloride relationships for waters from Upper Basin, Yellowstone National Park: ○, Geyser Hill type waters; ●, Black Sand type waters.

type waters possess $Cl^-/(HCO_3^- + CO_3^{2-})$ ratios near to 0·9 and chlorides ranging from 242 to 312 mg/kg. The minimum temperature of the water from the reservoir which feeds the Geyser Hill hot springs is determinable by initially drawing a straight line from the spring with maximum chloride (point A) to the enthalpy of steam at 100°C and thereafter extending a vertical line from the spring with least chloride (point B). The intersection of that vertical line with the previous line, point C, gives the minimum enthalpy of the water in the reservoir, 936 J/g, indicating a temperature of 218°C.[17] If the silica (quartz) geothermometer is applied to water A (with the assumption of maximum steam loss), a calculated reservoir temperature of 216°C is obtained; in the case of water B (assuming no steam loss), 217°C results. In fact, the close agreement between the calculated reservoir temperatures employing silica and chloride relations is evidence that the reservoir supplying Geyser Hill has a temperature near 218°C. The spectrum of chloride concentrations in the Black Sand type waters, E to D, suggests a reservoir temperature of approximately 209°C (point F); the silica geothermometer yielded 205°C.

Point F represents a more dilute and somewhat cooler water than point C, but F water cannot be derived from C water by a simple mixing process between hot and cold water (point N) because any such mixture would be on or near the line CN. Waters C and F probably both relate to higher-enthalpy water, e.g. G or H. Water F could relate to water G or H by mixing, in different proportions, with N. Water C could be related to G by boiling (adiabatic cooling resulting from a decrease in hydrostatic head) and the evaporative concentration of chloride of a slowly moving solution. Route G to C is more probable because boiling and loss of carbon dioxide into the steam phase may explain the different $Cl^-/(HCO_3^- + CO_3^{2-})$ ratios occurring in waters C and F. On the basis of further hot spring data from Midway and Lower Basins, R. O. Fournier *et al.* in 1976 concluded that a reservoir at about 270°C underlies the Upper and Lower Basins, and that an even hotter reservoir exists at greater depth.[18]

As for the effects of subterranean boiling (due to decreasing hydrostatic head) on the various geothermometers, adiabatic cooling has various results. The sodium–potassium geothermometer is unaffected, but the sodium–potassium–calcium one is affected significantly if loss of carbon dioxide causes calcium carbonate to precipitate, resultant calculated temperatures being too high. The silica geothermometer must be corrected for steam loss, as must the $\Delta^{18}O$ ($SO_4^{2-} - H_2O$) geothermometer must be corrected because of the temperature-dependent fractionation of oxygen isotopes between liquid water and steam.

8.8. GEOTHERMAL FIELDS

These contain exploitable heat resources associated with permeable strata and may be classified as shown in Table 8.2.

TABLE 8.2
DISTRIBUTION AND POTENTIAL OF THERMAL REGIONS

Type	Frequency	Economic value	Future
Vapour-dominated hyperthermal	Rare	High	Good
Water-dominated hyperthermal	Rare	High	Good
Semi-thermal	Rare	Fair	Fair
As above, but without associated permeable beds	Common	None	?

Semi-thermal areas have geothermal gradients up to about 70°C/km depth, while hyperthermal ones possess geothermal gradients many times greater than that.

The model of a vapour-dominated system which was presented by D. E. White et al. in 1971 is generally accepted.[19] In it, relatively impermeable rock and shallow groundwater cap a reservoir of considerable thickness in which the steam is the continuous phase in relatively open channels, with liquid water filling most of the intergranular pore spaces. Fluid temperature and pressure increase with increasing depth above the steam zone, in which the fluid pressure remains relatively constant (because steam is light compared with water); hence there is practically no change in hydrostatic head. Temperature also is relatively constant throughout the steam zone because the temperature of a steam/water mixture is pressure-dependent. Under the zone of steam, liquid fills open channels, both temperature and pressure again increasing with increasing depth. By contrast, in a hot water system, hot water is the continuous phase in such open channels (although there may be gas or steam bubbles in the water). The hydrostatic pressure continuously increases with increasing depth and the maximum temperature is limited by a boiling point curve given by J. L. Haas, Jr, in 1971.[20] In a vapour-dominated system it is probable that steam, together with relatively volatile components such as ammonia, carbon dioxide, hydrogen sulphide, mercury and boron, condenses in the cap rock region. Less volatile components will redissolve in this condensate, but the more volatile ones will probably continue to ascend into colder and overlying groundwater. In consequence, groundwater and springs over such vapour-dominated systems are usually rich in such volatiles or their reaction products and, of course, correspondingly poor in non-volatile components such as chloride (which remains in the residual brine at depth).

Surface expressions of vapour-dominated systems include fumaroles, mud pots, acid-sulphate springs with low discharge rates, acid-charged hot ground and sodium bicarbonate spring waters. However, these are not restricted to such vapour-dominated systems, and may be found also in areas over hot-water systems where there is subterranean boiling. It is often the case that chloride-rich, neutral to alkaline springs occur over hot-water systems at topographically low places, acid-sulphate springs and mud pots emerging at higher elevations.

8.9. THERMAL WATERS FROM SPRINGS AND WELLS

These were compared in 1981 by R. O. Fournier, who compiled data from the Yellowstone National Park, USA, Mexico El Salvador, Chile, Turkey, Taiwan and New Zealand.[12] His main generalization was that waters tend to react chemically with wall rocks after leaving deep reservoirs and prior to emerging at the surface. Such reactions may occur in intermediate reservoirs or in channels which lead to the surface. As temperatures decrease, both HCO_3^- and SO_4^{2-} usually increase, K^+ decreases relatively to Na^+, and Ca^{2+} increases relative to Na^+ unless calcium carbonate precipitates. The Mg^{2+} data are incomplete, but there may be significant increases in Mg^{2+} concentration in the spring waters compared with the deep waters as hot waters cool which causes a dilemma in the application of the magnesium correction for the sodium–potassium–calcium geothermometer which was devised by R. O. Fournier and R. W. Potter II in 1979 (see Section 8.6).[14] Waters which start hot and then react with wall rocks during cooling may give sodium–potassium–calcium temperatures which are closer to the temperature deep in the system than the Mg^{2+}-corrected temperatures.

Continued water–rock reactions as the solutions rise from the depths and cool are not invariably adverse in regard to securing valuable data about a particular hydrothermal system: the geographical distribution of springs with different water compositions and different contents of gas can show directions of subterranean hot-water movement and successive chemical re-equilibrations at lower temperatures.

8.10. RESERVOIR ENGINEERING

Geothermometers can be utilized in the estimation of aquifer temperatures in wells from weeks to months before the subterranean water temperatures return to normal after drilling. Flow testing may possibly accelerate the temperature recovery in the actual production zone, but it interferes with the acquisition of pre-drilling temperature data elsewhere in the well. In addition, extensive flow testing immediately after the cessation of drilling may be impossible because of limited brine disposal facilities. However, collecting a small quantity of fluid at the wellhead may suffice to give a reasonable indication of the temperature of the aquifer.

S. Mercado in 1970 noted that the variations in the sodium–potassium ratio in well waters at Cerro Prieto in Mexico give extremely valuable data for interpretation of the behaviour of wells.[21] In particular he referred to well M-20, on completion of which water commenced to flow spontaneously through a small, 1·27 cm diameter drain line. Later, the rate of discharge was increased from time to time by permitting the fluid to escape through larger and larger orifice plates up to 14·6 cm diameter. As the diameter of the orifice plate was enlarged, the enthalpy of the discharge decreased and the sodium–potassium ratio in the discharged fluid increased. Variations in this ratio demonstrated that more than one aquifer supplied water to the well, and also that the proportion of water from the cooler aquifers increased as the total rate of production from the well in question was permitted to increase and the downhole pressure decreased.

Flashing or boiling in the reservoir before the liquid enters the well may result from drawdown; the consequence is that the enthalpy of the discharged fluid may be higher or lower than the initial enthalpy of the liquid in the reservoir concerned.

Where a flashing front enters the rock away from a well, the fluid pressure in the formation dropping below the initial vapour pressure of the solution, the fluid temperatures decrease practically at once due to vaporization of the liquid. However, rock temperatures fall more slowly than fluid temperatures and transfer of heat from the aquifer rock to the fluid causes more steam to form than could occur through simple adiabatic expansion of the fluid. If this extra steam enters the well together with the residual liquid, the enthalpy of the discharged fluid will exceed that of the initial fluid in the reservoir. Excess steam of this type has been described in fluids obtained from wells at Cerro Prieto, and also at Wairakei and Broadlands in New Zealand.

Relatively low enthalpy in the discharged fluid is a concomitant of the process if the steam, or a part of it, fails to enter the well together with the parental water. The steam may be lost through porous rock or even form a steam cap above the inlet to the well. If a steam cap forms, enlargement of the steam zone could cause wells to switch from producing fluids with relatively low enthalpy to producing fluids with relatively high enthalpy or even dry steam.

Clearly, therefore, reservoir temperatures based on wellhead enthalpy can be too high or too low, even where only a single aquifer contributes fluid to the well. Geochemical investigations can indicate whether or not flashing is taking place in the reservoir, and whether excess or deficient steam accompanies the liquid produced.

Evaluation of flashing in the system is best approached by using sodium–potassium geothermometer temperatures in conjunction with silica and enthalpy temperatures. Aquifer temperatures calculated from these can be in agreement (without indicating flashing), partially agree or all differ. According to R. O. Fournier, there are 13 possible combinations.[12] He interpreted the likeliest ones. Where silica and sodium–potassium temperatures are in good agreement and either higher or lower than the enthalpy temperature, it is probable that the latter is erroneous due to flashing in the formation with a disproportionate quantity of steam entering the well. Where sodium–potassium and enthalpy agree, silica giving a lower temperature, some silica probably precipitated before the sample was collected. Where silica and enthalpy agree and sodium–potassium gives a lower temperature, there could be an outwardly moving flashing front in the formation without segregation of steam and residual liquid, which enter the well. Finally, where silica and enthalpy agree and the sodium–potassium geothermometer gives a higher temperature, this latter value may be residual from a time during which the water equilibrated with rock at a higher temperature than that existing in the nearby aquifer which supplied fluid to the well. In fact, A. J. Ellis and W. A. J. Mahon in 1967 stated that there are indications at Wairakei and Broadlands in New Zealand that sodium–

potassium does not respond as quickly as silica to a change in subterranean temperature.[22]

8.11. GEOCHEMICAL EVIDENCE OF HIGHER TEMPERATURE ELSEWHERE

Temperature measurements in a well may not correspond to deeper, higher-temperature waters. R. O. Fournier gave an excellent example of this:[12] at Coso Springs, a hot water source in California, the US Navy drilled a 114 m deep geothermal exploratory well in June 1967 and, in March 1968, the well was produced by bailing after the bottom hole temperature stabilized at 142°C. A wellhead sample collected towards the end of the bailing activity contained 3042 mg chloride/kg after flashing at atmospheric pressure. Cation geothermometer temperatures applied to that water showed a reservoir temperature of approximately 240–250°C. The water which flows into the well in response to the bailing could have ascended rapidly from the high-temperature reservoir, either (a) boiling as it moved with little conductive heat loss (CBA), or (b) cooling wholly through conduction prior to entering the well (DBA). In 1980, a downhole sample was collected and analyzed by R. O. Fournier et al., who showed that the chloride was 2370 mg/kg and geothermometer temperatures were (Na–K–Ca) 234°C, (Na–K) 231°C, and $(\Delta^{18}O(SO_4^{2-} - H_2O))$ 243°C.[23] The difference in chloride between the wellhead sample after flashing and without flashing indicates that flashing occurred in the formation in response to the bailing and that the minimum temperature in the aquifer supplying water to the well cannot be less than 215°C. The close agreement between all the geothermometer temperatures applied to the wellhead and downhole samples implies that the real temperature of the aquifer supplying water to the well must be near 240°C, also that the water concerned cooled partly adiabatically and partly by conduction when the well was bailed.[12]

Chloride contents have decreased in well waters from the southeastern part of the hot-water system at Cerro Prieto in Mexico and there has been an increase in the Na–K ratio as well; both of these changes are indicative of drawdown of fluids from an overlying and lower-temperature aquifer. At Larderello in Italy, movement of recharge water into the vapour-dominated system is shown by the appearance of tritium in a steam produced from wells at the margins of the zone of production, as R. Celati et al. reported in 1973.[24]

8.12. HYPERTHERMAL GEOTHERMAL FIELDS

Instances of vapour-dominated and water-dominated hyperthermal fields have been given. A model of either must incorporate information regarding the heat source, formations, permeability, etc., which can provide an aquifer containing steam and water or either of these separately. Additionally, the source of replenishment of fluid loss, the underlying bedrock and the overlying cap rock must be included. A combination of all these factors is rare in nature, so such potential energy concentrations as geothermal areas are uncommon. As noted earlier (Section 8.1), they are to be found at places of crustal weakness where adjacent plates are in relative motion (here, magma is most likely to intrude into the crust). High, outwardly directed heat flow in such hyperthermal fields may arise from localized hot spots in the mantle of the Earth, from localized higher concentrations of radioactive minerals in the crust or through direct ingress of intensely hot magmatic gases pressuring their way through bedrock faults into the aquifer, etc.

In the context of geothermal waters, the word aquifer is applicable not only to water-bearing formations, but also to steam-bearing ones. It is interesting to observe that, as might be expected, waters of magmatic origin differ isotopically from meteoric water; thus, the ratio H/D is about 6800:1 in precipitation, but it is 6400:1 in magmatic (juvenile) steam. Fortunately, the content of magmatic water in known hyperthermal fields is small (probably never in excess of 10%), nearly all of the water involved in the groundwater circulation system being meteoric in origin. This penetrates down to considerable depths at which heating occurs, perhaps mainly due to conduction through bedrock from underlying magma. The magmatic water component, when present, probably arises from ascending magmatic vapours. In the actual aquifer, convection current flow occurs and, in rising to the underside of the cap rock, hotter water from the depths tends to boil and release steam. Under certain circumstances this may accumulate, or it may migrate along the underside of the cap rock to reach cooler regions where condensation takes place. A number of such 'circulation cells' may exist within an aquifer. In the case where the aquifer is dome-shaped, hot water may accumulate below an overlying steam layer. Of course, where the cap rock is fractured, geothermal fluids reach the surface of the ground and hence produce surface manifestations.

REFERENCES

1. ARMSTEAD, H. Christopher H., 1978. *Geothermal Energy*. E. and F. N. Spon, London.
2. FOURNIER, R. O. and TRUESDELL, A. H., 1974. Geochemical indicators of subsurface temperature — Part 2, Estimation of temperature and fraction of hot water mixed with cold water. *J. Res., US Geol. Surv.*, **2**, 3, 263–70.
3. PANICHI, C., 1976. *International Post-graduate Course in Geothermics*. Inst. Intle per le Ricerche Geotermiche, 23 pp.
4. MAHON, W. A. J., 1966. Chemistry in the exploration and exploitation of hydrothermal systems. *Geothermics, Sp. Issue 2*, 1310–22.
5. ELLIS, A. J. and MAHON, W. A. J., 1967. Natural hydrothermal systems and experimental hot water/rock interactions. *Geochim. Cosmochim. Acta*, **28**, 519.
6. WHITE, D. E., 1970. Geochemistry applied to the discovery, evaluation and exploitation of geothermal energy resources. *Geothermics, Sp. Issue 2*, 58–80.
7. FOURNIER, R. O. and TRUESDELL, A. H., 1973. An empirical Na–K–Ca geothermometer for natural waters. *Geochim. Cosmochim. Acta*, **37**, 1255–75.
8. PACES, T., 1975. A systematic deviation from Na–K–Ca geothermometer below 75°C and above 10^{-4} atm p_{CO_2}. *Geochim. Cosmochim. Acta*, **39**, 541–4.
9. BOTTINGA, Y., 1969. Calculated fractionation factors for carbon and hydrogen isotope exchange in the system calcite carbon dioxide–graphite–methane–hydrogen–water vapour. *Geochim. Cosmochim. Acta*, **33**, 49.
10. CORTECCI, G., 1974. Oxygen isotopic ratios of sulfate ions–water pairs as a possible geothermometer. *Geothermics*, **3**, 3, 60–4.
11. BROWNE, P. R. L., 1978. Hydrothermal alteration in active geothermal fields. *Ann. Rev., Earth Sci. Planet.*, **6**, 229–50.
12. FOURNIER, R. O., 1981. Application of water geochemistry to geothermal exploration and reservoir engineering. In: *Geothermal Systems: Principles and Case Histories*, Ed. L. Rybach and L. J. P. Muffler. J. Wiley, New York, 359 pp., pp. 109–143.
13. ARNORSSON, S., 1975. Major element chemistry of the geothermal sea-water at Reykjanes and Svartsevgi, Iceland. *Miner. Mag.*, **42**, 209–20.
14. FOURNIER, R. O. and POTTER, R. W., II, 1979. Magnesium correction to the Na–K–Ca chemical geothermometer. *Geochim. Cosmochim. Acta*, **43**, 1543–50.
15. LLOYD, R. M., 1968. Oxygen isotope behaviour in the sulfate–water system. *J. Geophys. Res.*, **73**, 6099–110.
16. MCKENZIE, W. F. and TRUESDELL, A. H., 1977. Geothermal reservoir temperatures estimated from the oxygen isotope compositions of dissolved sulfate and water from hot springs and shallow drillholes. *Geothermics*, **5**, 51–61.
17. KEENAN, J. H., KEYES, F. G., HILL, P. G. and MOORE, J. G., 1969. *Steam Tables* (intl edn, metric units). J. Wiley, New York, 162 pp.
18. FOURNIER, R. O., WHITE, D. E. and TRUESDELL, A. H., 1976. Convective heat flow in Yellowstone National Park. *2nd UN Symp., Development and Use of Geothermal Resources, San Francisco, May 1975*, **1**, 731–9.
19. WHITE, D. E., MUFFLER, L. J. P. and TRUESDELL, A. H., 1971. Vapor-

dominated hydrothermal systems compared with hot-water systems. *Econ. Geol.*, **66**, 75–97.
20. HAAS, J. L., Jr, 1971. Effect of salinity on the maximum thermal gradient of a hydrothermal system at hydrostatic pressure. *Econ. Geol.*, **66**, 940–6.
21. MERCADO, S., 1970. High activity hydrothermal zones detected by Na/K, Cerro Prieto, Mexico. *Geothermics, Sp. Issue* **2**, 1367–76.
22. ELLIS, A. J. and MAHON, W. A. J., 1967. Natural hydrothermal systems and experimental hot water/rock interactions, Pt 2. *Geochim. Cosmochim. Acta*, **31**, 519–39.
23. FOURNIER, R. O., THOMPSON, J. M. and AUSTIN, C. L., 1980. Interpretation of chemical analyses of water collected from two geothermal wells at Coso, California. *J. Geophys. Res.*, **85**, 2405–10.
24. CELATI, R., NOTO, P., PANICHI, C., SQUARCI, P. and TAFFI, L., 1973. Interactions between the steam reservoir and surrounding aquifers in the Larderello geothermal field. *Geothermics*, **2**, 3–4, 174–85.

CHAPTER 9

Groundwater Quality

9.1. GROUNDWATER SALINITY

Salinity always exists in groundwater, but in very variable amounts — thus, reported salt contents range from under 25 mg/litre in a quartzite spring to over 300 000 mg/litre in brines.[1] The type and quantity of salts reflect the environment, movement and source of the groundwater. Since groundwater often occurs in association with soluble geological materials, higher concentrations of dissolved components are found in it than in surface water. Such soluble salts arise mainly from the solution of geological materials, but bicarbonate, normally the principal anion in groundwater, derives from carbon dioxide released as a result of organic decomposition in the soil. Factors which influence salinity include the specific surface area of the relevant aquifer, the solubility of its contained minerals and the duration of the contact. It is logical that the salt content should increase with decreasing mobility of the groundwater and, indeed, this is the case. The most common geochemical situation in groundwater is probably to have bicarbonate-waters near to the surface, with chloride-waters in the deeper portions of formations.

Precipitation which reaches the terrestrial surface is poor in dissolved mineral matter, but, on landing, it reacts with the multitudinous soil and rock minerals which it contacts. Naturally, the types and amounts of mineral matter which are dissolved depend upon the chemical composition and physical structure of the rocks involved, and upon the pH (hydrogen ion concentration) and the Eh (redox potential); see, for instance, W. Back and B. B. Hanshaw in 1965.[2]

Obviously, where large-scale recharge is occurring, the quality of the infiltrating surface water from rivers or artificial recharge areas has a great influence on the quality of the receiving groundwater. Soluble products of soil weathering and erosion resulting from rainfall and flowing water may add salts, and excess irrigation water percolating down to the water table can contribute large amounts of salt. However, water which passes through root zones in cultivated regions possesses salt concentrations which are several times greater than that of applied irrigation water, mainly because of evapotranspirative processes. Plants can absorb salts selectively and this, coupled with introduction of fertilizers, also alters the salt contents of percolating water.

Increase in salt contents is related to a series of factors such as the permeability of the soils, drainage facilities, the quantity of water which is applied, crops in the area and climate in general. High salinities are to be found in both soils and groundwaters in arid climates where rainwater leaching is ineffective in diluting the saline solutions. In a parallel manner, areas of poor drainage frequently have high salt concentrations. Some areas may possess remnants of earlier sedimentary deposition under saline waters and the excess salt content in soil and water which results has led to the use of the word 'badlands' for such features.

Where groundwater passes through igneous rocks, only minute quantities of mineral matter are dissolved because such rocks comprise mostly insoluble constituents. Of course, the carbon dioxide added to rainwater in its descent through the atmosphere increases the subterranean solvent power of this water when it percolates downwards. It is the silicate minerals which add silica to groundwaters.

Where groundwater passes through sedimentary rocks, the situation is rather different because many soluble constituents are present. They are also very abundant in the terrestrial crust, so they contribute most of the soluble materials to groundwaters. The most commonly contributed cations are sodium and calcium, while the most common anions to be introduced are bicarbonate and sulphate. Usually, there is not very much chloride, but it can be derived from sewage, from connate water and from seawater intrusion. Nitrate is an important natural constituent and, where it is found in unusually high concentrations, may indicate sources of former or present pollution. In limestone terrains, calcium and bicarbonate ions are added to groundwaters by solution. Table 9.1, from S. N. Davis and R. J. M. DeWiest in 1966, provides appropriate data on relative abundances of dissolved solids in potable water.[3]

The constituents listed in Table 9.1 have various effects on water. Thus, where calcium and magnesium occur, silica forms a scale in boilers and on steam turbines which both retards heat and is hard to remove. However, silica can be added to soft water in order to inhibit corrosion

TABLE 9.1
RELATIVE ABUNDANCE OF DISSOLVED CONSTITUENTS IN POTABLE WATER[3]

Major (1–1 000 mg/litre)	Secondary (0·01–10 mg/litre)	Minor 0·0001–0·1 mg/litre	Trace (<0·001 mg/litre)
Na^+	Fe^{2+}, Fe^{3+}	Al^{3+}	Be^{2+}
Ca^{2+}	Sr^{2+}	Sb^{3+a}	Bi^{3+}
Mg^{2+}	K^+	As^{3+}	Cerium
Bicarbonate	Carbonate	Ba^{2+}	Caesium[a]
Sulphate	Nitrate	Bromide	Gallium
Chloride	Fluoride	Cadmium[a]	Gold
Silica	Boron	Chromium[a]	Indium
		Cobalt	Lanthanum
		Copper	Niobium[a]
		Germanium[a]	Platinum
		Iodide	Radium
		Lead	Ruthenium[a]
		Lithium	Scandium[a]
		Manganese	Silver
		Molybdenum	Thallium[a]
		Nickel	Thorium[a]
		Phosphate	Tin
		Rubidium[a]	Tungsten[a]
		Selenium	Ytterbium
		Titanium[a]	Yttrium[a]
		Uranium	Zirconium[a]
		Vanadium	
		Zinc	

[a] These occupy an uncertain position in the list.

of iron pipes. If present in groundwater, iron may cause turbidity and staining as well as imparting an unpleasant taste and colour to food and drink; amounts exceeding 0·2 mg/litre are objectionable for most industrial uses. In the case of manganese, 0·2 mg/litre precipitates on oxidation and again is unacceptable in industry; this metal also imparts an undesirable taste, stains and promotes growths in reservoirs and filters. Calcium and magnesium combine with bicarbonate, carbonate, sulphate and silica, forming heat-retarding and pipe-clogging scale in boilers, etc. Also, they combine with ions of fatty acid in soap to make soapsuds; the higher their content is, the more soap is required to form suds. It is interesting that magnesium in high concentration is laxative in its effects. Sodium and potassium are deleterious if present in amounts exceeding 50 mg/litre in the presence of suspended matter because they cause foaming which accelerates scale formation and corrosion in boilers. Carbonate combines readily with calcium and magnesium forming a crust-like scale which retards heat flow through pipe walls and the flow of fluids in pipes. When bicarbonate is heated, it decomposes to steam, carbon dioxide and carbonate. Water with large quantities of bicarbonate is unacceptable in many industries. Sulphate combines with calcium and forms an adherent, heat-retarding scale: more than 200 mg/litre is objectionable in water employed in some industries. Water containing 500 mg sulphate/litre has a bitter taste and that containing 1000 mg/litre may be cathartic. A salty taste is given by chloride in water in excess of 100 mg/litre and higher concentrations can cause physiological damage — which is why the food processing industries normally require under 250 mg/litre. Some industries need lower values, e.g. paper manufacturing and synthetic rubber manufacturing require less than 100 mg/litre. It is asserted that fluoride concentration between 0·6 and 1·7 mg/litre in drinking water benefits both the structure and the decay resistance of the teeth of children. However, fluoride in excess of 1·5 mg/litre may promote 'mottled enamel' in children's teeth. In excess of 6 mg/litre, fluoride causes marked mottling and disfiguration of teeth. Water containing more than 100 mg nitrate/litre tastes bitter and can cause physiological distress. As noted earlier, water from shallow wells containing over 45 mg nitrate/litre are reported to cause methaemoglobinemia in infants (see Section 3.3). In industry, however, small quantities of nitrate assist in reducing cracking of high-pressure boiler steel. Dissolved solids, i.e. minerals dissolved in water, are not desirable in amounts exceeding 500 mg/litre in drinking water. In industry, also, less than 300 mg/litre is optimum for dyeing textiles and manufacturing plastics, pulp paper and rayon. Dissolved solids may cause foaming in steam boilers, the maximum permissible content decreasing with increasing operating pressure.[4]

Major natural sources and concentrations in natural waters of the above-mentioned constituents are shown in Table 9.2.

Much salinity in the groundwater of coastal areas derives from airborne salts originating from the air–water interface over the sea. Studies by E. Eriksson in 1959 and by S. Loewengart in 1961 imply that salts deposit on land by precipitation and also by dry fallout.[6,7] Deposition decreases inland and varies exponentially with distance from the sea. Israeli water measurements gave the following expression:

$$N = 110 \exp(-0.0133d)$$

TABLE 9.2
NATURAL OCCURRENCE OF COMMON SOLUTES IN WATER[a]

Constituent	Natural sources	Concentration in natural water
Silica	Feldspars, ferromagnesian and clay minerals, amorphous silica, chert and opal.	Ranges from 1 to 30 mg/litre but as much as 100 mg/litre can occur and, in brines, amounts may reach 4 000 mg/litre.
Iron	Igneous rocks: amphiboles, ferromagnesian micas, FeS_2, and magnetite, Fe_3O_4. Sandstone rocks: oxides, carbonates, sulphides or iron clay minerals.	Usually under 0·5 mg/litre in fully aerated water. Groundwater with pH less than 8 can contain 10 mg/litre; infrequently, 50 mg/litre may be present.
Manganese	Arises from soils and sediments. Metamorphic and sedimentary rocks and mica biotite and amphibole hornblende minerals contain large quantities of Mn.	Usually under 0·2 mg/litre; groundwater possesses over 10 mg/litre.
Calcium	Amphiboles, feldspars, gypsum, pyroxenes, dolomite, aragonite, calcite, clay minerals.	Usually under 100 mg/litre, but brines may contain up to 75 000 mg/litre.
Magnesium	Amphiboles, olivine, pyroxenes, dolomite, magnesite, clay minerals.	Usually under 50 mg/litre; about 1 000 mg/litre in ocean water; brines may have 57 000 mg/litre.
Sodium	Feldspars (albite), clay minerals, evaporites such as halite, NaCl, industrial wastes	Generally under 200 mg/litre; about 10 000 mg/litre in seawater; about 25 000 mg/litre in brines.

Potassium	Feldspars (orthoclase, microcline), feldspathoids, some micas, clay minerals.	Usually under 10 mg/litre, but up to 100 mg/litre in hot springs and 25 000 mg/litre in brines.
Carbonate	Limestone, dolomite.	Usually under 10 mg/litre, but can exceed 50 mg/litre in water highly charged with sodium.
Bicarbonate	Limestone, dolomite.	Usually under 500 mg/litre, but can exceed 1 000 mg/litre in water highly charged with CO_2.
Sulphate	Oxidation of sulphide ores, gypsum, anhydrite.	Usually under 300 mg/litre, except in wells influenced by acid mine drainage. Up to 200 000 mg/litre in some brines.
Chloride	Sedimentary rock (evaporites), a little from igneous rocks.	Usually under 10 mg/litre in humid areas; up to 1 000 mg/litre in more arid regions. Roughly 19 300 mg/litre in seawater and up to 200 000 mg/litre in brines.
Fluoride	Amphiboles (hornblende), apatite, fluorite, mica.	Usually under 10 mg/litre, but up to 1600 mg/litre in brines.
Nitrate	Atmosphere, legumes, plant debris, animal excrement.	Usually under 10 mg/litre.
Dissolved solids	Mineral constituents dissolved in water.	Usually under 5000 mg/litre, but some brines contain as much as 300 000 mg/litre.

[a] Modified table derived from C. N. Durfer and E. Baker in 1964[4] by D. K. Todd in 1980.[5]

where N is the annual quantity of chloride precipitation in kg/ha and d is the distance from the sea in kilometres. In arid regions, where surface runoff is small and evapotranspiration is large, the deposition of airborne salts is intensified considerably in groundwater.

9.2. QUALITY ANALYSES

These may be chemical, physical or biological.

A chemical analysis involves determination of the concentrations of inorganic constituents which are present. Organic and radiological components are usually not significant except where pollution occurs. In groundwater, where the salinity is normal, the dissolved salts are ionized, other minor constituents being reported in elemental form. The chemical analysis includes measurement of pH and specific electrical conductance.

Physical analysis provides information regarding temperature, colour, turbidity, smell and taste.

Biological analyses test for coliform bacteria; if these occur, the water is not fit for human consumption. They are found in the intestines of human beings and animals and become introduced into groundwater through sewage.

After chemical analysis of a groundwater sample has been effected, it is reported and related to established standards determining utility of the source in question in relation to its ultimate function.

Concentrations of the ions commonly occurring in groundwater used to be expressed as parts per million (ppm), but this unit has been replaced by mg/litre, which is numerically equivalent up to a concentration of dissolved solids of around 7000 mg/litre. The old method was a weight-per-weight approach (1 ppm = 1 mg/kg of solution) and the newer one is on a weight-per-volume basis. The equivalent to the metric mg/litre in Imperial measure is grains per gallon, 1 grain per Imperial gallon being 14.25 mg/litre; 1 grain per US gallon is 17·1 mg/litre. In irrigation engineering, the unit tons of dissolved solids per acre-foot of water (taf) has been utilized: 1 taf = 735 ppm.

Another system, chemical equivalence, can be employed. Positively charged ions (cations) and negatively charged ones (anions) combine and dissociate in definite weight ratios and, by expressing ion concentrations in equivalent weights, such ratios are easily determined. This is because exactly one equivalent weight of a cation combines with one equivalent weight of an anion. The combining weight of an ion is obtained by dividing its formula weight by its charge and, when the concentration in mg/litre is divided by the combining weight, equivalent concentrations expressed in milliequivalents per litre (meq/litre) result. Table 9.3, derived from J. D. Hem, lists conversion factors for chemical equivalence.[8]

TABLE 9.3
CONVERSION FACTORS FOR CHEMICAL EQUIVALENCE[8]
Concentration in mg/litre × conversion factor = concentration in meq/litre

Chemical constituent	Conversion factor
Aluminium, Al^{3+}	0·11119
Ammonium, NH_4^+	0·05544
Barium, Ba^{2+}	0·01456
Beryllium, Be^{3+}	0·33288
Bicarbonate, HCO_3^-	0·01639
Bromide, Br^-	0·01251
Cadmium, Cd^{2+}	0·01779
Calcium, Ca^{2+}	0·04990
Carbonate, CO_3^{2-}	0·03333
Chloride, Cl^-	0·02821
Cobalt, Co^{2+}	0·03394
Copper, Cu^{2+}	0·03148
Fluoride, F^-	0·05264
Hydrogen, H^+	0·99209
Hydroxide, OH^-	0·05880
Iodide, I^-	0·00788
Iron, Fe^{2+}	0·03581
Iron, Fe^{3+}	0·05372
Lithium, Li^+	0·14411
Magnesium, Mg^{2+}	0·08226
Manganese, Mn^{2+}	0·03640
Nitrate, NO_3^-	0·01613
Nitrate, NO_2^-	0·02174
Phosphate, PO_4^{3-}	0·03159
Phosphate, HPO_4^{2-}	0·02084
Phosphate, $H_2PO_4^-$	0·01031
Potassium, K^+	0·02557
Rubidium, Rb^+	0·01170
Sodium, Na^+	0·04350
Strontium, Sr^{2+}	0·02283
Sulphate, SO_4^{2-}	0·02082
Sulphide, S^{2-}	0·06238
Zinc, Zn^{2+}	0·03060

In assessing groundwater quality it is not possible to calculate an equivalent weight in the same way for undissociated species with zero charge, such as silica.

Now, hydrogen has an equivalent weight of unity and the equivalent weights of other elements are relative to it. In practice, the equivalent weight of an ion or compound in grams, i.e. the gram equivalent weight, is that weight in grams which combines with or is replaced by one gram of hydrogen.

Formerly, concentrations of substances in solution in terms of their chemical equivalence was expressed as equivalents per million (epm). For instance, if 30·6 ppm of calcium is present in a water sample, then:

$$\text{Atomic weight, Ca} = 40·08$$
$$\text{Valency (ionic charge)} = 2$$
$$\text{Equivalent weight} = 40·08/2 = 20·04$$
$$\therefore \text{Concentration of Ca} = 30·6/20·04 = 1·52 \text{ epm}$$

Some additional details regarding important chemical constituents of groundwater are given in Table 9.4.

TABLE 9.4
EQUIVALENT WEIGHTS OF COMMON IONIC CONSTITUENTS IN GROUNDWATER

Ion	Atomic or formula wt	Charge	Equivalent wt
Cations			
Calcium, Ca^{2+}	40·08	+2	20·04
Magnesium, Mg^{2+}	24·32	+2	12·16
Sodium, Na^+	23·00	+1	23·00
Potassium, K^+	39·10	+1	39·1
Anions			
Carbonate, CO_3^{2-}	60·01	−2	30·00
Bicarbonate, HCO_3^-	61·01	−1	61·01
Sulphate, SO_4^{2-}	96·06	−2	48·03
Chloride Cl^-	35·46	−1	35·46
Nitrate NO_3^-	62·01	−1	62·01

Chemical equivalence implies that the sum of the cations and the sum of the anions totalling the dissolved solids content in a sample of groundwater will equal each other when expressed in milliequivalents per litre (meq/litre). In the event that a chemical analysis finds to the contrary, it may be inferred that either there is an analytical error, or else there are other, undetermined constituents present.

Measuring the electrical conductance of a sample of groundwater is a useful means of assessing the total dissolved solids content. Conductance is preferred to its reciprocal, resistance, because it increases with salinity. Specific electrical conductance defines the conductance of a millilitre of water at a standard temperature, 25°C; an increase of 1°C increases conductance by approximately 2%. It is measured in microsiemens/cm (μS/cm), equivalent to micromhos/cm, but the results cannot be regarded as necessarily providing the value of the total dissolved solids content because of the great variety of ionic and undissociated species which may be present. However, it is easily measured and provides data which give some indication of this total. J. Logan in 1961 gave an approximate relation for most natural waters in the range 100–5000 μS/cm, leading to the equivalences 1 meq of cations/litre $\equiv$ 100 μS/cm and 1 mg/litre $\equiv$ 1·56 μS/cm.

The hardness of water is the consequence of divalent metallic cations in the groundwater, principally those of calcium and magnesium, which react with soap to form unwelcome precipitates and, if certain anions are present, even more unwelcome scale. As they have this characteristic, hard waters are most unsatisfactory for household cleansing and water softeners may be necessary.

The ultimate origin of hardness is traceable to the solution of carbon dioxide, liberated by bacterial activity in soils, in percolating rainwater, as C. N. Sawyer and P. L. McCarty indicated in 1967.[9] Conditions of low pH arise and these lead to the solution of previously insoluble carbonates in the soils and in limestones as they are converted into soluble bicarbonates. As the carbonates are removed in this way, impurities in the limestone, e.g. chlorides and silicates, are exposed to the solvent action of the water and may dissolve as well. It is not surprising, therefore, that hard water is common in regions of limestone overlaid by thick topsoils.

Normally, hardness is expressed as the equivalent of calcium carbonate, thus:

$$H_r = Ca^{2+} \times \frac{CaCO_3}{Ca^{2+}} + Mg^{2+} \times \frac{CaCO_3}{Mg^{2+}}$$

where H_r is the hardness; H_r, Ca^{2+} and Mg^{2+} are measured in mg/litre and the ratios in formula weights. The above equation reduces to:

$$H_r = 2·5 Ca^{2+} + 4·1 Mg^{2+}$$

or, more exactly:

$$H_r = 2·497 Ca^{2+} + 4·115 Mg^{2+}$$

An appropriate hardness classification appears in Table 9.5.[9]

TABLE 9.5
CLASSIFICATION OF HARDNESSES OF WATER[9]

Hardness, mg/litre as $CaCO_3$	Water category
0–75	Soft
75–150	Moderately hard
150–300	Hard
300	Very hard

In the examination of chemical analyses, it is often valuable to present graphical representations in order that they may be compared and also to illustrate both similarities and differences. Various types of graphs are available. One is the vertical bar graph in which an analysis appears as a vertical bar with a height proportional to the total concentration of anions or cations given as milliequivalents per litre, with anions on the right half of the bar, cations on the left. The two segments so produced are divided horizontally to show the concentrations of major ions and identified by patterns of shading. If required, extra segments can be added to cover hardness and silica content also.

Another approach is to employ radiating vectors of which the lengths represent ionic concentrations in milliequivalents per litre.

H. A. Stiff, Jr, in 1951 proposed pattern diagrams as a means of representing chemical analyses using four parallel axes.[10] Cation concentrations are plotted to the left of a vertical zero axis and anions to its right, all values being in milliequivalents per litre. Resultant points are connected and produce irregular, polygonal patterns; a given shape characterizes waters of a similar quality. Figure 9.1 shows a typical pattern.

A. M. Piper in 1944 suggested the trilinear diagram both for representing and comparing water quality analyses.[11] Cations, expressed as percentages of total cations in milliequivalents per litre, plot as a single point in the left triangle. Anions, similarly expressed as percentages of total anions, appear as a point in the right triangle. The two points so formed are projected into a central, diamond-shaped area parallel to the upper edges of the central area. A single point then is uniquely related to the total ionic distribution and a circle can be drawn centred on it with an area proportional to the total dissolved solids.

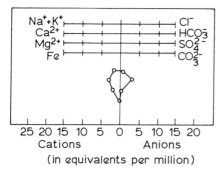 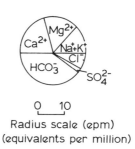

FIG. 9.1. Shape or pattern diagram for representing analyses of groundwater quality.

FIG. 9.2. Circular diagram for representing analyses of groundwater quality.

A simpler technique is to draw circles of water quality with a special scale for the radii so that the area of each is proportional to the total ionic concentration of the analysis. Sectors within each such circle show the fractions of the various ions given in milliequivalents per litre (Fig. 9.2).

H. Schoeller in 1961 devised a semi-logarithmic diagram which is widely used in Europe and termed a Schoeller diagram.[12] An example is given in Fig. 9.3. The main ionic concentrations (as equivalents per million) are plotted on six equally spaced logarithmic scales, and the points so plotted are then joined by straight lines. This diagram gives the absolute concentration of each ion and, in addition, the concentration differences among various analyses of groundwater. As a result of the semi-logarithmic arrangement, if a straight line joining two points of two ions in a sample of water is parallel to another straight line joining two points of the same two ions in a second sample of water, the ratio of the ions in the two analyses is equal. Finally, the Schoeller diagram can be so adapted that it may be utilized to determine the degree of saturation of calcium carbonate and calcium sulphate in groundwater.

Physical analyses include the reporting of the temperature of groundwater in degrees Celsius which must be measured immediately upon completion of sampling. Colour, which is caused by organic or mineral matter is also noted. It is given in terms of mg/litre in comparison with standard solutions. Turbidity measures suspended and colloidal matter content in the water sample, including for instance clay, silt and microorganisms. It is determined on the basis of the length of a path of light

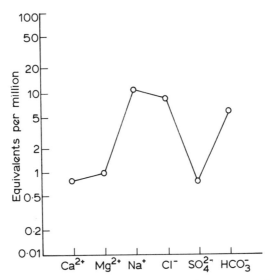

FIG. 9.3. Schoeller semi-logarithmic diagram for the representation of groundwater quality analyses.

through the water which causes the image of a flame from a standard candle to disappear. Usually, turbidity is eliminated naturally in aquifers as a result of their filtration potential in unconsolidated sediments. However, in other types of aquifer, it is not so easily removed.

Smells and tastes arise from a variety of agents such as dissolved gases, bacteria, phenols and mineral matter. The impact they make is subjective matter, but quantitative determinations are possible, based on the maximum degree of dilution which can be distinguished from odourless and tasteless water.[13]

Biological analyses are important in detecting pathogenic bacterial pollution; the easiest bacteria to isolate and identify from groundwater belong to the coliform group. Tests for these exist and results are given as the most probable number, MPN, of coliform group micro-organisms in a given volume of water. The MPN is computed from probability tables after analyzing a number of separate portions of a groundwater sample. For drinking water, low bacterial content is mandatory and the MPN should not exceed one per 100 ml of water.

The chemical characteristics of drinking water conforming to international standards of hygiene and safety are given in Table 9.6. For organic chemicals and radionuclides the data in Tables 9.7 and

TABLE 9.6
INTERNATIONAL STANDARDS FOR DRINKING WATER SUPPLIES[14]

Constituent	Maximum permissible level (mg/litre)	
	Recommended limit[a]	Tolerance limit[b]
ABS (alkylbenzene sulphonate)	0·5	—
Arsenic	0·01	0·05
Barium	—	1·00
Cadmium	—	0·01
CCE (carbon chloroform extract)	0·2	—
Chloride	250·0	—
Chromium, hexavalent	—	0·05
Copper	1·0	—
Cyanide	0·01	0·2
Fluoride	0·8–1·7[c,d]	1·4–2·4[d]
Iron	0·3	—
Lead	—	0·05
Manganese	0.05	—
Mercury	—	0·002
Nitrate, as N	10·0	—
Phenolic compounds, as phenol	0.001	—
Selenium	—	0.01
Silver	—	0.05
Sulphate	250·0	—
Zinc	5·0	—

[a] Concentrations not to be exceeded where more suitable water supplies are available.
[b] Higher concentrations than these are reasons to reject the supply.
[c] Dependent upon annual maximum daily air temperature.
[d] Where fluoridation is practised, minimum recommended limits are specified.

9.8 are appropriate. Physical requirements for drinking water standards in the USA are listed in Table 9.9, and biological standards are given in Table 9.10. Tables 9.6–9.10 all relate to drinking water supplies and embody the standards requirements set up by the US Environmental Protection Agency.[14]

In the case of water used industrially, there is a variety of requirements. The most significant parameters involved are salinity, hardness and silica content, but other important considerations relate not merely to these (among others), but to the relative constancy of the components as

TABLE 9.7
US STANDARDS FOR ORGANIC CHEMICALS IN DRINKING WATER SUPPLIES[14]

Constituent	Tolerance limit (mg/litre)
Chlorinated hydrocarbons	
Endrin	0·0002
Lindane	0·004
Methoxychlor	0·1
Toxaphene	0·005
Chlorophenoxys	
2,4-D	0·1
2,4,5-TP Silvex	0·01

TABLE 9.8
US STANDARDS FOR RADIOACTIVITY IN DRINKING WATER SUPPLIES[14]

Constituent	Recommended radioactivity limit (c/litre)
Radium-226	3
Strontium-90	10
Gross beta activity	1 000[a]

[a] In the absence of strontium-90 and alpha-emitters.

TABLE 9.9
PHYSICAL REQUIREMENTS FOR DRINKING WATER SUPPLIES IN THE USA[14]

Characteristic	Recommended limit[a]
Colour, units	15
Smell, threshold number	3, inoffensive
Residue: filtrable, mg/litre	500
Taste	Inoffensive
Turbidity, units	5

[a] Concentrations not to be exceeded where more suitable water supplies are available.

well as to fluctuations in the temperature of the water involved. From this standpoint, groundwater is preferable to a surface water supply, which will reflect seasonal variations in chemical and physical quality.

The American Water Works Association in 1971 supplied details of

TABLE 9.10
BIOLOGICAL STANDARDS FOR DRINKING WATER SUPPLIES IN THE USA[14]

Substance	Maximum permissible limit
Standard 10 ml portions	Not more than 10% in one month shall show coliforms.[a]
Standard 100 ml portions	Not more than 60% in one month shall show coliforms.[a]

[a] Subject to additional, specified restrictions.

the ranges in recommended limiting concentrations for industrial process waters.[15]

9.3. WATER IN IRRIGATION

The mineral constituents in groundwater govern its suitability for the purpose of irrigation. In particular, salts can be highly deleterious because they can damage the growth of plants physically, by restricting the taking-up of water by modification of osmotic processes, and chemically, by the effects of toxic substances upon metabolic processes. Salts promote changes in the structure, permeability and extent of aeration in soils; this may have an indirect, but adverse, result in terms of the growth of plants.

Unfortunately, there are such great differences in tolerance to salts among the various plants that no specific limits of permissible salt concentrations in irrigation water can be presented. Table 9.11 provides some pertinent data and is derived from L. A. Richards in 1954.[16] In this Table, the criterion utilized is the relative yield of the crop on a saline soil compared with its yield on a non-saline soil: under similar conditions of growth and, in each group, the crops are listed in order of increasing tolerance to salt with the electrical conductance values at the top and bottom of each column representing that range of salinity level at which a 50% decrease in yield is to be anticipated. The concentrations in question refer to soil water, which can contain much higher concentrations than applied irrigation water.

Drainage is of great importance and, where it is adequate, crops may be grown even if saline water is employed; however, if it is not, even water of good quality may not promote a satisfactory crop.

TABLE 9.11
THE RELATIVE TOLERANCES OF CROPS TO CONCENTRATIONS OF SALTS[a]

Type of crop	Low tolerance	Medium tolerance	High tolerance
Fruit	Avocado Lemon Strawberry Peach Apricot Almond Plum Prune Grapefruit Orange Apple Pear	Cantaloupe Date Olive Fig Pomegranate	Date palm
Vegetable	3 000 μS/cm Green bean Celery Radish 4 000 μS/cm	4 000 μS/cm Cucumber Squash Pea Onion Carrot Potato Sweetcorn Lettuce Cauliflower Bell pepper Cabbage Broccoli Tomato 10 000 μS/cm	10 000 μS/cm Spinach Asparagus Kale Garden beet 12 000 μS/cm
Forage	2 000 μS/cm Burnet Ladino clover Red clover Alsike clover Meadow foxtail White Dutch clover 4 000 μS/cm	4 000 μS/cm Sickle milkvetch Sour clover Cicer milkvetch Tall meadow oat grass Smooth brome Big trefoil Reed canary Meadow fescue Blue grame Orchard grass Oat (hay)	12 000 μS/cm Bird's-foot trefoil Barley (hay) Western wheat grass Canada wild rye Rescue grass Rhodes grass Bermuda grass Nattall alkali grass Salt grass Alkali sacaton 18 000 μS/cm

Table 9.11 — *contd.*

Type of crop	Low tolerance	Medium tolerance	High tolerance
Forage (*contd*)		Wheat (hay) Rye (hay) Tall fescue Alfalfa Hubam clover Sudan grass Dallis grass Strawberry clover Mountain brome Perennial rye grass Yellow sweet clover White sweet clover 12 000 μS/cm	
Field	4 000 μS/cm Field bean	6 000 μS/cm Castor bean Sunflower Flax Corn (field) Sorghum (grain) Rice Oat (grain) Wheat (grain) Rye (grain) 10 000 μS/cm	10 000 μS/cm Cotton Rape Sugar beet Barley (grain) 16 000 μS/cm

[a] The electrical conductance values represent the salinity levels of the saturation extract at which a 50% decrease in yield can be anticipated, compared with yields on non-saline soil under comparable growing conditions. The saturation extract is the solution extracted from a soil at its saturation percentage.

Quality in irrigation water is usually expressed in terms of classes of relative suitability, rather than salinity limits. The concentration of sodium is significant because this element reacts with soils to reduce permeability. Soils containing sodium and, as the dominant anion, carbonate are called alkali soils, as compared with saline soils in which chloride or sulphate is the dominant anion. Sodium content is expressed as the % sodium value, the sodium percentage or soluble-sodium percentage, defined as follows:

$$\%Na^+ = \frac{(Na^+ + K^+)100}{Ca^{2+} + Mg^{2+} + Na^+ + K^+}$$

in which all the ionic concentrations are given in milliequivalents per litre. Another parameter which is recommended by the US Department of Agriculture is the sodium absorption ratio, SAR, because it relates directly to the adsorption of sodium by soil.[15] This is:

$$SAR = \frac{Na}{\sqrt{(Ca^{2+} - Mg^{2+})/2}}$$

the concentrations of the components being given as in the previous equation. Boron is another element of considerable significance to plant growth. In minute quantities it is a necessity, but higher concentrations may be toxic. Some plants are extremely sensitive to it (e.g. lemon, grapefruit, orange, peach, cherry, grape, apple, pear and plum), some less so (e.g. corn, wheat, barley, radish, sweet pea, tomato, cotton and potato), and there is a group which is tolerant, this including the carrot, lettuce, cabbage, turnip, onion, broad bean and asparagus. The plants are listed in each category in the order of increasing tolerance.

Finally, a quality classification of waters for irrigation purposes derived from L. V. Wilcox in 1955[17] is shown in Table 9.12.

TABLE 9.12
QUALITY CLASSIFICATION OF IRRIGATION WATER

Class	$Na(\%)$	Specific conductance (S/cm)	Boron (mg/litre)		
			Sensitive crops	Semi-sensitive crops	Tolerant crops
Excellent	20	250	0·33	0·67	1·00
Good	20–40	250–750	0·33–0·67	0·67–1·33	1·00–2·00
Permissible	40–60	750–2 000	0·67–1·00	1·33–2·00	2·00–3·00
Doubtful	60–80	2 000–3 000	1·00–1·25	2·00–2·50	3·00–3·75
Unsuitable	80	3 000	1·25	2·50	3·75

9.4. SUBTERRANEAN GEOCHEMICAL VARIATIONS

Groundwater flowing underground always attempts to attain chemical equilibrium through reactions with the aquifer rocks, and the effects of some equilibria are relevant to well clogging, pollutant movement, etc. Ion exchange is one of the important phenomena. This depends upon the ability of many naturally occurring minerals to exchange one ion for another, ions adsorbed on the surface of fine-grained aquifer materials being replaced by ions in solution. As cations are commonly involved, principally sodium, calcium and magnesium, this process is termed base (cation) exchange and its direction is towards an equilibrium of bases actually present in the water and on the fine materials of the aquifer. Base exchange may soften groundwater naturally and also it may promote changes in the physical properties of soils. As noted above (see Section 9.3), sodium in high concentration can reduce soil permeability. In such a case, the number of sodium ions combined with the soil rises, displacing an equivalent amount of calcium or other ions. The result is deflocculation and the concomitant reduction of permeability. It may be noted that adding gypsum to a soil is beneficial because, by base exchange, it can improve both the texture and drainability of a soil.

Chemical reduction of oxidized sulphur to sulphate ions or to sulphide may occur quite often in groundwater and is probably effected in the presence of certain bacteria.[8] Usually, waters which undergo sulphate reduction possess high bicarbonate and carbon dioxide contents and they also contain hydrogen sulphide.

Once equilibrium is attained, it may persist because of the normally very slow movement of groundwater and its long residence time in a given geological formation. In fact, variations in quality are more pronounced in shallow aquifers in which the seasonal variations in recharge and discharge cause parallel fluctuations in salinity. It has been observed that the freezing of shallow groundwater in arctic areas promotes seasonal changes in chemical quality. In the winter, there is a noticeable increase in salinity resulting from freezing because this lowers the dilution effect which might otherwise spring from infiltration of recharge from precipitation. Dissolved gases also occur in groundwater; these have been discussed already (see Section 3.3).

9.5. TEMPERATURE

This parameter is important. On the terrestrial surface, the variation in solar radiation received creates daily and yearly periodicities in temperature underneath. However, the crust possesses insulating ability which serves to diminish rather rapidly the wide range of temperature occurring at its surface; hence, groundwater has a fairly constant temperature except when it is very shallow. In this case, the type of surface environment overlying it is extremely significant. For instance, E. J. Pluhowski and I. H. Kantrowitz in 1963 demonstrated that, on Long Island, USA, the mean annual temperature and the annual range in the temperature of groundwater were greater under clear areas than beneath those which are wooded.[18] This difference is probably due to the absence of shade and lack of an insulating layer of organic matter on the ground in the areas which have been cleared. It must be remembered also that, due to the geothermal gradient, groundwater from deep wells is warmer than shallow groundwater and, in this connection, W. D. Collins stated that groundwater from 10 or 20 m depth is usually some 1–2°C warmer than the local mean annual air temperature in the USA.[19]

9.6. POLLUTION

This has been discussed earlier in some detail, but here it may be useful to summarize some additional data.

Leakage from sewers may be extremely adverse in introducing high concentrations of biochemical oxygen demand (BOD), chemical oxygen

TABLE 9.13

NORMAL RANGE OF INCREASE IN THE MINERAL CONSTITUENTS FOUND IN DOMESTIC SEWAGE[20]

Mineral	Range of increase (mg/litre)
Dissolved solids	100–300
Boron	0·1–0·4
Sodium	40–70
Potassium	7–15
Magnesium	3–6
Calcium	6–16
Total nitrogen (NO_3^-)	20–40
Phosphate	20–40
Sulphate	15–30
Chloride	20–50
Alkalinity, as $CaCO_3$	100–150

demand (COD), nitrate, organic chemicals and perhaps bacteria into groundwater. D. W. Miller in 1980 gave interesting information, which is embodied in Table 9.13.[20]

Of course, sewers are intended not to leak, but they do so because of a variety of factors such as defective piping, breaking by tree roots, seismic-induced fracturing, etc. Suspended solids in the sewage tend to clog cracks, however, and this may serve to decelerate leakage. A parallel case is leakage from tanks and pipelines, from which most of the pollution constitutes petroleum and petroleum products; see, for instance, J. O. Osgood in 1974.[21] As might be expected, the problem often arises with petrol station and domestic fuel oil tanks.

Mining may produce many groundwater pollution problems. The nature of the pollutant depends upon the material actually being mined and also upon the milling process. Among very important contributors are coal, phosphate and uranium mines, ore bodies producing iron, copper, zinc and lead, etc., and, since surface and subterranean mines usually extend below the water table, expansion of mining activities necessitates de-watering. The water so removed is very highly mineralized and is referred to as acid mine drainage. Its properties include low pH and high iron, aluminium and sulphate contents. Coal accumulations are usually associated with pyrite, FeS_2, which is stable for sub-water table conditions, but which oxidizes if the water table is lowered. Oxidation succeeded by contact with water produces ferrous sulphate ($FeSO_4$) and sulphuric acid (H_2SO_4) in solution and, of course, if they reach groundwater its pH will be reduced and its iron and sulphate contents will increase.

Oil and gas production is usually accompanied by the discharge of brine of which the constituents include sodium, calcium, ammonia, boron, chloride, sulphate, trace metals and a high total dissolved solids content. Formerly, such waste brine was discharged into streams or 'evaporation ponds'. This practice tended to pollute aquifers in the area, so such disposal is now disallowed by most regulatory bodies. Now, most brine is injected through deep wells into deep formations which are geologically isolated from the overlying, freshwater aquifers.

Pollutants may originate from agricultural activities, particularly return flows from irrigation. These represent the approximate one-half to one-third of water supplied for irrigation which does not disappear through evapotranspiration and drains off to surface channels and/or contributes to the groundwater system. They are much more saline than the applied water because salts are released by dissolution during irrigation and subsequently incorporated in them. Much of this salt contribution arises from fertilizers. As irrigation is so important in semi-arid and arid regions, return flow waters often promote most of the groundwater pollution.

Animal wastes also may add to pollution, especially where too many beasts are concentrated in one area. A consequence of this is that the soil cannot cope and the excess manure may be removed by storm runoff and contribute both to surface and subterranean waters, often together with salts and bacteria.

Fertilizers have been mentioned above: whenever they are applied, some are likely to leak through the soils to reach the water table. Mostly, they comprise nitrogen, phosphorus and potassium compounds, of which the last two usually become adsorbed on particles of soil and hence are quite innocuous. However, nitrogen in solution is only in part utilized by plants or adsorbed by the soil, the balance being available as a source of pollution.

In the same way, pesticides can contribute to groundwater pollution, a problem which continually increases because of their ever-widening employment. Most of them are rather insoluble and others are easily adsorbed by soil particles or may undergo microbial degradation; see for instance the 'state of the art' review of the health aspects of waste water reclamation for groundwater recharge issued by the California State Water Resources Control Board in 1978.[22] In 1979, analytical tests of water in California showed that about 100 water supply wells possessed trace quantities of DBCP (dibromochloropropane), which was widely used earlier and is a suspected carcinogenic agent.[5]

Pollution may originate in spills and surface discharges since these may penetrate into and degrade groundwater. Septic tanks and cesspools also make a major contribution to this phenomenon and it is interesting to note that, in the USA, most septic tanks are to be found in residential suburban areas which were constructed after the Second World War. A cesspool comprises a large and buried chamber which has porous walls; it can receive and percolate raw sewage.

There remains the matter of salt-water intrusion, the invasion of fresh water aquifers, where these are coastally located, intrusion is due to the sea, or where they are inland, it is due to underlying saline water. This whole subject is discussed in detail in Chapter 10.

The pollution in an aquifer may slowly dissipate with time and transit distance. Various mechanisms may play a role, for example filtration, sorption, microbiological decomposition and dilution. Filtration is most effective near the surface of the ground. Sorptive materials include clays, metallic oxides and hydroxides together with organic matter and they can take up most pollutants except for chloride, nitrate and sulphate. Since many pathogenic micro-organisms do not flourish in soils, they tend ultimately to attenuate and pathogens such as bacteria and viruses react as explained earlier (see Section 3.2).

In order to evaluate the potential pollution from any given source, an empirical point-count system was developed by H. E. LeGrand in 1964.[23] This may be used in respect of waste disposal sites and wells; a water table aquifer is assumed. Factors include depth to water table, sorption above the water table, permeability, water table gradient and horizontal distance, numerical values being read off above the lines for the five factors. Total point values are then interpreted in possible pollution terms ranging from imminent, through probable, possible and very improbable to impossible.

A pollution watch is desirable. This entails sampling and analyzing groundwater for quality, also determining groundwater levels and directions of flow and degree of moisture in the unsaturated zone; geophysical surveys, investigations of wastes which might be contributing to pollution and aerial surveillance should also be involved (see L. C. Everett et al. in 1976[24]). Additionally, hardness should be monitored carefully because, as H. Mitchell Perry, Jr, pointed out in 1971, comparisons, between drinking water and death rates established a statistically significant negative correlation involving hardness and deaths due to cardiovascular diseases.[25]

9.7. RECENT DEVELOPMENTS REGARDING POLLUTANTS

New European Economic Community standards for potability of water came into effect in July 1985 and the British Water Authorities have indicated that 600 of their source supplies cannot meet them. The most controversial pollutants involved are nitrates (which were discussed in detail in Section 3.3), described by an Opposition politician as perhaps the greatest environmental time-bomb ticking in the United Kingdom, lead and aluminium. Lead may retard brain development in children and it is estimated that over seven million people live in areas at risk such as Glasgow, Birmingham and Manchester. In Glasgow, about one-third of children examined showed contamination above safety levels laid down by the British Government. Unfortunately, it is not known how many of the relevant populations actually drink water which, being acidic, has dissolved lead from pipes and water tanks because this process only occurs after the water supply enters consumers' houses. Aluminium discolours water if present in adequate quantities. The metal is believed to be a possible cause of senile dementia, although this is denied by some experts. About two million people are exposed to potential aluminium hazards. In comparison, about 1·5 million are thought to be at risk from nitrates which, as mentioned in Section 3.3, can be linked with the production of stomach nitrosamines which are known to be carcinogens. The geographic areas most at risk are at Scunthorpe, Wisbech, and in parts of Leamington, Kenilworth, Warwick, Lichfield and Sutton Coldfield.

It is interesting to note that another hazardous pollutant occurs in Canada, where arseniferous groundwaters have been found in Nova Scotia and New Brunswick, as reported by D. J. Bottomley in 1984.[26] In Nova Scotia, a Carboniferous volcanic tuff contains iron oxide aggregations with arsenic and New Brunswick has bedrock composed of early Paleozoic metasediments, the Meguma Group, with arsenic concentrated in arsenopyrite in quartz veins.

REFERENCES

1. WHITE, D. E. et al., 1963. Data of Geochemistry — Chemical Composition of Subsurface Water, 6th edn. US Geol. Surv. Prof. Paper 440-F, 67 pp.
2. BACK, W. and HANSHAW, B. B., 1965. Chemical geohydrology. In: Advances in Hydroscience, Ed. V. T. Chow, Vol. 2. Academic Press, New York, pp. 49–109.
3. DAVIS, S. N. and DEWIEST, R. J. M., 1966. Hydrogeology. J. Wiley, New York, 463 pp.
4. DURFER, C. N. and BAKER, E., 1964. Principal Chemical Constituents in Groundwater, Their Sources, Concentrations and Effect on Usability. US Geol. Surv. Water Supply Paper 1812.
5. TODD, D. K., 1980. Groundwater Hydrology, 2nd edn. J. Wiley, New York, 535 pp., pp. 272–6, 334.
6. ERIKSSON, E., 1959. Atmospheric transport of oceanic constituents in their circulation in nature. Tellus, **11**, 1–72.
7. LOEWENGART, S., 1961. The major source of the salinity of waters in Israel. Bull. Res. Council Israel, **10G**, 183–206.
8. HEM, J. D., 1970. Study and Interpretation of the Chemical Characteristics of Natural Waters, 2nd edn. US Geol. Surv. Water Supply Paper 1473, 363 pp.
9. SAWYER, C. N. and MCCARTY, P. L., 1967. Chemistry for Sanitary Engineers, 2nd edn. McGraw-Hill, New York, 518 pp.

10. STIFF, H. A., Jr, 1951. The interpretation of chemical water analysis by means of patterns. *J. Petr. Technology*, **3**, 10, 15–7.
11. PIPER, A. M., 1944. A graphic procedure in the geochemical interpretation of water analyses. *Trans. Amer. Geophys. Union*, **25**, 914–28.
12. SCHOELLER, H., 1961. *Les Eaux Souterraines*. Masson et Cie, Paris, 642 pp.
13. AMERICAN PUBLIC HEALTH ASSN, AMERICAN WATER WORKS ASSN and WATER POLLUTION CONTROL FEDN, 1975. *Standard Methods for the Examination of Water and Wastewater*, 14 edn. American Public Health Association, Washington, DC, 1200 pp.
14. US ENVIRONMENTAL PROTECTION AGENCY, 1975. Federal Register, Vol. 40, No. 248 59566–59588.
15. AMERICAN WATER WORKS ASSN, 1971. *Water Quality and Treatment*. McGraw-Hill, New York, 654 pp.
16. RICHARDS, L. A., Ed., 1954. *Diagnosis and Improvement of Saline and Alkali Soils*. Agric. Handbook 60, US Dept Agric., Washington, DC, 60 pp.
17. WILCOX, L. V., 1955. *Classification and Use of Irrigation Waters*. US Dept Agric., Circ. 969. Washington, DC, 19 pp.
18. PLUHOWSKI, E. J. and KANTROWITZ, I. H., 1963. *Influence of Land Surface Conditions on Ground-water Temperatures in Southwestern Suffolk County, Long Island, New York*, US Geol. Surv. Prof. Paper 475-B, 186–8.
19. COLLINS, W. D., 1965. *Temperature of Water Available for Industrial Use in the United States*. US Geol. Surv. Water Supply Paper 520-F, 97-104.
20. MILLER, D. W., Ed., 1980. *Waste Disposal Effects on Ground Water*. Premier Press, Berkeley, California, 512 pp.
21. OSGOOD, J. O., 1974. Hydrocarbon dispersion in ground water: significance and characteristics. *Ground Water*, **12**, 6, 427–38.
22. CALIF. STATE WATER RESOURCES CONTROL BD, DEPT WATER RES. and DEPT HEALTH, 1978. *A 'State of the Art' Review of Health Aspects of Wastewater Reclamation for Groundwater Recharge*. Water Info. Center, Huntington, New York, 240 pp.
23. LEGRAND, H. E., 1964. System for evaluation of contamination potential of some waste disposal sites. *J. Amer. Water Works Assn*, **56**, 959–74.
24. EVERETT, L. C. *et al.*, 1976. *Monitoring Groundwater Quality: Methods and Costs*. Rept EPA-600/4-76-023. US Env. Prot. Agency, Las Vegas, 140 pp.
25. MITCHELL PERRY, H., Jr, 1971. Trace elements related to cardiovascular disease. Geol. Soc. Amer., Memoir 123, *Environmental Geochemistry in Health and Disease*, Amer. Assn for Advancement of Science Symposium, Dallas, Texas, December 1968, Ed. H. L. Cannon, H. C. Hopps, pp. 179–95.
26. BOTTOMLEY, D. J., 1984. Origins of some arseniferous groundwaters in Nova Scotia and New Brunswick, Canada. *J. Hydrol.*, **69**, 223–57.

CHAPTER 10

Saline Intrusion

10.1. OCCURRENCE

Investigation of saline intrusion is extremely important because it constitutes probably the commonest of all pollutants in fresh water, which it may displace or with which it may mix. In shallow inland aquifers it usually arises from waste discharges on the surface but in the case of coastal aquifers it is caused by a seawater invasion. In deep aquifers, it can be caused by the ascent of saline waters from a deep-seated geological origin. As implied, human activities are responsible for some saline intrusion. The phenomenon is not invariably adverse, however; in arid inland areas, it may become a useful resource and, of course, such water may be employed in industrial processes such as cooling and desalination.

Often, shallow fresh water may overlie saline water when recent flushing has removed salts from antique marine deposits; hence this may be considered as a desalinated, connate water. The author encountered many such situations when working in the Punjab, India, in 1980, and in addition there were instances of saline water overlying fresh water.[1] In the lower regions the movement of groundwater is slower and, in these circumstances, saline water displacement is also slower. At depths of some few thousands of metres it is common to encounter brines and it is interesting that the US Geological Survey estimated in 1965 that about two-thirds of the USA is underlain by aquifers which are known to produce waters containing over 1000 mg salt/litre.[2]

In summary, therefore, saline waters occurring in aquifers may derive from the following sources:

(a) seawater, in coastal areas;

(b) seawater which penetrated aquifers during past geological time;
(c) salt from salt domes or thin salt beds or disseminated in geological formations,
(d) evaporated water residue left over in tidal lagoons, playas, etc.;
(e) return flows from irrigated land to streams;
(f) saline wastewaters of human origin.

There are several ways in which the intrusion into fresh water aquifers may be effected. One is the reversal or reduction of groundwater gradients which can allow denser, saline water to displace fresh water, a state of affairs often found in coastal aquifers which are hydraulically continuous with the sea and in which excess well pumping has disturbed the hydrodynamic equilibrium. Another method is based upon the removal of natural barriers separating fresh and saline waters, and a third depends upon the subsurface disposal of waste saline waters.

Intrusion of saline water in coastal environments occurs all over the world. Instances have been recorded all round the USA where the most severely affected states include Connecticut, New York, Florida, Texas, California and Hawaii. In Europe, there are cases in England (the earliest being reported by F. Braithwaite from London and Liverpool in 1855[3]), Germany, the Netherlands (see, for instance, D. K. Todd and L. Huisman in 1959[4]) and elsewhere. In Israel and Japan there are also examples, and small oceanic islands are frequently underlain with aquifers which contain seawater.

10.2. THE GHYBEN–HERZBERG RELATION

In 1888–89, J. Drabbe and W. Baydon-Ghyben and, independently in 1901, B. Herzberg, who were investigating along the coast of Europe, discovered that saline water occurred underground at a depth approximately forty times the height of fresh water above sea level.[5,6] The phenomenon is attributed to a hydrostatic equilibrium existing between two fluids having different densities, and is described by the Ghyben–Herzberg equation, which is derived as follows.

The hydrostatic balance which obtains between fresh and saline waters can be demonstrated in a U-tube (Fig. 10.1). As the pressures on each side of this must be the same,

$$\rho_s g h_s = \rho_f g(z + h_f)$$

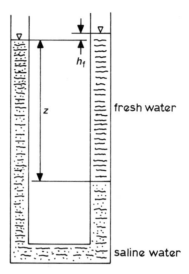

Fig. 10.1. The hydrostatic balance between fresh and saline water is demonstrated in a U-tube.

where ρ_s is the density of the saline water, ρ_f is the density of the fresh water, g is the acceleration due to gravity and z and h_f are as observed in the Figure. From this, it follows that

$$z = \frac{\rho_f}{\rho_s - \rho_f} h_f$$

and this is the Ghyben–Herzberg relation. In most cases, ρ_s has a value of about $1·025\text{ g/cm}^3$ and $\rho_f = 1·000\text{ g/cm}^3$. Therefore,

$$z = 40 h_f$$

Figure 10.2 demonstrates the situation in a coastal milieu, h_f being the elevation of the water table above sea level and h_s the depth to the fresh water/saline water interface below sea level, a hydrodynamic rather than a hydrostatic balance (since fresh water is flowing seawards). In this case:

$$h_s = \frac{\rho_f}{\rho_s - \rho_f} h_f$$

and:

$$h_s = 40 h_f$$

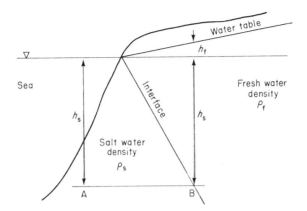

FIG. 10.2. Illustrating the Ghyben–Herzberg relation.

If density alone is considered, no flow occurring, the result would be the development of a horizontal interface with fresh water lying above saline water. Where the flow is almost horizontal, however, the Ghyben–Herzberg relation does provide quite acceptable results. Nevertheless, significant errors may be introduced near shorelines where components of vertical flow intrude.

The above argument applies to an unconfined, coastal aquifer, but it may be extended to a confined aquifer in a similar location by substituting the piezometric surface for the water table. The Ghyben–Herzberg relation requires that, in both cases, in order to satisfy fresh water/saline water equilibrium needs, the appropriate surface (either the water table of the piezometric surface) must be above sea level and must also incline towards the ocean.

M. K. Hubbert in 1940 discussed the theory of groundwater motion[7] and, from this basis, N. J. Lusczynski and W. V. Swarzenski[8,9] generalized the Ghyben–Herzberg relation to cover cases where underlying saline water is in motion with heads either above or below sea level. Their result, for non-equilibrium conditions, is:

$$z = \frac{\rho_f}{\rho_s - \rho_f} h_f - \frac{\rho_f}{\rho_s - \rho_f} h_s$$

where h_f is the altitude of the water level in a well filled with fresh water of density ρ_f and terminated at a depth z and h_s is the altitude of the water level in a well filled with saline water of density ρ_s which also terminates at depth z. When $h_s = 0$, then the saline water is in equilibrium with the sea, and:

$$z = \frac{\rho_f}{\rho_s - \rho_f} h_f$$

More precise solutions for the shape of the fresh water/saline water interface have been proposed from potential flow theory and one result is:

$$z^2 = \frac{2qx}{\Delta \rho K} + \left(\frac{q}{\Delta \rho K}\right)^2$$

where z and x are as illustrated in Fig. 10.3, $\Delta \rho = \rho_s - \rho_f$, K is the hydraulic conductivity of the aquifer and q is the fresh water flow per unit length of shoreline.

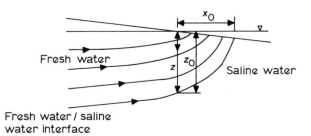

Fig. 10.3. The flow pattern developed by fresh water in an unconfined coastal aquifer.

The corresponding shape for the water table is:

$$h_f = \left(\frac{2\Delta \rho q x}{K}\right)^{1/2}$$

The width, x_0, of the submarine zone through which fresh water discharges into the sea can be derived for $z = 0$, giving:

$$x_0 = -\frac{q}{2\Delta \rho K}$$

The depth of the interface beneath the shoreline, z_0, occurs when $x = 0$, hence:

$$z_0 = \frac{q}{\Delta \rho K}$$

10.3. THE FRESH WATER/SALINE WATER INTERFACE

This is not a sharply defined boundary in nature. In fact, there is a brackish water transitional zone between the two fluids and this has a variable, but finite, thickness. It results from the flow dispersion of the fresh water accompanied by unsteady displacements of the interface as a consequence of influences such as tides and well pumping. The transition zone attains maximum thicknesses in very permeable coastal aquifers which are subjected to heavy pumping. Usually, these are around 100 m, but in the Honolulu–Pearl Harbor region of Hawaii local transition zones have been found to attain thicknesses exceeding 300 m. Minimal thicknesses, by contrast, may be as little as 1 m. The transition zone is flowing seawards and therefore conducts saline water (arising from underlying sources) in the same direction. As a result, from continuity, there is a landwards flow in the saline water region, as D. K. Todd indicated in 1960.[10] Inside the transition zone, the groundwater salinity increases progressively with depth from that of fresh water to that of saline water and, since the distribution of salinity with depth varies as an error function, the relative salinity, S_R, can be calculated as a percentage thus:

$$S_R = 100\left(\frac{c - c_f}{c_s - c_f}\right)$$

where c is the salinity at a particular depth in the transition zone and c_f, c_s the salinities of the fresh and saline waters respectively. If S_R values are plotted against depth on probability paper, a straight line usually results and this facilitates estimation of the transition zone from any two point measurements of salinity.

If there are heterogeneous conditions in coastal aquifers, stratification of fresh and saline waters may occur. Appropriate researches have been effected by a number of investigators; see, for instance, M. A. Collins and L. W. Gelhar in 1971 on seawater intrusion in layered aquifers.[11]

An interesting variant is manifested by the up-coning of saline water where a well is pumped from an aquifer which has an underlying layer of this fluid. The matter has been studied in recent years so that criteria may be formulated to design and operate wells in order to draw off fresh water lying above saline water. Whilst some of this work involved an assumption of an abrupt interface between the two fluids, this cannot be the case in nature because of their miscibility. Hence, once again a transition of variable, but finite, thickness must exist. There are analytic solutions for up-coning, but some apply only if the rise is limited. It has been found that up-coning can be minimized if wells and galleries are properly designed and operated and, in given aquifer conditions, wells must be separated vertically as far as is feasible from the zone of salinity; they should also be pumped at a low, uniform rate. Tests have demonstrated that wells which pump brackish or saline water from below the fresh water, the so-called scavenger wells, successfully counteract up-coning.

10.4. OCEANIC ISLANDS

Most oceanic islands are rather permeable and comprise limestone, coral, sand or occasionally lava; hence the seawater contacts groundwater on every side. The groundwater is limited in quantity and fresh groundwater is obtained through rainfall alone. Since fresh water movement is radial towards the sub-circular coast, a fresh water lens forms and floats on deeper-lying saline water, its thickness diminishing coastwards.

From the Dupuit assumptions and the Ghyben–Herzberg relation, an approximate fresh water boundary is determinable. If the oceanic island is circular in plan with a radius of R and there is an effective recharge from rainfall at a rate W, the outward flow, Q, at a radius r is:

$$Q = 2\pi r K (z+h) \frac{dh}{dr}$$

where K is the hydraulic conductivity, h and z being defined in Fig. 10.4. Since $h = (\Delta\rho/\rho)z$ and, from continuity, $Q = \pi r^2 W$, therefore:

$$z\,dz = \frac{Wr\,dr}{2K\left[1+\dfrac{\Delta\rho}{\rho}\right]\left[\dfrac{\Delta\rho}{\rho}\right]}$$

and, integrating and applying a boundary condition that $h = 0$ when $r = R$:

$$z^2 = \frac{W(R^2 - r^2)}{2K\left[1+\dfrac{\Delta\rho}{\rho}\right]\left[\dfrac{\Delta\rho}{\rho}\right]}$$

Hence, the depth to saline water at any location is a function of the

SALINE INTRUSION

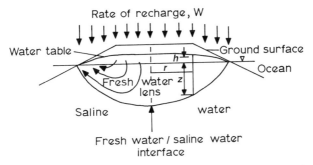

FIG. 10.4. A fresh water lens in an oceanic island under natural conditions with a recharge rate of W.

actual rainfall recharge, the size of the island involved and the hydraulic conductivity.

Fluctuations in rainfall and tides, coupled with the effects of dispersion, cause a transition zone along the interface in an oceanic island. This boundary is close to the water table and as it can contribute saline water in a well by up-coning, care must be taken to avoid disturbing the fresh water/salt water equilibrium by pumping; this activity is usually operated so as to produce as small a drawdown as possible. Sometimes, a horizontal infiltration gallery is an alternative approach. There are a number of these, for instance on Bermuda and in the Bahamas in the Atlantic Ocean and on the Gilbert and Marianas islands in the Pacific Ocean. In cases where the water table is very deep, shafts may be sunk to it, and from them adits (horizontal tunnels) can be extended outwards so as to intercept the overlying layer of fresh water; many exist in Barbados, Guam and Hawaii. Barbados, in fact, has a thin, very permeable coral limestone as an aquifer; rainfall percolates through solution channels along the base of this limestone until it reaches sea level. From there to the coast, fresh water floats in a layer which is locally termed sheet water, lying above saline water. Water is taken from large-diameter dug wells which connect to horizontal adits at the water table. In Honolulu, the aquifer is in permeable basalt, impermeable cap rock acting as a groundwater dam. Wells were installed first in 1880. Prior to that date, groundwater discharged as springs at inland or submarine cap rock boundaries. At that time, the transition zone was sub-horizontal and narrow, but, after well pumping developed widely, the water table level declined and the transition zone expanded. Thus, the actual volume of fresh water has been reduced greatly and up-coning has occurred, as a result of which, the depths of production wells have decreased from 450 m to 85 m as the phenomenon has progressed through the years.

10.5. KARSTIC TERRAINS

Coastal aquifers in karstic limestones are especially liable to seawater intrusion, the multitudinous fissures, fractures and solution openings facilitating the admission of this fluid, as discussed by V. T. Stringfield and H. E. LeGrand in 1971.[12] Among unique concomitant features are sporadic brackish springs possibly resulting from channels which connect with the sea. Solution channels are, in fact, very common in circum-Mediterranean karstic regions, and discharge fresh water as submarine springs. However, pumping of such channels in an attempt to avoid loss of this fresh water resource frequently yields saline water because it can enter these channels easily if fresh water flow is reduced.

10.6. CONTROLLING SALINE WATER INTRUSION

The most common methods utilized are listed in Table 10.1.

TABLE 10.1
CONTROL OF SALINE WATER INTRUSION

Cause of intrusion	Control techniques
Seawater in a coastal aquifer	Alteration of the pumping pattern.
	Artificial recharge.[a]
	Extraction barrier.
	Injection barrier.
	Subsurface barrier.
Up-coning	Alteration of the pumping pattern.
	Saline scavenger wells.
Oil field brine	Eliminating surface disposal.
	Injection wells.
	Plugging abandoned wells.
Defective well casing	Plugging defective wells.
Surface infiltration	Eliminating the surface source.
Saline water zones in fresh water aquifers	Relocating and re-designing wells.

[a] See Chapter 11.

The pumping pattern can be varied by changing the location of the wells concerned by dispersal and, if this is optimally effected, it may result in the establishing of a more powerful seaward hydraulic gradient. Reducing the degree of pumping in existing wells can have the same consequences.

As noted in the Table, artificial recharge will be discussed in Chapter 11.

Extraction barriers are created by maintaining a continuous pumping trough with a line of wells adjacent to the sea. Seawater will flow inland from the latter to the trough and fresh water within the basin will flow seawards towards the trough. Water so pumped is brackish and usually it is discharged into the sea.

The injection barrier concept entails maintaining a pressure ridge along the coast using a line of recharge wells through which fresh water is introduced and flows both landwards and seawards. Of course, very good-quality water must be imported to the relevant area for this purpose. A better solution may be to operate both extraction and injection barriers simultaneously, but this requires a larger number of wells.

A subsurface barrier is a physical one constructed parallel to the coast throughout the vertical thickness of an aquifer in order to prevent seawater flowing into a basin. Various materials are applicable, for instance cement grout, bentonite, silica gel, calcium acrylate and plastics as well as clay, emulsified asphalt and even sheet piling. The main problems arising from this approach are the great expense involved and the resistance of the installed structure to earthquakes and chemical erosion.

Many examples of seawater intrusion exist and the case of Los Angeles, California, may be cited.[13] Here, saline intrusion was noted in the 1930s along the west coast of the county and the problem accelerated in the next decade. In consequence, an injection barrier was constructed along 11 km of this particular shoreline and 94 recharge wells create a pressure ridge in the relevant confined aquifers, seawater thus being separated from overpumped inland basin supplies. Profiles of the piezometric surface perpendicular to the barrier taken while this part of it was established show how the landward gradient in the injected Silverado aquifer altered into a pressure ridge which extends above sea level. No less than 267 observation wells are monitored in order to assess the effects of this injection barrier. Filtered water is delivered to the distribution line with an average chlorine residual of 0·5 mg/litre and, under normal conditions of operation, 1·5 mg chlorine/litre is added before injection takes place. The elevations in the piezometric head in the barrier are kept 1–3 m above sea level in the main aquifer. Actually, the heads at mid-points between the injection wells are measured every week and the flow rates in the wells adjusted as proves necessary to maintain the injection barrier. The injection rate per well in the past few years has averaged 1500 m^3 daily.

10.7. ORIGIN AND MOVEMENT OF SALINE CONTAMINATION

Sometimes, groundwater analyses from samples collected in areas of seawater intrusion have a chemical composition which differs from that obtained by a simple proportional mixing of seawater and groundwater. There are several ways in which seawater composition may be modified when it enters an aquifer. Thus, base exchange may take place between the water and the minerals in the aquifer, sulphate reduction may occur together with substitution of carbonic or other weak acid radicals, or solution and precipitation may occur. Solution and precipitation can change the total content of salt, but the first two processes, which require the maintenance of ionic balance, can change the percentage by weight of the different salt components and thus the total dissolved solids in mg/litre.

R. Revelle proposed that the chloride–bicarbonate ratio be used to evaluate saline intrusion; chloride, the dominant anion in seawater, is unaffected by the above-mentioned processes and it occurs only in very small quantities in groundwater.[14] Bicarbonate, by contrast, is the most abundant anion in groundwater and occurs in very small amounts in seawater. Usually, the ratio $Cl^-/(CO_3^{2-} + HCO_3^-)$ is employed in such evaluations.

One problem which may arise is that of distinguishing seawater intrusion from other sources of salinity in coastal regions. For instance, in Long Beach, California, there were no less than four possible local saline sources, namely the sea, oil field waste brines, irrigation wastewaters and shallow connate waters. A useful approach in such cases is to utilize isotopic analyses because isotopic contrasts may be expected among the various possible sources of salinity. Stable-isotope investigations of deuterium and oxygen-18 contents are valuable because marine waters have completely different ratios from meteoric waters. Saline water which is derived from lakes and other surface water bodies enriched in these heavy stable isotopes by evaporation processes is of course distinguishable from both seawater and precipitation, which are relatively depleted.

Oil field brines and deep fossil waters usually show oxygen-18 exchange when compared with seawater and precipitation.

Tritium and radiocarbon are also valuable because modern precipitation is labelled by these isotopes. Many tritium-dead, saline waters may be 'dated' by their radiocarbon content because carbon-14 has a much longer half-life than tritium (5730 years as opposed to 12·323 years). Near-surface seawaters possess a low, but consistent, tritium content and a rather high, but also consistent, radiocarbon content. It is apparent, therefore, that a distinction can be made between the possible sources of saline contamination on the basis of the isotopic content of the waters. Reconnaissance sampling may produce satisfactory results. The following generalizations may be applicable.

(a) Terrestrial waters concentrated by evaporation can be distinguished easily from waters of marine origin, the former being relatively enriched in the heavy, stable isotopes.
(b) Modern seawater can sometimes be distinguished from connate waters, although often there is little stable-isotopic contrast.
(c) Seawater can be distinguished from deep brines because the latter are relatively enriched in oxygen-18 with respect to deuterium due to the exchange of oxygen with rock minerals.

It is interesting to observe also that if connate water entered an aquifer while Pleistocene glacial episodes prevailed, deuterium and oxygen-18 are depleted relative to modern seawater and, of course, lack of tritium and radiocarbon is in sharp contrast to the presence of these labels in seawater. In principle, $^{13}C/^{12}C$ and $^{34}S/^{32}S$ ratios might be utilized in distinguishing seawater from meteoric waters.

R. N. Clayton and others in 1966 used deuterium and oxygen-18 ratios in a study of 96 brines from oil fields in the Illinois and Michigan basins, and also on the Gulf Coast and in the Alberta Basin.[15] They found that variation in D content among basins was much greater than that noted in individual basins, and it was related to geographic location; also, ^{18}O contents covered a wide range in each basin and this correlated with salinity and formation temperature. Their main conclusions were that the water is not marine in origin, but arises from local meteoric sources, the deuterium contents not having undergone much exchange or fractionation. However, there has been extensive exchange between the water and the reservoir rocks. Some of the samples may have originated as Pleistocene glacial precipitation.

REFERENCES

1. BOWEN, Robert, 1985. Hydrogeology of the Bist Doab and adjacent areas, Punjab, India. *Nordic Hydrology*, **16**, 33–44.
2. FETH, J. H. *et al.*, 1965. Preliminary Map of the Conterminous United States Showing Depth to and Quality of Shallowest Ground Water Containing More Than 1000 ppm Dissolved Solids. U.S. Geol. Surv. Hydrol. Inv. Atlas HA-199, 31 pp.
3. BRAITHWAITE, F., 1855. On the infiltration of salt water into the springs of wells under London and Liverpool. *Proc. Inst. Civ. Engrs*, **14**, 507–23.
4. TODD, D. K. and HUISMAN, L., 1959. Ground water flow in the Netherlands coastal dunes. *J. Hydraulics Divn, Am. Soc. Civ. Engrs*, **83**, HY75, 63–81.
5. DRABBE, J. and BAYDON-GHYBEN, W., 1888–89. Nota in verband met de voorgenomen putborung nabij Amsterdam. *Tijdschrift van het Koninklijk Instituut van Ingenieurs*, The Hague, Netherlands, 8–22.
6. HERZBERG, B., 1901. Die Wasserversorgung einiger Nordseebader. *J. Gasbeleuchtung und Wasserversorgung*, München, **44**, 815–9, 842–4.
7. HUBBERT, M. K., 1940. The theory of ground-water motion. *J. Geol.*, **48**, 785–944.
8. LUSCZYNSKI, N. J., 1961. Head and flow of ground water of variable density. *J. Geophys. Res.*, **66**, 4247–56.
9. LUSCZYNSKI, N. J. and SWARZENSKI, W. V., 1966. *Salt-water Encroachment in Southern Nassau and Southeastern Queens Counties, New York*. US Geol. Surv. Water Supply Paper 1613-F, 76 pp.
10. TODD, D. K., 1960. *Salt Water Intrusion of Coastal Aquifers in the United States*. Intl. Assn. Sci. Hydrol., Publ. 52, pp. 452–61.
11. COLLINS, M. A. and GELHAR, L. W., 1971. Seawater intrusion in layered aquifers. *Water Res. Res.*, **7**, 971–9.
12. STRINGFIELD, V. T. and LEGRAND, H. E., 1971. Effects of karst features on circulation of water in carbonate rocks in coastal areas. *J. Hydrol.*, **14**, 139–57.
13. BRUINGTON, A. E. and SEARES, F. D., 1965. Operating a sea water barrier project. *J. Irrig. Drain. Divn, Amer. Soc. Civ. Engrs*, **91**, IRI, 117–40.
14. REVELLE, R., 1941. Criteria for recognition of sea water in groundwaters. *Trans. Amer. Geophys. Union*, **22**, 593–7.
15. CLAYTON, R. N., FRIEDMAN, I., GRAF, D. L., MAYEDA, T. K., MEENTS, W. F. and SHIMP, N. F., 1966. The origin of saline formation waters. I: Isotopic composition. *J. Geophys. Res.*, **71**, 16, 3869–82.

CHAPTER 11

Artificial Recharge

11.1. INCREASING NATURAL GROUNDWATER SUPPLY

The artificial recharge of groundwater basins involves augmenting the natural infiltration of precipitation or other surface waters into subterranean formations in other words assisting their natural movement by various means.

Methods employed include water spreading on the ground surface, pumping in order to induce recharge from water bodies on the surface, and recharging through wells and boreholes. The technique selected for any particular location depends upon factors such as topography, soil state and geology as well as the quantity of water to be recharged and the end use of the water.

Artificial recharge has been employed in the USA since the end of the nineteenth century, and almost 30 years ago the state of California recharged as much as 375 million gallons daily using surface water. At that time, this represented more than half of all water artificially recharged in the United States, although all artificial recharge water then only amounted to approximately 1·5% of total groundwater use there.

Similar activities have been effected in other parts of the world; for instance, pumping water from the surface down wells has proved useful in the Netherlands, where fine sand beds are recharged with treated water from the Rhine River, both to provide water storage for supply and to act as a barrier against the inwards seepage of seawater from the North Sea.

11.2. PROJECT OBJECTIVES

These may be grouped as follows:

(a) to maintain or augment natural groundwater as a resource;
(b) to coordinate the operation of surface and groundwater reservoirs;
(c) to ameliorate unfavourable conditions, for instance saline water intrusion;
(d) to afford subsurface storage for local or imported surface waters;
(e) to minimize or halt totally subsidence of land;
(f) to provide a localized subsurface system of distribution for wells that are already established;
(g) to offer facilities for treatment and storage for reclaimed waste waters for later re-use;
(h) to conserve or extract energy as hot or cold water.

For water from the surface to be stored underground, enough of it must be available and, in order to ensure this, collection may be effected in ditches, reservoirs, etc., from which recharge may occur later. Recharge basins are common in Swedish municipal water supply systems and also in the Netherlands in the water supply systems for the cities of Amsterdam, Leiden and The Hague. In this decade, there are approximately 276 artificial recharge projects in operation in areas of extensive groundwater exploitation.

11.3. THE METHODS OF RECHARGE

As noted above, there are many methods of recharge, of which the most commonly used are variants of the water spreading approach, i.e. releasing water over the surface of the ground so as to increase the amount of water infiltrating into it and thereafter percolating down to the water table. Of the factors which affect the rate of water entry into the soil, the most important are the actual area of recharge and the length of time during which the water is in contact with the soil. The efficiency of the spreading is measured in terms of the recharge rate, given as the velocity of downward movement of water over the relevant wetted area. Spreading is accomplished through the following techniques: basin; stream channel; ditch and furrow; flooding and irrigation; and through pits and recharge wells. Appropriate data from 1959 in California are given in Table 11.1.[1]

TABLE 11.1
DISTRIBUTION OF SPREADING TECHNIQUES IN CALIFORNIA, USA[1]

Technique	Recharge projects (%)	Recharged water (%)
Basin	54	58·4
Stream channel	15	29·5
Ditch and furrow	8	9·4
Pit	7	1·3
Well	12	1·0
Flooding	4	0·4

The basin approach to artificial recharge entails introducing water into appropriate basins which may be excavated or impounded by barriers, their geometry integrating with the local geomorphology. In order for water to remain free of silt, regular inspections must be made and attention must be given to the basins, including bottom scraping when they are dry.[2] The number of basins required depends upon the method of recharge. Thus, if storm runoff is utilized, one basin is usually sufficient. On the other hand, diversion of streamflow for the same purpose necessitates several basins located more or less parallel to the stream channel involved. Water from the stream is conducted through a ditch into the highest of the basins which, when full, discharges into a second lower one, and so on, down a series, from the lowest of which surplus water may be returned to the stream channel. This technique allows water to contact more than three-quarters of the total area. An advantage of the multiple basin system is that, if one is taken out of service for attention, the rest can still be employed. Also, the upper basins can be utilized for settling out of undesirable silt. The rates of recharge achieved in various regions show considerable variability; D. K. Todd in 1959 cited informative data which are set out in Table 11.2.[3]

It seems that the natural slope of the ground is indicative of the long-time rates of recharge which may be expected.[1] In the case of alluvial soils with a range of slope of 0·1–10%, the long-time rate of infiltration, W (m/day), is expressed thus:

$$W = 0·65 + 0·56i$$

where i is the natural slope of the ground as a percentage. Individual rates have been found to vary within a factor of two of this estimate.

In the stream channel method of water spreading, the approach entails

TABLE 11.2
SELECTED SPREADING RATES USING BASIN RECHARGE
IN THE USA[3]

Location	Rate (m/day)
Arizona	
Santa Cruz River	0·3–1·2
California	
Los Angeles County	0·7–1·9
Madera County	0·3–1·2
San Gabriel River	0·6–1·6
San Joaquin Valley	0·1–0·5
Santa Ana River	0·5–2·9
Santa Clara Valley	0·4–2·2
Tulare County	0·1
Ventura County	0·4–0·5
Iowa	
Des Moines	0·5
Massachussetts	
Newton	1·3
New Jersey	
East Orange	0·1
Princeton	0·1
New York	
Long Island	0·2–0·9
Washington	
Richland	2·3

widening the area and increasing the time over which water is recharged from a naturally losing channel. Upstream streamflow management is required and there may have to be modifications of the channel in order to increase the infiltration. Ideally, reservoirs sited upstream facilitate the control of erratic runoff water and limit rates of streamflow to those which conform to the absorptive capacity of the downstream channels. Remedial work on stream channels sometimes involves constructing low check dams across streams where these are wide-bottomed. These dams function as weirs and distribute water into shallow ponds occupying the whole of the stream bed. Such dams are temporary in nature, but, if permanent ones are needed, these must be made in such a way that no flood hazard can arise from them. The approach has been extensively employed in California, where channel spreading with rock and wire check dams is effected in Cucamonga Creek, for instance. Most of the spreading works in and near Los Angeles county form part of an integrated water conservation and flood protection plan.

In the ditch and furrow method of water spreading, water is distributed by a set of ditches or furrows which are shallow, flat-bottomed and closely spaced in order to provide a maximum area of contact for the water. There are several basic patterns: contour in which the ditch follows the contour of the ground surface, tree-shaped in which it successively branches into smaller and smaller canals, and lateral in which a set of small ditches extends laterally from a main one. The width of ditches varies from about 0·3 m to about 1·8 m and their gradient should be adequate to permit suspended material to pass through the whole system because, otherwise, finer-grained sediment may well block some of the surface openings. At the lower end of the relevant site, a collecting ditch is installed so as to convey excess water back into the main stream channel.

A rather flat area is required for flooding so that water can be diverted and is able to spread more or less evenly over it. The practice is to release the water through canals on a periodic basis at the upper end of the area of flooding. The great advantage of this technique, where it can be applied, is that it is the cheapest method as regards the preparation of the land.

In irrigation, water is used for crop land during the winter or dry season, and the spreading effected for this purpose serves a dual function; its application to recharge by spreading requires no special work because the delivering system of distribution already exists. If the irrigation canals are maintained full, a contribution to recharge will be made anyway through seepage from them.

The pit method entails excavation of a pit into a permeable formation. To save money, abandoned gravel pits, etc., are used. In areas possessing shallow, impermeable subsurface strata such as clays and hard pans, the pits enable recharge water to reach geological materials with higher rates of infiltration. Pits are valuable in that their steep sides afford a high tolerance for silt, which often settles out to their bottoms and thus leaves the walls almost unblocked so that water may continue successfully to infiltrate.

It may be noted that recharge pits of this type have been utilized in Peoria, Illinois, USA, since 1951.[4] They penetrate a shallow sand and gravel aquifer which supplied well fields there and they recharge chlorinated river water. Depths range up to 9 m and sides and bottoms are covered with natural gravel having a permeability about 19 times greater than that of the aquifer itself. This gravel takes up silt and when the content reaches 0·97 kg silt/litre of gravel, the gravel is replaced: this occurs roughly every 3–4 years. The addition of chlorine to the recharge water is at a rate of 3–5 mg/litre. The local groundwater here is utilized for industrial cooling, so the river water recharge is introduced only when the temperature of the river is under 18°C, which happens for about six months of every year. At first, the rates of recharge were about 23 m/day over the filtering surface, but these declined until, after just over a decade, they were only just over half of this (actually 12 m/day), as a result of penetration of silt into the upper levels of the aquifer. The Illinois State Water Survey made laboratory studies of the filtration efficiency of coarse media and produced the following expression:[5]

$$SS_0 = 13 \cdot 1 H^{-0.25} d^{0.5} Q_0^{0.33} SS_i^{1.33}$$

where SS_0 is the suspended solids concentration (mg/litre) transmitted through the filter layer, H is the filter layer thickness (cm), d is the mean diameter (mm) of the particles forming the filter layer, Q is the rate of recharge (m/day) and SS_i is the suspended solids concentration of the recharged water. The relation can be valuable in designing and operating a recharge pit.

The recharge well method involves using a well which transfers water from the surface to fresh water aquifers; alternative terms for them include disposal wells and drain wells. The recharge well must be distinguished from an injection well which recharges brines and toxic waste products of industry to deep, saline water aquifers.

Flow in a recharge well reverses that in a pumping well so that a cone of recharge replaces the cone of depression and, if these are equal, the recharge capacity, equals the pumping capacity of that well.

An equation for the curve of the cone of depression in a recharging well can be derived similarly to that for a pumping well and, for a confined aquifer with water being recharged into a totally penetrating well at a rate of Q_r, an approximate steady-state expression is:

$$Q_r = \frac{2Kb(h_w - h_0)}{\ln(r_0/r_w)}$$

where the symbols are those shown on Fig. 11.1. In the case of an unconfined aquifer, the appropriate equation is:

$$Q_r = \frac{K(h_w^2 - h_0^2)}{\ln(r_0/r_w)}$$

for which Fig. 11.2 is applicable.

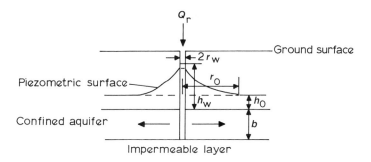

Fig. 11.1. Radial flow from a recharge well penetrating a confined aquifer.

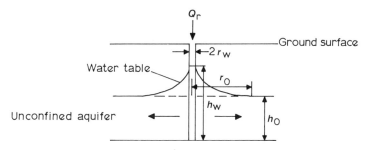

Fig. 11.2. Radial flow from a recharge well penetrating an unconfined aquifer.

The above-mentioned equality between a pumping and a recharging well hardly ever occurs in nature because more factors are involved than just a reversal of the direction of flow. When water is withdrawn from a well undergoing pumping, fine-grained material filters through coarser sediment surrounding the well and enters it. However, silt carried by water into a well which is recharging an aquifer filters out and tends to clog the aquifer in the immediate vicinity of the well. Also, water entering into a recharge well transports quite large quantities of dissolved air which, by air binding, tends to reduce the permeability of the receiving aquifer. Finally, recharge water usually contains bacteria and these may form growths on the well screens which may reduce their effective flow areas. All of these factors are adverse to the amount of recharge water which can be introduced into an aquifer. Nevertheless, recharging activities have been carried out in a number of localities with a fair degree of success; see for instance the reports of H. O. Reeder *et al.* in 1976 on fissured carbonate rock at West St. Paul, Minnesota, USA, and B. L. Foxworthy in 1970 on the Salem Heights area of Salem, Oregon,[6,7] both concerned with recharge wells. Of course, the maximum rate of recharge is achieved in highly porous formations much as limestone and lava.

Supply wells may alternate as recharge wells. In the Second World War, distilleries in Louisville, Kentucky, recharged municipal water into their pumping wells in order to alleviate overdraft of the groundwater. In some places, recharge wells may be an alternative to sewer systems to dispose of storm runoff; one of these is at Orlando, Florida, where wells penetrate from 35 m to 300 m into limestone and are cased to depths of 20–120 m. They have diameters which vary from 12 to 45 cm. It is noteworthy that, although much waste material enters them, the limestone is hardly ever clogged because it is so cavernous. The actual water table occurs between 10 and 15 m depth. There is a danger of pollution of water supply wells in the vicinity because the formation does not allow adequate filtration to take place.

Other wastes which are disposed of in recharge wells include the effluent of septic tanks, excess irrigation water and surface runoff from permeable volcanic terrains in the north-western USA; see for instance J. E. Sceva in 1968.[8]

W. G. Graham *et al.* in 1977 stated that there are more than 2000 recharge wells in operation over 130 000 ha of agricultural land in the Snake River plain in southern Idaho, most of these being 10–30 cm diameter, 30–50 m in depth and able to take flows as great as 20 000 m^3/day.[9] Many recharge wells are to be found also in the coastal areas of Los Angeles and Orange Counties in California; a fresh water pressure ridge has been created and is maintained which controls seawater intrusion (see Section 10.6). The wells were installed in 1953 and later; they demonstrate that gravel-packed wells are the most efficient, favourable recharge rates being kept up by chlorination and also by de-aerating the water supply. In addition, there is a first-class programme of maintenance which includes periodic pumping of the wells involved. One practical point which derived from this work is that, where well casings pass through an impermeable zone above a confined alluvial aquifer, the outside of such casing should have a concrete seal.

Extensive field experience and also laboratory investigations confirm that fresh water may be stored temporarily in saline water aquifers using wells which are recharged and then pumped, as F. F. Brown and W. D. Silvey indicated in 1977.[10]

11.4. UNPLANNED RECHARGE

This happens when water penetrates the ground as a by-product of human activities not aimed at artificial recharge and a large number of examples may be cited, such as water from irrigation, cesspools and septic tanks, water mains, sewers, landfills, canals and reservoirs and so on.

The interesting aspect is that water so introduced enters at a rate greatly exceeding anything which can be achieved by deliberate artificial recharging. Of course, the difficulty is that almost all such 'unintended' recharge is more or less polluted.

Water spreading for recharge proceeds simultaneously with irrigation. Both require a high rate of infiltration if they are to be efficient. Typical rate curves demonstrate that there is a trend towards decrease with time. There are several probable reasons for this. Initially, it can arise because of the effects of wetting upon particles of soil, which both swell and disperse; thereafter the decline in rate may give way to an increase caused by trapped air dissolving in transient water and so being removed. Later, a gradual rate decrease is the result of the growth of micro-organisms which clog soil pores.

It has also been observed that the rates of recharge usually decline as the mean size of particles of the soil in an area subject to spreading diminishes. Attempts to keep the soil pores open for the movement of water have been made by adding organic matter and various chemicals to the soil concerned, also to vegetation growing on the surface in the spreading area.

Other work shows that the rate of infiltration is directly proportional to the head of water.

H. I. Nightingale and W. C. Bianchi in 1977 in Fresno, California, examined the environmental impact of recharge basins. The recommendations they made for attaining maximum rates of infiltration in them[11] may be summarized. Firstly, soil for levees must be obtained from borrow pits and heavy construction equipment must be kept out of basins. Then recharge water should be brought into a basin at its lowest point in order to obviate intra-basin erosion and clogging of the soil surface. Trees and shrubs should be removed initially and turbid water should not be permitted to enter basins. Aquatic vegetation must be minimized so as to prevent any biological clogging of soils. Finally, recharge should be concentrated in summer, when the basin water temperature is maximal.

11.5. RE-USING MUNICIPAL WASTEWATER

Increasing efforts are being made to re-use this water for the purposes of industry, irrigation or supplementation of groundwater, although the common occurrence of pollutants might be adverse to this last application: see for instance a 'state of the art' review of health aspects of wastewater reclamation for groundwater recharge issued in 1978.[12] However, secondary treatment can aid in reducing any health hazards.

The previously described methods of recharge are applicable. Thus, surface sprinklers can introduce the water, but this is a slow process with rates ranging from about 0·05 to 0·2 m weekly and, even then, much water is lost through plant consumption. Higher rates are achieved if effluent is recharged onto bare ground whence infiltration and percolation can occur. Values up to 10 m weekly have been obtained. Wastewater can be utilized with surface irrigation or through spraying, but sometimes it is introduced through spreading basins. High-rate systems entail having deep, permeable soils and a water table which does not reach the surface of the ground; flooding is intermittent. Movement of effluent through soils is accompanied by much of the bacterial and viral content as well as the BOD (biochemical oxygen demand) and suspended solids, half of the nitrogen and most of the phosphate being removed. Municipal usage of water results in approximately 300 mg dissolved solids/litre being added to it; they cannot be removed by recharging, so wastewater can affect groundwater quality adversely unless there is sufficient subterranean dilution potential.

Recharge wells can be utilized in order to admit high-quality treated effluent. The optimum situation is BOD < 5 mg/litre, suspended solids < 1 mg/litre, phosphate < 1 mg/litre, iron < 0·5 mg/litre and turbidity < 0·3 turbidity units. However, to obtain effluent of this quality by purification is expensive.

Removal of pollutants from secondary effluent by recharging is related to the time and distance of movement underground and also to the type and properties of the soils and subsurface formations. Generally, for percolation through fine-grained alluvium, bacteria and viruses are removed, nitrogen is reduced and trace elements together with heavy metals may be reduced as well. An instance of a very successful wastewater reclamation plant is at Whittier Narrows in Los Angeles County, Southern California, where secondary treatment of effluent is provided: 57 000 m³/day are handled and discharged by means of spreading basins along the Rio Hondo and San Gabriel Rivers, where percolation through permeable alluvial deposits to the groundwater takes place. The investigations indicated that drinking water standards were complied with after percolation and subterranean dilution.[13]

11.6. RECHARGE MOUNDS

The recharge mounds that arise in the water table when water percolates underneath a spreading basin have dimensions related to the shapes and sizes of the basin, the rate of recharge, its duration and aquifer properties. The geometries of such mounds have been calculated by several investigators, for instance M. S. Hantush in 1967.[14] The calculations are based upon complex mathematical analyses arising from generalized non-steady groundwater flow equations. The same limitations occur, i.e. aquifers are assumed to be isotropic and homogeneous, vertical recharge is taken as occurring at a uniform rate, the top of the recharge mound should not be in contact with the spreading basin and its height must be low in relation to the initial saturated thickness.

Under uniform conditions of recharge, a recharge mound continues growing until halted either by a lateral or by a potential control. The former is found where the mound intersects a constant surface water elevation such as a lake and, if this is extensive, equilibrium may be approached, leading to a constant rate of recharge. Potential control occurs when the recharge mound reaches the recharge surface. With a fixed maximum height, the gradient and, consequently, the rate of recharge decrease with time.

11.7. LONG ISLAND, NEW YORK, USA

This is a very interesting case. In 1944, C. E. Jacob investigated the relation between groundwater levels and precipitation.[15] He demonstrated that, in the absence of rainfall, the decline in the level of the water table on a peninsula is expressible as:

$$h = h_0 \exp(-\pi^2 T t / 4 a^2 S)$$

where h_0 is the initial height of the water table above mean sea level, h is the height of the water table after a given time t, a is half the width of the peninsula and T is the product of the permeability and the thickness of the aquifer. A limiting condition is that the thickness of an aquifer must be large compared with the height of the water table above mean sea level. It has been found that the logarithm increases directly with time and, therefore, the plot of log h against t is a straight line. Plots of this type provide useful data for investigations into changes in groundwater storage.

By the early 1970s, groundwater had become the only fresh water source for over 2·5 million people on the island outside metropolitan New York, this occupying only the western end of it. The primary aquifer is made up of unconsolidated, coastal-plain deposits which are overlaid by a thin sequence of glacial materials. It has been utilized to the extent that, in coastal areas, there has been seawater intrusion so that remedial measures had to be put in hand. There are about 2100 recharge basins which dispose of some 231 000 m^3 daily of storm runoff according to G. E. Seaburn and D. A. Aronson in 1974.[16] Originally, abandoned gravel pits were used, but now it is necessary to develop them as places in which to accommodate increased runoff and also to conserve water. It is planned that, ultimately, there will be anything up to a total of 5000 such basins in Long Island (with a maximum density of three basins to 2 km^2 in urban areas). The overlying glacial materials are where these basins are located at depths of from 3 to 6 m with areas ranging from under 0·1 ha to more than 12 ha. Three basins gave average rates of infiltration of 1·4–6·6 m/day for storm runoff.

As well as basins for storm runoff, there are about 200 others which remove daily approximately 114 000 m^3 of industrial and commercial wastes into the ground. There is also incidental recharge from cesspools and septic tanks which amounts to 450 000 m^3 daily. In certain locations, there has been a marked degradation in the quality of the groundwater and, to counteract it, sewage systems were established. Unfortunately, this measure has lowered the water table which may trigger seawater intrusion.

Further incidental recharge arises from pressurized water supply pipe distribution systems and, if a leakage factor of 10% is assumed, about 378 000 m^3 per day is recharged.

There are more than 1000 recharge wells, which remit about 189 000 m^3 daily to the groundwater; this complies with government regulations which require that groundwater which is pumped for air-conditioning and industrial cooling has to be returned to the aquifer from which it is derived. Earlier, pits or wells were constructed for this purpose, but now drilled wells have been installed, with gravel packing and extending below the water table. Table 11.3 provides some details of artificial recharge on Long Island, New York, in the early 1970s.[16]

TABLE 11.3
ARTIFICIAL RECHARGE ON LONG ISLAND, NEW YORK

Water source	Recharge rate m^3/day
Recharge basins	
Storm runoff	231 000
Industrial wastewater	114 000
Cesspools and septic tanks	450 000
Leakage from water mains	378 000
Recharge wells	189 000
Total	1 362 000

Experiments are being undertaken at Riverhead and Bay Park with the object of finding out the feasibility of recharging treated municipal wastewater into wells as a means of controlling seawater intrusion; see the papers of J. J. Baffa in 1970,[17] J. Vecchioli in 1972,[18] E. Koch et al in 1973[19] and S. E. Ragone in 1977.[20]

11.8. RECHARGE INDUCTION

This technique is quite different from all methods discussed above because it does not require surface water to be conveyed to an appropriate location from which it can enter the ground. The approach is to withdraw groundwater near a river or lake so that the lowering of the level of the subterranean water promotes the ingress of water to the ground directly from the surface source involved. In the case where a perennial stream is utilized for supply, this supply becomes continuous even in areas near others suffering from overdraft. Induced infiltration has turned out to be very effective where unconsolidated formations of permeable alluvial materials are involved and hydraulically an interconnection exists between the river and the relevant aquifer. The actual quantities of water gaining admission to the aquifer depend upon factors such as the rate of pumping, permeability, the type of well, the distance from the surface stream and the natural movement of the groundwater. One requirement is that the water velocity in the stream should be great enough to preclude any silt from depositing and clogging the bed.

It has been shown that recharge so induced can provide water which is free of pathogenic bacteria and organic matter, as R. G. Kazmann pointed out in 1978.[21] Actually, surface water is usually less mineralized than groundwater, so when it is mixed with it through recharge induction it can improve the quality of the resultant water.

11.9. ARTIFICIAL RECHARGE ACTIVITIES IN EUROPE

These are mentioned as an attempt to broaden the picture, because most information regarding artificial recharge derives from the USA. Work of this type commenced early in the nineteenth century and, in fact, the very first infiltration basin for recharging water to the groundwater system was constructed at Göteborg in Sweden in 1897. In that country, such basins comprise only one aspect of many sources of municipal water supply. They are located mostly on glacial eskers which function very efficiently as conduits conveying recharge water to pumping installations. Adjacent stream or lake waters are transmitted through mechanical and rapid sand filters prior to recharging; most of the plants use rectangular basins with unprotected side slopes of 1:2 with a layer of uniform sand up to a metre thick on their bottoms. V. Jansa in 1952 gave some interesting data on several of the Swedish infiltration basins: they are quoted in Table 11.4.[22] The unusually high rates of infiltration result from the high permeability of the geological materials involved, coupled with the optimum distances between the basins and the pumping wells.

TABLE 11.4
INFILTRATION BASINS IN SWEDEN[22]

Location	Capacity (cfs)	Height of basin above water table (ft)[a]	Infiltration rate per day (ft)	Distance from basin to the pumping wells (ft)
Västerås	16·3	50	13	700–1 600
Hälsingborg	6·5	9	13	>1 000
Eskilstuna	5·3	72–82	13	1 000
Södertälje	4·1	30–56	16	1 300–5 600
Luleå	3·7	43	8	700
Landskrona	2·0	0–6	16	300–1 600
Malmö	2·0	16–33	7–10	1 600–3 300
Katrineholm	1·6	23	52	1 500
Kristinehamm	1·2	9	33	750
Eksjö	0·4	0–6	10–16	1 000

[a] 1 ft = 0·305 m.

In the Federal Republic of Germany, artificial recharge has been effected in a number of places, most notably on three rivers, the Ruhr, the Rhine and the Main, which were polluted.

In the Netherlands, water obtained by artificial recharge is used in

Amsterdam, Leiden and The Hague. Coastal sand dunes retain fresh water originating as rainfall, but overdraft has promoted penetration of underlying saline water. Introduction of recharge water stabilizes the latter, increases the volume of groundwater storage and provides natural filtration for the polluted surface water.

In England, artificial recharge was attempted in the valley of the River Lea in London where, before 1965, a scheme was proposed and later implemented. According to K. J. Edworthy and R. A. Downing in 1979, in 1970–71 a recharge borehole and 11 observation boreholes were installed in Bunter Sandstone (Triassic) near Clipstone, Nottinghamshire, experiments involving them being carried out from June, 1971 to March, 1974.[23] The main conclusions were that the technique there used is applicable to main aquifers in the United Kingdom, that an average rate of recharge of 0·3–0·5 m daily is to be expected where lagoons are utilized to recharge aquifers as in the Permo-Triassic sandstones, and the quality of a water is improved by recharge through an unfissured aquifer.

11.10. MAJOR PROBLEMS IN ARTIFICIAL RECHARGING

D. K. Todd in 1970 edited *The Water Encyclopedia*, which contains much information from various investigators on problems involved in artificial recharge.[24]

One of the most important is clogging due to silting, which may reduce the rate of infiltration of recharge water appreciably. Silt particles can block the pores in soils in superficial levels and it is then necessary to effect de-silting. This can be accomplished in various ways. One is by the employment of suitable flocculating agents such as 'Separan'. The water can be re-routed until the silt level drops to an appropriate level. Where water is recharged through ditches, its velocity may be sufficient to ensure satisfactory infiltration, even if high concentrations of silt exist.

Weed growth can actually increase the rate of percolation, but it can constitute a fire hazard. The solution is to remove weeds manually or by applying chemicals; they do not constitute a very significant impediment in operating pits or injection wells.

Animals can cause trouble, however, and rodents are particularly dangerous in causing leakage in, and frequently failure of, levees and dikes. The only remedies are traps and poisons. Eradication of rodents serves a dual purpose because they are in any case a menace to public health. However, fencing of the relevant area, patrolling and using warning signs may be necessary.

Bacteria and algae can be controlled by chlorination or introducing copper sulphate into the water; chemicals are also useful in preventing the deposition of calcium carbonate. Aeration of the water is undesirable and its head must be as great as possible. Dry ice, hydrochloric acid or sulphuric acid can be utilized in reconditioning of wells.

Water spreading in winter should be continuous in order to prevent water from freezing. Soils may be reconditioned using organic matter, e.g. cotton gin trash, or suitable chemical agents such as krillium.

The problem of maintenance of diversion structures may arise when spreading operations break down. Prevention is possible by systematic and routine checks coupled with patrolling during operations. Undercutting of structures should be examined, especially on the downstream end of the operation. Such a process can be arrested by applying riprap. In addition, accumulated debris in the vicinity of a structure can be removed by sluicing the channel.

11.11. HEAT STORAGE

A recent development is the possibility of utilizing hot storage wells for charging confined aquifers with surplus hot water obtained from industries generating both electricity and heat. C. F. Meyer in 1976 discussed this matter.[25] Hot water would radiate out from the well, override the groundwater and create an inverted cone. Heat would be lost by conduction into upper and lower layers which confine the aquifer, and by dispersion along the hot/cool interface, but it has been calculated that at least three-quarters of the heat which is stored would be recoverable. As R. G. Kazmann indicated in 1978, such storage wells are desirable because they will conserve energy and be rather economical.[26]

Geothermal energy may also be produced from hot, dry rock. The technique first developed at Los Alamos and later by the Camborne School of Mines is to drill two adjacent wells to depths exceeding 1000 m, and to introduce water into one of them. The water is heated by the hot, fractured rock and pumped back to the surface through the second well.

REFERENCES

1. RICHTER, R. C. and CHUN, R. Y. D., 1959. Artificial recharge of ground water reservoirs in California. *J. Irrig. Drain. Divn, Amer. Soc. Civ. Engrs*, **85**, IR4, 1–27.
2. TASK GROUP ON ARTIFICIAL GROUND WATER RECHARGE, 1963. Design and operation of recharge basins. *J. Amer. Water Works Assn*, **55**, 697–704.
3. TODD, D. K., 1959. *Annotated Bibliography on Artificial Recharge of Ground Water Through 1954*. US Geol. Surv. Water Supply Paper 1477, 115 pp.
4. SUTER, M. and HARMESON, R. H., 1960. *Artificial Ground-water Recharge at Peoria, Illinois*. Bull. 48, Illinois State Water Survey, Urbana, 48 pp.
5. HARMESON, R. H. *et al.*, 1968. Coarse media filtration for artificial recharge. *J. Amer. Water Works Assn*, **60**, 1396–1403
6. REEDER, H. O. *et al.*, 1976. *Artificial Recharge Through a Well in Fissured Carbonate Rock, West St Paul, Minnesota*. US Geol. Surv. Water Supply Paper 2004, 80 pp.
7. FOXWORTHY, B. L., 1970. *Hydrologic Conditions and Artificial Recharge Through a Well in the Salem Heights Area of Salem, Oregon*. US Geol. Surv. Water Supply Paper, 1594-F, 56 pp.
8. SCEVA, J. E., 1968. *Liquid Waste Disposal in the Lava Terranes of Central Oregon*. US Fed. Water Pollution Control Admin., Corvallis, Oregon, 2 Vols.
9. GRAHAM, W. G. *et al.*, 1977. *Irrigation Waste Water Disposal Well Studies — Snake Plain Aquifer*. Rept EPA-600/3-77-071. US Env. Prot. Agency, Ada, Oklahoma, 51 pp.
10. BROWN, F. F. and SILVEY, W. D., 1977. *Artificial Recharge to a Freshwatersensitive Brackish-water Sand Aquifer, Norfolk, Virginia*. US Geol. Surv. Prof. Paper 939. 53 pp.
11. NIGHTINGALE, H. I. and BIANCHI, W. C., 1977. *Environmental Aspects of Water Spreading for Ground-water Recharge*. Tech. Bull. 1568. Agric. Res. Service, US Dept of Agriculture, 21 pp.
12. STATE WATER RESOURCES CONTROL BOARD, DEPARTMENT OF WATER RESOURCES AND DEPARTMENT OF HEALTH, 1978. *A 'state of the Art' Review of Health Aspects of Wastewater Reclamation for Groundwater Recharge*. Water Information Center, Huntington, New York, 240 pp.
13. MCMICHAEL, F. C. and MCKEE, J. E., 1966. *Wastewater Reclamation at Whittier Narrows*. Calif. State Water Quality Control Board, Publn 33, Sacramento, California, 100 pp.
14. HANTUSH, M. S., 1967. Growth and decay of groundwater-mounds in response to uniform percolation. *Water Res. Res.*, **3**, 227–34.
15. JACOB, C. E., 1944. On the flow of water in an elastic artesian aquifer. *Trans. Amer. Geophys. Union*, **21**, 574–86.
16. SEABURN, G. E. and ARONSON, D. A., 1974. *Influence of Recharge Basins on the Hydrology of Nassau and Suffolk Counties, Long Island, New York*. US Geol. Surv. Water Supply Paper 2031, 68 pp.
17. BAFFA, J. J., 1970. Injection well experience at Riverhead, New York. *J. Amer. Water Works Assn*, **62**, 41–6.
18. VECCHIOLI, J., 1972. Experimental injection of tertiary-treated sewage in a

deep well at Bay Park, Long Island, N.Y. — a summary of early results, *J. New England Water Works Assn*, **86**, 67–103.
19. KOCH, E. et al., 1973. *Design and Operation of the Artificial Recharge Plant at Bay Park, New York.* US Geol. Surv. Prof. Paper 751-B, 14 pp.
20. RAGONE, S. E., 1977. *Geochemical Effects of Recharging the Magothy Aquifer, Bay Park, New York, with Tertiary-treated Sewage.* US Geol. Surv. Prof. Paper 751-D, 22 pp.
21. KAZMANN, R. G., 1948. River infiltration as a source of ground-water supply. *Trans. Amer. Soc. Civ. Engrs*, **113**, 404–24.
22. JANSA, V., 1952. Artificial replenishment of underground water. *Intl Water Supply Assn*, 2nd Congress, Paris.
23. EDWORTHY, K. J. and DOWNING, R. A., 1979. Artificial groundwater recharge and its relevance in Britain. *J. Inst. Water Engrs and Scientists*, **33**, 2, 151–72.
24. TODD, D. K., Ed., 1970. *The Water Encyclopedia.* Water Information Center, Water Research Building, Manhasset Isle, Port Washington, New York.
25. MEYER, C. F., 1976. Status report on heat storage wells. *Water Res. Bull.*, **12**, 237–52.
26. KAZMANN, R. G., 1978. Underground hot water storage could cut national fuel needs 10%. *Civil Engrng*, **48**, 5, 57–60.

CHAPTER 12

Isotope Hydrology and Groundwater

12.1. ISOTOPE HYDROLOGY

This is a field of scientific investigation to which several allusions have been made earlier in this book and which has proved to be valuable in its applicability to hydrological studies in general and groundwater in particular, using both environmental and artificially introduced isotopes of which some constitute part of the actual water molecule. In the past twenty years, there has been a flood of research papers on the subject, many appearing through the International Atomic Energy Agency in the form of technical reports, symposia, etc., and, in a book of limited size covering the subject of groundwater as completely as possible it is not possible to discuss this subject as adequately as would be desirable. Thus, the approach to be used will be essentially to attempt a summary of the theoretical background and available knowledge.

12.2. ENVIRONMENTAL ISOTOPE HYDROLOGY

Naturally occurring isotopic variations provide some labelling of the natural waters in which they occur. This labelling can be observed and, using knowledge of the chemical isotopic fractionations involved in such natural processes, it can be utilized in interpreting some hydrologic problems, such as the identification of the recharge area of an aquifer, the determination of the age of any particular water, the study of mixing phenomena between different waters, water–rock and water–gas interactions and so on. In addition, the study of isotopic equilibria can result in the creating of new geothermometers. Such isotopic tracing is of great value, but, unlike the case of artificially introduced isotopes, it cannot be controlled.

Isotopes may be defined as two or more nuclides which have the same atomic number, i.e. constitute the same element, but differ in mass number. Hence, isotopes of a particular element have the same number of protons in the nucleus, but differing numbers of neutrons. Naturally occurring chemical elements are mostly mixtures of isotopes so that the non-integer atomic weights in the Periodic Table represent average values for these mixtures.

The number of protons in the nucleus, i.e. the atomic number Z, determines the number and arrangement of the outer electrons of the atom and hence the physico-chemical properties of the element. As a result, isotopes of the same element cannot be distinguished by normal chemical methods.

The element with atomic number 1 is hydrogen and has three isotopes, namely ^1_1H (99·985% average in nature, stable), ^2_1H (D, deuterium, 0·015%, stable) and ^3_1H (T, tritium, 10^{-15}%, radioactive, half-life $T_{1/2} =$ 12·323 years). The upper index to the left of the chemical symbol indicates the mass of the nuclide and the lower one gives the atomic number. Among other elements of interest in hydrology are carbon and oxygen. The former has three isotopes of significance, namely, $^{12}_6C$ (98·89%, stable), $^{13}_6C$ (1·11%, stable) and $^{14}_6C$ ($1·2 \times 10^{-10}$%, radioactive, half-life $T_{1/2} = 5730$ years). The latter has also three important isotopes, namely $^{16}_8O$ (99·759%, stable), $^{17}_8O$ (0·037%, stable) and $^{18}_8O$ (0·204%, stable). Usually, the chemical number of the element is omitted and the isotopes are identified only by their chemical symbol and the mass number.

Isotopes can be stable or radioactive. In fact, only 260 of nearly 1700 known nuclides are stable so that nuclear stability is the exception, not the rule. Most of the stable nuclides possess even numbers of protons and neutrons. Stable isotopes do not change their concentrations with time in a closed system unless they are produced by some radioactive element also present in the system. Radioactive isotopes have unstable nuclei and decay into different isotopes with time. For instance, there is the reaction: $^{14}_6C \xrightarrow{\beta} {}^{14}_7N$. As a result, their concentrations in a closed system decrease with time following the law:

$$N = N_0 \exp(-\lambda t)$$

where N is the number of atoms at time t, N_0 at time 0 and λ is the rate constant which is characteristic of the decay of the isotope.

The isotope ratio, R, is the ratio in numbers of atoms between a given isotope and the most abundant isotope of that element. In the case of deuterium, for instance, the ratio is:

$$RD = \frac{^2H}{^1H}$$

and its average value in nature is usually:

$$R = 1.5 \times 10^{-4}$$

For carbon-13, $R^{13}C$ is given by $^{13}C/^{12}C = 1.12 \times 10^{-2}$ and, for oxygen-18, it is $R^{18}O = {^{18}O}/{^{16}O} = 2.04 \times 10^{-3}$.

Although isotopes of the same element have practically identical chemical properties, because of their differences in mass they have different rates of reaction, and also different distributions in two chemical compounds or phases undergoing mutual isotopic exchange. Physical processes such as diffusion produce isotopic differentiation. All such variations in isotopic compositions resulting from physico-chemical processes in compounds or phases occurring in the same system are termed isotopic fractionations.

In 1931, Harold C. Urey predicted, on theoretical grounds, that there should be a difference in the vapour pressures of the isotopes of hydrogen. This interest in hydrogen stemmed from a suggestion that this element might have naturally occurring isotopes. An experiment was carried out to detect ^2_1H and ^3_1H by spectroscopic methods in the residual volume of gas produced by evaporating approximately six litres of liquid hydrogen. The results showed the presence of deuterium, but tritium was not found. Urey actually called ^2_1H deuterium because it has a mass almost twice that of 'ordinary' hydrogen. At the time it was not known why the mass was greater because neutrons were not discovered until later, in 1932. This work led to Urey winning the Nobel Prize for Chemistry in 1934. During the Second World War, he applied this knowledge of isotopic fractionation to developing methods for separating uranium-235 by gaseous diffusion. After the War, he considered the possibility that stable isotopes of oxygen may be fractionated by natural processes. He suggested that such fractionation may take place during the formation of calcium carbonate in oceans, also that the degree of fractionation depends upon the temperature. This idea led to the development of the oxygen isotope method of measuring the temperature of deposition of skeletal calcium carbonate.

In later years, the original research investigations of Urey and his associates became an important branch of isotope geology. Fractionation processes entail fractionation factors. The fractionation factor α is the relationship between the isotopic ratios in the different phases of a system that is in equilibrium. For instance, the following equation is relevant to the water–vapour equilibrium at 0°C:

$$\alpha_{wv}{}^{18}O = \frac{R(^{18}O)_{water}}{R(^{18}O)_{vapour}} = 1.0115$$

so that, if a water with an isotopic ratio of for example 2.0449×10^{-3} attains equilibrium with its vapour at 0°C, then the isotopic ratio in the vapour is:

$$R(^{18}O)_{vapour} = \frac{2.0449 \times 10^{-3}}{1.0115} = 2.0217 \times 10^{-3}$$

It may be observed that the vapour is depleted in the heavier isotope relative to its parental water and this fractionation has been very thoroughly investigated, for instance by J. Bigeleisen in 1965.[1]

The fractionation factor for deuterium under the same conditions of water–vapour equilibrium at 0°C is given by:

$$\alpha(D)_{wv} = \frac{R(D)_{water}}{R(D)_{vapour}} = 1.1085$$

The α for D is higher than for oxygen-18 and isotopic fractionations are higher for light elements because of greater relative mass differences. Equations correlating fractionation factor α with temperature are:

$$1000 \ln \alpha\,^{18}O = 2.644 - 3.206 \times 10^3/T + 1.534 \times 10^6/T^2$$

for oxygen, according to Y. Bottinga and H. Craig in 1969,[2] and:

$$\ln \alpha D = -0.0771 + 13.436 \times 10^3/T^2$$

for hydrogen, according to R. Gonfiantini in 1971.[3] Temperatures are expressed in K. Table 12.1 contains appropriate data.

In isotopic hydrology, the isotopic abundances are expressed in the delta, δ, notation, where:

$$\delta = \frac{R_{sample} - R_{standard}}{R_{standard}} \times 1000$$

The standard adopted until 1976 for both $\delta^{18}O$ and δD was SMOW (Standard Mean Ocean Water), of which the isotopic composition represented an average of that of oceanic water.[4] This standard was selected because the oceans represent the start and end of any significant

TABLE 12.1

FRACTIONATION FACTORS FOR ^{18}O AND D AT VARIOUS TEMPERATURES

$T(°C)$	$\alpha^{18}O$	αD
−5	1·0121	1·116
0	1·0115	1·108
5	1·0110	1·101
10	1·0105	1·095
15	1·0100	1·088
20	1·0096	1·082
25	1·0092	1·077
30	1·0088	1·072
35	1·0084	1·067
40	1·0081	1·062

hydrologic circuit and also contain approximately 97% of all water on the crust of the Earth, this having a rather uniform composition, (see the 1953 papers of I. Friedman and S. Epstein and T. Mayeda[5,6]). Clearly, SMOW has null values, so that $\delta^{18}O = 0$ and $\delta D = 0$.

In September 1976 at a Consultants' Meeting on Stable Isotope Standards and Intercalibration in Hydrology and Geochemistry held in Vienna, it was decided to substitute Vienna-SMOW, V-SMOW, which has the following isotopic composition relative to SMOW:

$$\delta D = +0·2\permil \qquad \delta^{18}O = +0·04\permil$$

The D/H absolute ratio of V-SMOW was determined by R. Hagemann et al. in 1970[7] and the $^{18}O/^{16}O$ absolute ratio was determined by P. Baertschi in 1976.[8] They are as follows:

$$\frac{D}{H}\text{(V-SMOW)} = (155·76 \pm 0·05) \times 10^{-6}$$

$$\frac{^{18}O}{^{16}O}\text{(V-SMOW)} = (2005·20 \pm 0·45) \times 10^{-6}$$

R. Gonfiantini in 1978 explained that V-SMOW was obtained by mixing distilled ocean water with small amounts of other waters so as to bring the isotopic composition as close as possible to the defined SMOW.[9] A further standard which must be mentioned is SLAP, Standard Light Antarctic Precipitation, which has the following values relative to

V-SMOW:

$$^{18}O(SLAP) = -55\cdot 5\text{\textperthousand} \qquad D(SLAP) = -428\text{\textperthousand}$$

The original standard used by Harold C. Urey and his group for measurements in paleotemperature analyses was a Belemnite fossil, *Belemnitella americana*, from the Peedee Formation of the Cretaceous of South Carolina, USA, PDB-1 Chicago, of which the value on the SMOW scale was $+0\cdot 22\%$. This no longer exists. Another pair of standards are NBS-20 (Solenhofen limestone) and TS (marble); also, in 1983, a new carbonate standard, NBS-19, was proposed. This has the following values relative to PDB:

$$^{18}O = -2\cdot 20\text{\textperthousand} \qquad ^{13}C = +1\cdot 95\text{\textperthousand}$$

Full details are given in the Dictionary of Isotope Terms, Appendix 2, under PDB.

In general terms, fractionation between two substances A and B accompanying some specific process is given by:

$$\alpha_{A-B} = \frac{1+10^{-3}R_A}{1+10^{-3}R_B} = \frac{1000+R_A}{1000+R_B}$$

and, in terms of δ values:

$$\alpha_{A-B} = \frac{1+10^{-3}\delta_A}{1+10^{-3}\delta_B} = \frac{1000+\delta_A}{1000+\delta_B}$$

and α-values are applicable to processes involving either kinetic or equilibrium isotope effects, but more commonly the latter. In the case of equilibrium isotope fractionation between compounds A and B, the fractionation factor, α, is related to the normal equilibrium constant, K, for the isotope exchange reaction by:

$$\alpha = K^{1/n}$$

where n is the number of atoms exchanged, provided that the isotopes are distributed randomly over all positions in the equilibrating compounds. Because of this relationship, isotope exchange reactions are usually written so that only one atom is exchanged.

Because isotope fractionation factors are close to unity (cf. Table 12.1), they can be expressed in per mille terms by introducing the ε value which is defined as follows:

$$\varepsilon_{A-B} = (\alpha_{A-B} - 1) \times 1000$$

Theoretical considerations of the temperature dependence of isotopic fractionation demonstrate that $\ln \alpha$ varies as $1/T^2$ and $1/T$ in high and low temperature limits, respectively. The $1/T^2$ dependence exists for many important equilibrium fractionations over a wide temperature range; see the paper by I. Friedman and J. R. O'Neil in 1977.[10]

It is noteworthy that:

$$10^3 \ln(1 \cdot 00x) \cong x$$

so, for typical values of α ($\cong 1 \cdot 00x$), $10^3 \ln \alpha$ expresses the fractionation (‰) and this value is well approximated by the difference in δ values:

$$\delta(A) - \delta(B) = \Delta_{A-B} = 10^3 \ln \alpha_{A-B}$$

It is relevant to examine the laboratory methods for determining the isotopic composition of oxygen and hydrogen in water.

There are several possible approaches in the case of oxygen, of which the most common is to equilibrate the water with carbon dioxide and then to measure the isotopic composition of the CO_2. The isotopic abundance in the water is obtained from appropriate equations. The experimental procedure is as follows: 2–3 ml of water in a pyrex glass vessel is frozen by a mixture of acetone and dry ice prior to pumping off the air from the system through a vacuum line. The frozen water is allowed to melt so as to release dissolved air trapped in the ice during freezing. The water is re-frozen and the second package of air is pumped away. At this point, carbon dioxide is introduced into the system from a cylinder of ordinary commercial gas which has been purified earlier in the same line at a pressure of about 10 cm mercury below atmospheric pressure. The pyrex tube and connecting tap joint (in a closed position) are placed in a thermostatic bath at 25°C where the following isotopic exchange reaction takes place:

$$C^{16}O_2 + H_2{}^{18}O \rightleftharpoons C^{16}O^{18}O + H_2{}^{16}O$$

The mixture reaches equilibrium for that temperature after a day or two. The test tube is thereafter returned to the vacuum line and the carbon dioxide is separated from the water using liquid-nitrogen traps and acetone–dry ice. It is collected in a sample tube and is then ready for mass spectrometric analysis.

It may be noted that the Isotopic Geophysical facility at the University of Copenhagen has designed an automatized system for determination of the oxygen-18 concentrations in natural waters and is able to handle several hundred samples daily.

For hydrogen, the only reliable method is to separate the water in the form of vapour at 800°C on metallic uranium, following this reaction:

$$3H_2O + 2U \xrightarrow{800°C} 3H_2 + U_2O_3$$

Argon is blown into a small, clean, dry pyrex finger with cone to remove as much air as possible. A small drop of water is then inserted in the finger using a capillary pipette. The finger is connected to the vacuum line, the water frozen by a liquid-air trap and the air pumped away. The water is transferred to a trap beside the uranium furnace. At this stage, the water is passed over the heated uranium and separation takes place. Should any undecomposed water pass through, the process is repeated in the opposite direction. During this latter stage, the resulting hydrogen can be removed to a sample tube employing a Toepler mercury pump. When the vacuum gauge ceases moving, the sample is ready to be analyzed.

After the gases have been obtained, mass spectrometric methods are utilized to measure the relevant isotope abundance ratios. Essentially, the mass spectrometer comprises four sections which are an inlet system, an ion source, a mass separation system and an ion detector. The instrument separates and detects ions through the motions of charged particles with different masses in magnetic or electrical fields. The inlet system is so constructed that the gas is admitted by viscous flow, i.e. the mean free path of molecules is short compared with the tubing through which the gas flows. This prevents mass discrimination. Also, the inlet system is 'symmetrical' which allows rapid alternation and comparison between the sample gas and a standard reference gas. This comparison must be made in order to attain a precision adequate to measure small differences in isotopic abundance. Hence, dual inlet systems are essential components of all isotope-ratio mass spectrometers.

In the stainless-steel capillary tube admitting the gas by viscous flow, there is a constriction ('leak'), after which the gas flow becomes molecular and the gas enters the ion source. In this, a tungsten filament emits electrons in a beam which is first ribbon-shaped and then twisted by a coaxial magnetic field to increase the efficiency of ionization. As a consequence of the magnetic force lines, positive ions produced by electron impact on the gas which have been accelerated across an electric field into the magnetic field of the mass separation system follow a circular orbit of which the radius is given by:

$$r = \frac{1}{H}\left(\frac{2mV}{e}\right)^{1/2}$$

where H is the magnetic field strength, m is the mass of the ion, V is the accelerating voltage and e is the electrical charge of the electron. Thus, ions of different mass will follow different trajectories and can be collected separately. The mass spectrometer has a series of physical and electrostatic collimating slits through which the ions are accelerated by the above-mentioned accelerating voltage which can be adjusted up to 10 kV. When emerging from the source, ions of each isotopic mass have the same kinetic energy. The monoenergetic ion beam separates into the isotopic mass constituents in the magnetic field (the analyzer).

Each of the resulting ion beams is characterized by a particular mass-to-charge relationship. The ion or mass beams are subsequently collected in a metal cup (Faraday cage) or on a metal plate which is connected to earth through a very high resistance (10^9–10^{11} ohms). The potential drop is directly proportional to the ion current. All isotope-ratio instruments collect at least two ion beams simultaneously. The direct measurement of the ratio of ion beams gives a much higher precision than that obtainable by single ion beam collection. The precision is further increased by comparing the ion intensity ratio of the sample to that of the reference standard gas measured under identical conditions. In the older instruments, double collectors are installed and a Wheatstone bridge-type voltage divider is used for ratio measurements. In newer ones, multiple Faraday cage collectors are connected to voltage-to-frequency convertors and counters. These detector systems are valuable in the measurement of oxygen and carbon isotopic ratios from a single set of sample-reference comparisons on carbon dioxide, also in measuring D/H ratios on hydrogen gas and isotopic ratios on other gases in the same instrument.

Of all the isotopic species of water, only three are of interest in isotopic hydrology; these are $H_2^{16}O$ (99·73% of the mean ocean water), $HD^{16}O$ (0·031%) and $H_2^{18}O$ (0·199%). It may be noted that, when water evaporates from the oceans, the water vapour is depleted in heavy isotopes because $H_2^{16}O$ is more volatile than $HD^{16}O$ and $H_2^{18}O$. Atmospheric water vapour over oceans has values of about:

$$\delta(^{18}O) = -12·0\text{\textperthousand} \qquad \delta(D) = -85\text{\textperthousand}$$

relative to SMOW. When the relative humidity decreases, lower values are observed in the water vapour. Evaporation does not occur in nature in isotopic equilibrium; this is attained only when the process takes place in thermodynamic equilibrium, i.e. at relative humidity of 100% which is saturation. However, these conditions can occur only in closed systems or in systems from which the water vapour is removed very slowly and they do not occur very often, if at all, in nature. The above δ values, which were derived by H. Craig and L. I. Gordon in 1965,[11] differ from values found in North Atlantic mists by R. Gonfiantini and A. Longinelli in 1962.[12]

Since the fractionation factor is temperature-dependent, the isotopic composition of any precipitation is also dependent upon its temperature of formation. The following effects are observed. Firstly, summer rainfall has higher ^{18}O and D contents than winter rainfall. Secondly, there is an altitude effect, rainfall on mountains having lower ^{18}O and D contents than rainfall on lowlands. This may be due to the longer time the rain takes to reach lower levels so that rainfall originally having the same isotopic composition may be partially re-evaporated while falling, thus undergoing enrichment in the heavier isotopes. Thirdly, there is a latitude effect, rainfall at higher latitudes having lower ^{18}O and D contents than rainfall at lower latitudes.

Continental and altitude effects in groundwaters in the Federal Republic of Germany may be mentioned as examples. H. Förstel and H. Hützen in 1982 gave results of environmental oxygen-18 analyses of 900 groundwater samples collected from 480 municipal water factories.[13] The $^{18}O/^{16}O$ ratios decrease from north ($-7‰$) to south (under $-11‰$) as a consequence of increasing distance from the sea, the so-called continental effect, and an increasing altitude above sea level, the altitude effect. The decrease of the $^{18}O/^{16}O$ ratio per 1000 m altitude above sea level is 2·8‰ and per 1000 km from the sea, it is 2·4‰. Generally, the groundwater samples represent the mean oxygen isotope ratio of the local precipitation; this is to be expected also in samples from the conductive tissues of trees.

Seasonal variation is paralleled by similar variations in the hydrogen and oxygen isotopes, with the empirically observed relationship:

$$\delta D = 8\delta^{18}O + B$$

where B is usually 10.

$$B = \delta D - 8\delta^{18}O$$

The deuterium excess was termed B by W. Dansgaard in 1964.[14] However, values of B as high as $+22$ have been noted in the eastern Mediterranean area and also in Romania (see A. Tenu et al. in 1975[15]). This is attributed to the fact that the precipitation there is produced from water vapour of Mediterranean origin formed by an evaporation process at a relatively high rate. The line with $B = +22$ is called the Levant meteoric line. Values of B under 10 occur in rainfall of low intensity and/or short duration which is partially re-evaporated during its descent. Where water is evaporated naturally, the residue is enriched in the heavy stable isotopes partly due to their lower volatility and partly because of isotopic exchange with atmospheric water vapour. As isotopic equilibrium between water vapour and the liquid does not occur in evaporation, the fractionation factors are higher than the corresponding equilibrium ones. Hence, evaporated waters give a $\delta D - \delta^{18}O$ slope which is lower than that of meteoric waters (4 to 6). Consequently, the deuterium excess is also lower. Where evaporation affects a lake, enrichment in heavy isotopes is limited by the ratio between outflow and inflow (not including the evaporation in the outflow, of course). But when evaporation is not negligible for the balance of the lake, then its waters will be enriched in heavy isotopes and isotopically labelled with respect to waters not evaporated in the environment.

Another process affecting the isotopic composition of water is that of isotopic exchange with rocks during the circulation of water. Such exchange takes place only for oxygen because rocks do not contain sufficient hydrogen to modify the isotopic composition of water appreciably. The exchange kinetics with rocks are very slow; therefore a high temperature is mandatory if the isotopic composition of water is to undergo substantial modification. Actually, the characteristic $\delta^{18}O$ shift due to isotopic exchange with rocks was first seen in geothermal steam having a temperature of 300°C.

Radioactive isotopes are characterized by having an unstable nucleus which decays into another element in time. Radioactive decay is often accompanied by the emission of γ-rays, which are electromagnetic radiations produced by the de-excitation of the newly-formed nucleus. There are three classes of radionuclides, namely

(a) primary, which have half-life times exceeding 10^8 years and emit α- or β-particles;
(b) secondary, formed in radioactive transformations commencing with ^{238}U, ^{235}U or ^{232}Th;

(c) induced, with geologically short lifetimes, formed by induced nuclear reactions occurring in nature.

The forms of decay are listed below.

(a) α-decay, the nucleus of the element emitting an α-particle and thus losing two protons and two neutrons:

$$^{232}_{90}\text{Th} \xrightarrow{\alpha} {}^{228}_{88}\text{Ra}$$

(b) β-decay where the nucleus of the element emits a β-particle (electron) to produce an element with the same mass, but with the atomic number higher by one unit:

$$^{228}_{88}\text{Ra} \xrightarrow{\beta^-} {}^{228}_{89}\text{Ac}$$

(c) electron capture (EC) in which the nucleus captures an electron from the 'inner' layer producing an element with the same mass, but with the atomic number lower by one unit:

$$^{7}_{4}\text{Be} \xrightarrow{\text{EC}} {}^{7}_{3}\text{Li}$$

It was noted earlier that the decay law for radionuclides in a closed system is:

$$N = N_0 \exp(-\lambda t)$$

from which may be derived:

$$t = \frac{1}{\lambda} \ln \frac{N_0}{N}$$

Hence, if N_0, the initial concentration of the radioactive isotope, and N, the present concentration, are known, it is possible to calculate the age of the system under investigation.

When the number of radioactive atoms initially present is reduced by half, then $N = N_0/2$ and the time t represents the half-life, $T_{1/2}$:

$$T_{1/2} = \frac{1}{\lambda} \ln 2$$

where $1/\lambda$ is the so-called mean life.

The radioactive unit is the Curie unit, Ci, which corresponds to $3 \cdot 7 \times 10^{10}$ disintegrations per second, equivalent to the radioactivity of

1 g of radium. Tritium, designated T, is the radioactive isotope of hydrogen, with mass 3. It is formed in the atmosphere by the interaction of cosmic ray-produced neutrons with nitrogen:

$$^{14}_{7}N(n, T)^{12}_{6}C$$

and decays by β-emission:

$$^{3}_{1}T \longrightarrow {}^{3}_{2}He$$

The rate of production is about 0·25 atoms/cm² s, according to D. Lal and H. E. Suess in 1968.[16]

Large quantities of man-made tritium were released to the atmosphere by thermonuclear tests in the period 1953–62 and minor amounts are being released by industrial nuclear activities. Most tritium produced in the atmosphere is quickly oxidized to HTO and incorporated into the hydrologic cycle where it constitutes a useful label for water that has been in the atmosphere within the past 30 years or so. The great dilution by water of the H_2O species causes a very low concentration of tritium which can be measured only after isotopic enrichment.

The tritium content of natural waters is expressed in Tritium Units, T.U. One Tritium Unit is equivalent to a concentration of 1 tritium atom per 10^{18} hydrogen atoms, just about the lowest limit of feasible detection. Cosmic ray-produced tritium established a pre-bomb level of about 10 T.U. in temperate-zone continental meteoric waters, but after 1953, the tritium content of precipitation increased through thermonuclear testing and values up to 10 000 T.U. were attained in the northern hemisphere in 1963 as a result of the extensive tests of 1961–62; see the paper of L. L. Thatcher and B. R. Payne which appeared in 1965.[17] From 1963, the tritium content in precipitation decreased until 1968 because of the moratorium on atmospheric thermonuclear testing. Since 1968, the tritium content has more or less stabilized. Most of the bomb tritium was deposited in the stratosphere and about half of the stratospheric inventory is transferred to the troposphere in the spring and summer months of the year, so maxima in the stratosphere occur in winters, and minima in summers.

The tritium content of precipitation varies greatly in relation to latitude. Lower values occur at oceanic and coastal sites than at continental ones. This is because the ocean serves as a sink for HTO through isotopic exchange between atmospheric water vapour and ocean water, which has a low tritium content. Low concentrations are found in equatorial and southern areas. This is because most tritium was released in the northern hemisphere and is transferred from the stratosphere to the troposphere preferentially at high latitudes. Also, the higher proportion of ocean surface in the southern hemisphere and the high vapour pressure in equatorial regions provide more of a washout and dilution effect: the tritium content in the southern hemisphere is roughly a tenth of that in the northern.

The tritium content of precipitation is used for estimating the input of this radionuclide to groundwater systems. Allowance is made for the seasonal nature of much recharge and for the fact that, in vegetated areas, most precipitation (and tritium) returns to the atmosphere by evapotranspiration. In semi-arid locations, light precipitation is evaporated from the soil before it is able to infiltrate. In analyzing the tritium content of groundwater, a number of situations may be encountered. Absence of tritium indicates that the water of recent recharge has not reached that point where sampling is effected and, in that aquifer, more than 20 years are necessary for water to get there. Presence of an appreciable quantity of tritium, which varies with time, indicates that water younger than 20 years is present, the variations implying a short circulation time of a few years. Another possibility is that water from two different sources is present, perhaps involving a mixture of old tritium-free water and young water containing tritium. Variations in time of the relative amount of the two waters determine a change in the tritium content. On the other hand, there may be an appreciable tritium content which is constant in time. The inference is that the young water is well mixed in the aquifer with old water, the size of the reservoir masking the fluctuation in the recharge. Thus, in groundwater studies, tritium measurements give data on the time of recharge to the system.

The above argument depends, to a degree, on discrete segregation of groundwater of different ages during transit, except in the cases mentioned of incomplete or complete mixing. Occurrence of these cases reflects the dispersive effect of the granular materials in the aquifer which smooths out tritium valleys and peaks. The well-mixed reservoir shows exponential discharge and a mean residence of water within the system can be estimated; see T. Dinçer and G. H. Davis in their paper on tritium dating of 1967.[18]

Tritium has been utilized in studying the downward progress of water in the unsaturated zone, the usual means of recharging a groundwater system. Water artificially enriched in tritium and deuterium was used to label a specific horizon of soil moisture and its rate of downward movement as well as the dispersion of the tracer were monitored in core samples subsequently taken from the test plot, according to U. Zimmerman and his associates in 1967.[19] The downward progress of environmental tritium demonstrates the infiltration over many years in a completely undisturbed system, as reported by D. B. Smith and his associates in 1970.[20] This percolation of water through overlying soil material is an important component of recharge to unconfined aquifers in humid and semi-humid regions and G. B. Allison and M. W. Hughes in 1976 made important contributions to its study.[21] The area of investigation was the Gambier Plain in Australia which is underlain by an unconfined aquifer of great economic importance receiving most of its recharge by percolation through the soil. This recharge varies considerably and 11 cores were taken from three different sites, all within 30 km of each other. They were 'undisturbed', with 100 mm diameters, involving depths of up to 8 m. Water was extracted from each core section by vacuum distillation at 200°C. The tritium concentrations were obtained by 50–80-fold electrolytic enrichment of the soil–water samples in constant feed cells, followed by conversion to ethane and gas-proportional counting. The tritium concentration of rainfall in southern Australia in general and at Mt Gambier in particular is well known since late 1965, one monitoring station (that at the mountain) being within 20 km of the sampling stations. For earlier years, tritium concentrations were estimated by analyzing wine samples dating back to 1950. There is a good correlation between wine and rainfall tritium concentrations. The mean annual rainfall at the three sites selected is ~ 750 mm annually. Comparison of observed and model tritium profiles shows that piston-type flow of soil water dominates under the root zone (~ 1.5 m), but above this, direct input to beneath the soil surface becomes significant. Estimation of the recharge to groundwater from the tritium profiles obtained, utilizing both their shapes and the total quantities of tritium held in them, demonstrated that the mean annual recharge varies between 40 and 140 mm per year for the three selected sites. Multiple coring at two of the three sites also established that a recharge estimate made from a single core has a coefficient of variation of 15–20%.

Radiocarbon, carbon-14, emits β radiation as tritium does. However, the maximum energy of tritium is 18.1 keV compared with 156 keV for carbon-14. As with tritium, ^{14}C occurs as the result of natural and man-made processes. The natural production results from the interaction between cosmic ray-produced neutrons and ^{14}N atoms in the total atmosphere:

$$^{14}_{7}N(n,p)^{14}_{6}C$$

and decays with β-emission:

$$^{14}_{6}C \xrightarrow{\beta^-} {}^{14}_{7}N$$

It has a rate estimated at 2·5 atoms/cm^2 s by D. Lal and H. E. Suess in 1968.[16] The carbon-14 is oxidized to carbon dioxide, mixes with the carbon dioxide of the atmosphere and enters the global carbon cycle. Important characteristics of the radionuclide are: $T_{1/2} = 5730$ years, $1/\lambda = 8267$ years, $\lambda = 1·21 \times 10^{-4}$ years^{-1}. The ^{14}C concentration is expressed as a percentage of the modern ^{14}C content of atmospheric carbon dioxide, which is about 1·2 ^{14}C atoms per 10^{12} carbon atoms. Variation during the past 7000 years or so of the natural carbon-14 concentration has been demonstrated by measurements of the ^{14}C content of tree rings of the *Sequoia gigantea* and Bristle-cone pine, as reported by H. E. Suess in 1967.[22] It must be noted, however, that the uncertainty which this causes in applying ^{14}C in hydrology is much smaller than errors which may arise from other sources. The modern ^{14}C content on which measurements are based refers to pre-Bomb times, because the production of ^{14}C by detonation of thermonuclear explosives caused the atmospheric ^{14}C in the northern hemisphere almost to double by 1963. Subsequently it fell, but it is still above the pre-Bomb level. Similar, but lower, increases were observed in the southern hemisphere also, the peak value there attaining a level of approximately 60% above the pre-Bomb one. However, these variations are only significant when measuring relatively young waters for their ^{14}C content.

The application of ^{14}C to the dating of groundwater was proposed first by K. O. Münnich in 1957 on the basis that soil-zone carbon dioxide is of biogenic origin, resulting from the respiration of plant roots and plant decay and therefore containing ^{14}C derived by plants from the atmosphere.[23] The biogenic carbon dioxide dissolves in infiltrating water and is transferred to the groundwater reservoir. The ^{14}C content decreases by radioactive decay and the fraction of the original which remains is a measure of the time which has elapsed since its removal from the soil zone, i.e. the time since the associated water infiltrated, Thus,

$$t_{(years)} = 8270 \ln (C_0/C)$$

where 8270 is the mean life of ^{14}C in years, C_0 is the initial ^{14}C concentration and C is the ^{14}C concentration in the sample. ^{14}C is measured relative to the total carbon content of the sample, so the origins of both the ^{14}C and the stable carbon in the sample must be examined. Not all the stable carbon in groundwater is of the same origin as the ^{14}C. Infiltrating water which contains carbon dioxide dissolved from the soil zone dissolves carbonate minerals in the soil:

$$H_2CO_3 \rightleftharpoons HCO_3^- + H^+$$
$$CaCO_3 + H^+ \rightleftharpoons HCO_3^- + Ca^{2+}$$

However, carbon deriving from limestone in general does not contain ^{14}C, so the water reaching the water table contains dissolved carbon in the chemical forms H_2CO_3, HCO_3^- and CO_3^{2-} with a ^{14}C content lower than that occurring in the soil, biogenic carbon dioxide. The evaluation of the dilution of soil carbon dioxide originally containing 100% of modern ^{14}C with ^{14}C-free carbonate, to estimate the initial ^{14}C concentration in recharge water reaching the water table, is one of the most difficult problems of ^{14}C age determination of water.

Correcting for mineral carbonate entails developing various methods for the evaluation of the initial ^{14}C content. Many ^{14}C analyses of groundwaters from Europe and Africa suggest that only a few samples have more than 90% (relative to modern) ^{14}C and some of these have been contaminated by thermonuclear ^{14}C. On the other hand, many occur with 80 to 90% of ^{14}C and it is thought that a value of 85% would be a reasonable average for the ^{14}C content of recent, but pre-Bomb, waters; in fact, this was proposed as the initial ^{14}C content of carbon dissolved in groundwater by J. C. Vogel in 1970.[24]

It is also possible to evaluate the initial ^{14}C content of groundwater by utilizing the different $^{13}C/^{12}C$ ratios of biogenic carbon dioxide and limestone (mean abundance ^{12}C, 98·9%; ^{13}C, 1·1%). The ^{13}C content is measured by mass spectrometry and expressed as per mil relative deviation [$\delta^{13}C‰$] from the standard PDB which has a $^{13}C/^{12}C$ ratio close to that of marine limestone. In temperate climates, soil biogenic carbon dioxide is derived from plant respiration and organic matter decay and has a $\delta^{13}C$ value of $-25 \pm 3‰$, corresponding to that noted in local plants. Isotopic fractionation on dissolution of this carbon dioxide in water and conversion to bicarbonate raises the value to $-17 \pm 3‰$. Usually, limestone has a $\delta^{13}C$ value of $0 \pm 2‰$. The $\delta^{13}C$ values of carbon species dissolved in groundwaters mostly range from -20 to -5. A simple proportion gives the fraction of biogenic carbon present in a groundwater which corresponds to the fraction of modern ^{14}C present at the time of recharge. Of course, such treatments are of limited significance because the overall geochemistry of the system is not fully understood. Thus, for instance, it is important to know the pH if the relative quantities of H_2CO_3 HCO_3^- present are to be determined. It is also necessary to consider isotopic fractionation taking place between them. In addition, the effects of other ions such as sodium and magnesium on the carbonate equilibria are significant. With sodium bicarbonate waters, calcium ions may have been removed from solution by exchange with the sodium present in clay minerals, hence allowing extra calcium carbonate to be dissolved, which would give more positive values for $\delta^{13}C$. Another problem, arising in arid areas, is due to the lack of knowledge of the effects of climate on the isotopic composition of vegetation: soil carbon dioxide. $\delta^{13}C$ values of ca $-12‰$ have been recorded and may influence the evaluation of the initial ^{14}C content based on ^{13}C values. There is too the possibility of exchange of ^{14}C between the dissolved bicarbonate and the carbonates of the matrix of a limestone aquifer; this could be regarded as yet another problem, but it appears that the effect may be rather small in waters not subjected to high temperatures. In geothermal waters there is evidence of exchange which necessitates caution in interpreting ^{14}C data.

The radiocarbon method is applicable to waters younger than about 30 000 years and can be used to study groundwater movement in confined aquifers. Where recharge occurs in an outcrop area, the water chemistry and isotopic composition of dissolved carbon species being relatively uniform, the age differences between different locations are unaffected by the uncertainties which affect the determination of the absolute age of water. Hence, it is possible to determine the velocity of flow in groundwater by determining the age differences between two sampling points a known distance apart, providing that the water body involved is not too large. The matter was discussed earlier; see Section 4.3.

If satisfactory results are obtained, an estimate of the mean regional permeability can be made. ^{14}C measurements can also provide information on mixing processes of waters of different ages within a given aquifer. Recently, radioisotope dating has been attempted by direct counting of radioactive isotopes. Such single-atom detection is feasible utilizing an electrostatic accelerator as part of an ultra-sensitive mass spectrometer. At the Swiss Federal Institute of Technology in Zürich, the EN-Tandem accelerator has been employed. A new sputter source was developed with a caesium beam which impinges on the surface of the sputter sample at an angle of 60° to produce negative ions. The source is combined with a computer controlled sample exchange mechanism with a magazine having 25 samples exchangeable through a vacuum lock. Negative ions are accelerated to energies of 60 keV and mass-analyzed by a double-focusing 90° magnet. Rapid switching from one isotope to another is possible by varying the beam energy through this magnet. Ions are accelerated to the high-voltage terminal (6 MV), where their outer electrons are stripped in a gas stripper. Multiply-charged positive ions are then accelerated to ground potential. Ions with a charge state of +3 and the correct energy are selected in the electrostatic accelerator and analyzed by a double-focusing 90° magnet with a momentum resolution of $p/\Delta p = 500$. Rare isotopes are identified with a ΔE–E–gas counter.

Early measurements on geological samples used a carbon beam injected through a 20° magnet with low mass resolution. Ratios of the various carbon isotopes were derived by injecting and sequentially analyzing ^{12}C, ^{13}C and ^{14}C; to change the isotope being studied, the field of the 20° magnet was varied. The sequence was computer-controlled and time intervals of 15 s for ^{12}C and ^{13}C, 150 s for ^{14}C were selected. To avoid loading effects of the accelerator, the mean intensity was limited to 100 nA for ^{12}C of charge state 3+.

Four sample with known $^{14}C/^{12}C$ ratios were utilized for calibration. Alternating in 1 h intervals, samples and standards were measured. Calibration samples showed the correct relative dependence within the limit of statistical error, but fluctuations up to 4% in the $^{13}C/^{12}C$ ratios arose from variations in transmission through the accelerator, restricting the accuracy.

A contribution to the Europhysics Conference on Nuclear Physics Methods in Material Research at Darmstadt in 1980 referred to joint work effected by G. Bonani and his associates at the Nuclear Physics Laboratory of the Swiss Federal Institute of Technology at Zürich, where the accelerator is located, and J. Beer and colleagues of the Physics Institute of the University of Berne.[25] In this paper, an example of a spectrum taken with the heavy-ion counter showed the ^{14}C peak plus well-separated peaks of ^{12}C, ^{13}C and ^{14}N, the background peaks being partly ascribed to the low mass resolution of the 20° injection magnet, ^{14}N not forming negative ions (it was assumed that $^{14}NH^-$-ions of mass 15 derived from the source). All particles which reach the detector are stated to have the same momentum. Extra charge exchange and scattering processes were considered in an attempt to explain the observed background peaks.

The progress report of E. T. Hall, published in 1983, may be consulted.[26]

Another radioisotope used in isotype hydrology is ^{222}Rn, produced in all geofluids by the radioactive decay of radium:

$$^{226}_{88}\text{Ra} \xrightarrow{\alpha} {}^{222}_{86}\text{Rn}$$

the decay being with α emission.

$$^{222}_{86}\text{Rn} \xrightarrow{\alpha} {}^{218}_{84}\text{Po}$$

Radon-222 possesses the following characteristics; $T_{1/2} = 3\cdot823$ days, $1/\lambda = 5\cdot515$ days, $\lambda = 0\cdot18$ days^{-1}. The concentration of radon is expressed as nCi/litre or mCi/litre.

The gas radon emanating from rocks is conveyed convectively to the surface by flowing fluids, the velocity and flow rate of which are dependent upon the permeability of the rocks involved. The lifetime of radon is of the same order of magnitude as the transit times in the secondary reservoirs in the Italian geothermal fields (see a paper by F. D'Amore and his associates in 1976[27]). This is almost the same time as that needed to attain a stationary state in the new thermodynamic conditions of the fluid when operations are proceeding at wellhead. Hence, from the radon content, it is possible to obtain kinetic information which cannot be secured from traditional geochemical parameters. A hypothesis has been derived from radon measurements which envisages a separate front between vapour-dominated and hot water-dominated zones in Larderello, moving gradually northwards.

Some of the radon-222 in groundwater may ascend and escape to the atmosphere: indeed, concentrations of it in rain have been utilized to distinguish the source of monsoonal precipitation in India.

The last radionuclide of hydrological interest to be mentioned here is silicon-32, of which the half-life is not precisely known. It is usually taken to be about 600 years, but there is a variety of choices in the literature. The following may be cited:

(a) 60 years (A. Turkevich and A. Samuels in 1954[28]);
(b) 276 ± 32 years (D. J. Demaster in 1980[29]);
(c) 280 years (K. Jantsch in 1967[30]);

(d) 330 years (H. B. Clausen in 1973[31]);
(e) 500 years (M. Honda and D. Lal in 1964[32]);
(f) 650 years (D. Geithoff in 1962[33]);
(g) 710 years (M. Lindner in 1953[34]).

The radioisotope could be a useful 'dating' tool, falling between tritium and ^{14}C. It is produced by cosmic ray-induced spallation of argon and rapidly oxidizes to $^{32}SiO_2$ which is scavenged from the atmosphere by precipitation. The total annual production is extremely small, probably only a few grams, but an excess was injected artificially during the thermonculear tests of 1961–62.

Some applications of environmental isotope hydrology involving groundwater are cited below.

T. Dinçer and B. R. Payne in 1971 demonstrated interconnections between groundwater and lakes in Turkey.[35] Near Antalya, large karstic springs adjacent to the coast were thought to be supplied by inland lakes in the plateau north of the Taurus Mountains, which were known to leak important quantities of water through sinkholes and fractures. The fraction of water lost through such leakage was evaluated on the basis of an earlier isotopic study of the lake water by T. Dinçer in 1968.[36] The oxygen-18 and deuterium analyses of the spring waters demonstrated that the contribution from the lakes, Burdur, Eğridir and Beyşehir, is almost negligible. In fact, however, the Kirközler, Köprücay and Manavgat springs are recharged from the southern side of the Taurus Mountains and, crossing these, there is a change in the value of the intercept of the meteoric water line on the δ D-axis.

B. R. Payne in 1970, in a similar experiment, found that springs near Lake Chala in Kenya are not connected with it.[37]

J.-C. Fontes and his associates in 1970 showed that, around Lake Chad, Africa, groundwater samples collected at various depths near the shore line showed more or less regular variations with depth and distance from the lake as regards contributions of heavy-isotope-enriched lake water.[38] This makes feasible the evaluation of the proportions of lake water and direct infiltration water from precipitation at each point of sampling.

R. Gonfiantini and his co-workers in 1962 used oxygen-18 to test an hypothesis that Lake Bracciano in central Italy was leaking into a nearby water table with the phreatic surface 15–20 m lower than the surface of the lake.[39]

Investigations have been made on interconnections not only between groundwater and lakes, but also between groundwater and rivers. For example, in the Gorizia area of Italy two main rivers which descend from the eastern Alps, the Isonzo and the Vipacco, have waters with a $\delta^{18}O$ value of $-10.5\%_{00}$. The wells in an unconfined aquifer in the same area have values of $\delta^{18}O$ ranging from -7 to $-10\%_{00}$, values becoming heavier away from the rivers. S. Morgante and his associates in 1966 were able to define the areas of infiltration of river waters in the unconfined aquifer and also to assess its fraction with respect to local precipitation.[40]

Finally, there is the case of interconnections between aquifers. The famous and vast Early Cretaceous 'Continental Intercalaire' aquifer occupying the western Sahara contains water with a stable-isotope composition which is uniform throughout and devoid of ^{14}C. However, in the region adjacent to the unconfined aquifer of Grand Erg Occidental, the stable-isotope composition of the Continental Intercalaire alters and it contains ^{14}C. This demonstrates that there is a contribution from the Grand Erg Occidental aquifer, of which the water is rather recent and has a piezometric surface higher than that of the Continental Intercalaire; see G. Conrad and J.-C. Fontes in 1970 and 1972.[41,42]

In another area, Tunisia, the Continental Intercalaire discharges in the overlying aquifer which is confined and coastal at Djeffara, using a fault system near El Hamma, west of Gabès. The amount of water coming from the Continental Intercalaire decreases in the coastal aquifer from El Hamma to Gabès in the direction of water flow.

As noted earlier (Section 4.3 and above), flow velocity may be obtained from ^{14}C data. This was the case with the Floridan (Ocala limestone) aquifer investigated by B. B. Hanshaw and his associates in 1965.[43] Another experiment was made in Texas where a large confined aquifer has estimated ages adjusted by the ^{13}C approach ranging from zero in the area of recharge to approximately 30 000 years. Estimated flow velocities of 1·5–2 m annually agreed well with those derived from hydrological data (see F. J. Pearson Jr, and D. E. White in 1967[44]). Similar work was done in a confined aquifer near the coast in South Africa.[12]

Investigations have been made also on transit times and origins of water in aquifers. Stable isotope and tritium analyses were utilized in the volcanic island of Cheju in the Republic of Korea by G. H. Davis and his associates in 1970.[45] They characterized groundwaters as regards time and place of recharge and facilitated determination of the nature of mixing of the various groundwater sources and estimation of their residence times. The waters were classified in a number of flow regimes. Springs at medium and high altitudes showed a relatively high tritium content and a wide spectrum of stable-isotope composition. This was attributed to the water having a short transit time accompanied by poor mixing. Large coastal springs, however, had much lower tritium contents, but a similar stable-isotope spread, which suggests a similar source to that of the high-altitude springs, but involving a longer transit time. The period covered by tritium analyses was rather short, but an estimate of the mean transit time of the waters was made, assuming a well-mixed reservoir model with recharge effective only for monthly precipitation over 100 mm, and values of from 2–8·5 years were obtained, the longer times being connected with waters discharged from an extensive, well-mixed fresh water lens. A similar experiment was conducted by J.-C. Fontes and his associates in 1967: the recharge area and the turnover time (8 ± 3 years) were determined for the Evian aquifer in the Haute Savoie, France, using stable isotopes and tritium.[46]

Extensive regional environmental isotope projects have arisen in recent years, one of which in the Great Artesian Basin in Australia, has been mentioned earlier (Section 4.3). The point of interest is the employment of chlorine-36, half-life 301 000 years, in groundwater dating there. This radionuclide results from the spallation of atmospheric argon and the yields are latitude-dependent, attaining a maximum between 30 and 50°. The isotope is generated also in exposed rocks by spallation of potassium and calcium and the neutron activation of chlorine, and may be released locally in significant quantities by weathering. It appears, however, that the level of ^{36}Cl accumulated in groundwater by subsurface neutron activation is probably significant only if the age exceeds 10^6 years. The specific activity of the isotope at input is dependent not merely on its rate of generation, but also upon the rate of accession of 'dead' marine chloride. Estimates have been made from the rates of chloride deposition in rainfall, but these are approximate since no account could be taken of dry fall-out or of local recycling. In the Great Artesian Basin, observed values of the $^{36}Cl/Cl$ ratio at input (100×10^{-15}–150×10^{-15}) are greater than those calculated (3×10^{15}–8×10^{-15}) at coastal stations at about the same latitude. This may relate to the decrease in the rate of accession of 'dead' marine salt with distance from the coast. Chlorine-36 isochrones, assuming a constant value for the porosity/hydraulic conductivity ratio of 0·2 day/m over the basin, may be interpreted as groundwater ages, assuming $^{36}Cl/Cl$ levels at input of

150×10^{-15} and 100×10^{-15} respectively. There is a factor which requires careful examination in this context, i.e. the possibility of accession of subsurface 'dead' chloride in the region which may come from the underlying Devonian Adavale Basin. The significance of this work by P. L. Airey and his associates in 1983 lies in their having established a solid, experimental basis for this conclusion. Otherwise, mean $\delta^{18}O/\delta^{16}O$ values indicated conformity with the meteoric line $\delta D = 8\ ^{18}O + 10$ and the D/H ratios demonstrated a depletion westwards. The latter correlates with European and Middle Eastern studies in which data were correlated with the global average decrease in paleotemperatures over the past 50 000 years.[47]

12.3. ARTIFICIAL ISOTOPE HYDROLOGY

Artificial radioisotopes can be measured in minute concentrations and often *in situ*, thus facilitating convenient experiments in the field. They are expensive, however, due to the fact that there is a health hazard involved against which protective measures have to be taken. Radioactive isotopes are utilized as tracers to determine the local characteristics of aquifers.

It is thus possible to measure the following parameters of an aquifer:

(a) its characteristics, i.e. porosity, transmissivity and dispersivity;
(b) the direction and velocity of groundwater flow (see Section 4.1);
(c) the stratification.

The radiotracer chosen in any particular instance depends upon the nature of the problem involved. Generally, it is necessary to select one with the following points in mind.

(a) Its life should be comparable to the intended duration of the observation. This will prevent longer-lived isotopes from polluting the water and constituting a persistent hazard to health.
(b) It should not be adsorbed by the components of the soil.
(c) If possible, it should be measurable in the field; γ-emitters are preferable in this respect.
(d) It must be available at reasonable cost on site.

The most commonly utilized radioisotope tracers in groundwater investigations are tritium, ^{24}Na ($T_{1/2} = 15$ h), ^{51}Cr ($T_{1/2} = 27 \cdot 8$ d), ^{58}Co ($T_{1/2} = 71$ d), ^{82}Br ($T_{1/2} = 35 \cdot 7$ h), ^{110}Ag ($T_{1/2} = 249$ d), ^{131}I

($T_{1/2} = 8\cdot05$ d) and ^{198}Au ($T_{1/2} = 64\cdot8$ h) of which full details are given earlier (Section 1.2).

The chemical form of the radioactive solution utilized in water tracing work is important. In many cases cations act as tracers and, in the form in which they are usually introduced, they may be subject to strong adsorption in the ground that interferes with their tracing activity. To minimize this effect, resort is made to complex chemical forms, of which one of the best-known is a chelated metallic compound formed with ethylenediaminetetraacetic acid, EDTA.

Tritium is a good tracer because it comprises part of the water molecule, but it does not meet half-life and radiation requirements, nor can it be detected *in situ*. In some cases tritiated water, HTO, may be delayed in comparison with bulk water velocity because of ion exchange with clays (see W. J. Kaufman and G. T. Orlob in 1956[48] and G. Knutsson and H. G. Forsberg in 1967.[49] Artificial introduction of tritium into waters necessitates great care and the lowest activity levels possible in order to prevent rather long pollution and interference with environmental tritium contents.

Any radiotracer can be introduced to a borehole either by pouring it down a thin pipe or by crushing an ampoule at the required depth; alternatively, a special injection device can be utilized. Injection may be carried out at one or at several depths so as to facilitate mixing of the radiotracer with the standing water in the borehole. After the radiotracer has been released and mixed in the borehole, the radioactivity can be measured by an appropriate probe, e.g. a Geiger–Müller counter or a scintillation counter, inserted at the selected depth. Incidentally, the radiotracer being introduced must have the same temperature as the water in the borehole so as to preclude development of any density current; cf. Section 4.1.

In 1963 J. Guizerix and his associates gave details of specialized probes which seal off a defined volume of the borehole using inflatable bladders; radiotracer is then released into this volume and radioactivity is measured by detectors in or just above the release volume.[50]

M. I. Kaufman and D. K. Todd in 1962 circulated labelled water between an isolated segment of a borehole and the ground surface, using a small pump.[51] In this way they obtained gentle mixing and the injection and measurement of radioactivity was effected at the surface without the necessity of a special probe.

The single well dilution technique of E. Halevy and his associates in 1967 was described earlier (Section 4.1): the rate of dilution of the radiotracer by the natural flow of water through the borehole is observed.[52] Also, there is the single well pulse technique, described by M. Borowczyk and his associates in 1967, in which radiotracer is forced into an aquifer by pumping water into the borehole and then recalling it by pumping water out.[53]

There is a multiple well technique for observing the movement of water between boreholes, in which the observation well is usually pumped and the radioactivity of the pumped water is monitored at the surface; see E. Halevy and A. Nir in 1962.[54]

Some applications that may be considered are given below, related to the aquifer characteristics listed at the beginning of this section.

The effective porosity of an aquifer may be determined. The calculation is based upon the approximate equality of porosity, i.e. void volume/total volume and the partial volume of water, i.e. water volume/total volume. The radiotracer is introduced into a well and a second well at a distance r is pumped. If the possible effects of dispersion are ignored, the arrival of the radiotracer at the second well indicates that the volume of water pumped, V, is:

$$V = \pi r^2 b s$$

where b is the thickness of the aquifer, r is the distance between the two wells and s is the effective porosity, i.e. the partial volume of water mentioned earlier. It is necessary for the distance r to be large compared with the thickness of the aquifer, also for radial pumping velocities to be greater than the natural velocities and the cone of depression at the pumping well to be small compared with the volume of water pumped.

The coefficient of transmissibility, T, i.e. transmissivity, used in the Theis non-equilibrium equation (Section 5.2, eqn (5.29)) demonstrates the capacity of the aquifer to transmit water. T is related to the hydraulic conductivity, K, and the thickness of the aquifer, b, thus:

$$T = bK$$

and, for the porosity pumping experiment mentioned in the preceding paragraph, it has been found that the volume of water that passes before the arrival of the radiotracer peak is inversely proportional to the transmissivity. The transmissivity of two layers of the same aquifer has been determined by means of two injection wells and one observation well, from the following relationships:

$$V_1 = \pi r_1^2 b_1 s_1 T/T_1$$
$$V_2 = \pi r_2^2 b_2 s_2 T/T_2$$
$$T_1 + T_2 = T$$

where T is the total transmissivity, T_1, T_2 are the partial transmissivities of the layers 1 and 2, r_1 and r_2 are the distances between the injection and observation wells, b_1 and b_2 are the thicknesses of the layers 1 and 2 and s_1 and s_2 are the effective porosities of the layers 1 and 2. Precision of 5×10^{-6} m^2/s was obtained.

Dispersivity is an important parameter and its coefficient, D, occurs in the radiotracer transport equation.[52] It characterizes the mixing property of an aquifer and may be estimated by finding that theoretical curve which best fits the experimental radiotracer breakthrough curve in an observation well. The parameters on such theoretical curves are dispersion coefficients derived from the mathematical model of the radiotracer transport (see A. Lenda and A. Zuber in 1970[55]).

The velocity and direction of groundwater flow have already been discussed in detail in Section 4.1. As regards direction of flow, however, another method, proposed by I. B. Hazzaa in 1970, must be mentioned.[56] A radiotracer, ^{32}P, with a half-life of 14·3 days is adsorbed on two coaxial metallic screens and subsequent β-radiographic examination of these gives information on flow direction. Filtration velocity could be determined as well.

REFERENCES

1. BIGELEISEN, J., 1965. Chemistry of isotopes. *Science*, **147**, 463–71.
2. BOTTINGA Y. and CRAIG, H., 1969. Oxygen isotope fractionation between CO_2 and water and the isotopic composition of marine atmosphere. *Earth Sci. Planet. Lett.*, **5**, 285–95.
3. GONFIANTINI, R., 1971. *Notes on Isotope Hydrology*. IAEA, Vienna.
4. CRAIG, H., 1961. Standards for reporting concentrations of deuterium and oxygen-18 in natural waters. *Science*, **133**, 1833–4.
5. FRIEDMAN, I., 1953. Deuterium content of natural waters and other substances. *Geochim. Cosmochim. Acta*, **4**, 89–103.
6. EPSTEIN, S. and MAYEDA, T., 1953. Variation of ^{18}O content of waters from natural sources. *Geochim. Cosmochim Acta*, **4**, 213–24.
7. HAGEMANN, R., NIEF, G. and ROTH, E., 1970. Absolute isotopic scale for

deuterium analysis of natural waters. Absolute D/H ratios for SMOW. *Tellus*, **22**, 712.
8. BAERTSCHI, P., 1976. Absolute ^{18}O content of Standard Mean Ocean Water. *Earth Planet, Sci. Lett.*, **31**, 341.
9. GONFIANTINI, R., 1978. Standards for stable isotope measurements in natural compounds. *Nature*, **271**, 534–6.
10. FRIEDMAN, I. and O'NEIL, J. R., 1977. Compilation of stable isotope fractionation factors of geochemical interest. In: *Data of Geochemistry*, 6th edn, Ed. M. Fleischer. U.S. Geol. Surv. Prof. Paper 440-KK.
11. CRAIG, H. and GORDON, L. I., 1965. Deuterium and oxygen-18 variations in the ocean and marine atmosphere. In: *Stable Isotopes in Oceanographic Studies and Paleotemperatures*. CNR Lab. Geol. Nucl., Pisa, pp. 9–130.
12. GONFIANTINI, R. and LONGINELLI, A., 1962. Oxygen isotope composition of fogs and rains from the North Atlantic. *Experientia*, **18**, 222.
13. FÖRSTEL, H. and HÜTZEN, H. 1982. $^{18}O/^{16}O$-Verhältnis im Grundwasser der Bundesrepublik Deutschland. Kernforschungsanlage Jülich GmbH, Institut für Radioagronomie, ISSN 0366–0885, 57 pp.
14. DANSGAARD, W., 1964. Stable isotopes in precipitation. *Tellus*, **19**, 434–68.
15. TENU, A., NOTO, P., CORTECCI, G. and NUTI, S., 1975. Environmental isotope study of the Barremian–Jurassic aquifer in South Dobrogea (Rumania). *J. Hydrol.*, 26, 185–98.
16. LAL, D. and SUESS, H. E. 1968. The radioactivity of the atmosphere and hydrosphere. *Ann. Rev. Nucl. Sci.*, **18**, 407–37.
17. THATCHER, L. L. and PAYNE, B. R., 1965. The distribution of tritium in precipitation over continents and its significance to groundwater dating. *Proc. 6th Intl Conf. Radiocarbon and Tritium Dating*. Pullman, Washington, pp. 604–29.
18. DINÇER, T. and DAVIS, G. H., 1967. Some considerations on tritium dating and the estimates of tritium input function. *Proc. 8th Cong. Intl Assn Hydrogeol., Istanbul*, pp. 276–85.
19. ZIMMERMAN, U., MÜNNICH, K. O. and ROETHER, W., 1967. Downward movement of soil moisture traced by means of hydrogen isotopes. In: *Isotope Techniques in the Hydrologic Cycle*. Amer. Geophys. Union, Washington, DC, pp. 28–36.
20. SMITH, D. B., WEARN, P. L., RICHARDS, H. J. and ROWE, P. C., 1970. Water movement in the unsaturated zone of high and low permeability strata by measuring natural tritium. *Isotope Hydrology*, IAEA, Vienna, pp. 73–87.
21. ALLISON, G. B. and HUGHES, M. W., 1976. Environmental tritium in the unsaturated zone: estimation of recharge to an unconfined aquifer. *Interpretation of Environmental Isotopes and Hydrochemical Data, Proc.*, Vienna, IAEA, SM-182/4, pp. 57–72.
22. SUESS, H. E., 1967. Bristlecone pine calibration of the radiocarbon time scale from 4100 B.C. to 1500 B.C. *Radioactive Dating and Methods of Low-level Counting*. IAEA, Vienna, pp. 143–51.
23. MÜNNICH, K. O., 1957. Messung des ^{14}C Gehaltes von hartem Grundwasser. *Naturwiss.*, **44**, 32.
24. VOGEL, J. C., 1970. Investigation on groundwater flow with radiocarbon. *Isotope Hydrology*, IAEA, Vienna, pp. 355–69.

25. BONANI, G., BALZER, R., SUTER, M., WÖLFI, W., BEER, J., OESCHGER, H. and STAUFFER, B., 1980. Radioisotope dating using an EN-Tandem accelerator. *Proc. Europhysics Conf. on Nuclear Physics Methods in Material Research*, Darmstadt, pp. 357–9.
26. HALL, E. T., 1983. Radiocarbon dating by accelerator — a progress report. *Proc. 22nd Symp. on Archaeometry, Bradford University (30 March–3 April, 1983)*, pp. 130–4. Available from Schools of Physics and Archaeological Sciences, University of Bradford, Richmond Road, Bradford, W. Yorks BD7 1DP, UK. ISBN 0 9508–482 0 4.
27. D'AMORE, F., FERRARA, G. C., NUTI, S. and SABROUX, J. C., 1976. Variations in radon-222 content and its implications in a geothermal field. *Intl Cong. Thermal Waters, Geothermal Energy and Volcanism of the Mediterranean Area*, Athens.
28. TURKEVICH, A. and SAMUELS, A., 1954. Evidence for ^{32}Si, a long lived beta emitter. *Phys. Rev.*, **94**, 364.
29. DEMASTER, D. J., 1980. The half-life of ^{32}Si determined from a varved Gulf of California sediment core. *Earth Planet. Sci. Lett.*, **48**, 209–17.
30. JANTSCH, K., 1967. Kernreaktionen mit Tritonen beim ^{32}Si: Bestimmüng der Halbwertzeit von ^{32}Si. *Kernenergie*, **10**, 89.
31. CLAUSEN, H. B., 1973. Dating of polar ice by ^{32}Si. *J. Glaciol.*, **12**, 411.
32. HONDA, M. and LAL, D., 1954. Spallation cross sections for long lived radionuclides in iron and light nuclei. *Nucl. Phys.*, **51**, 363.
33. GEITHOFF, D., 1962. Über die Herstellung von ^{32}Si durch einen (t, p)-Prozess. *Radiochim. Acta*, **1**, 1, 3.
34. LINDNER, M., 1953. New nuclides produced in chlorine spallation reactions. *Phys. Rev.*, **91**, 642.
35. DINÇER, T. and PAYNE, B. R., 1971. An environmental isotope study of the southwestern karst region of Turkey. *J. Hydrol.*, **14**, 233–58.
36. DINÇER, T., 1968. The use of oxygen-18 and deuterium concentrations in the water balance of lakes. *Water Res. Res.*, **4**, 1289–1306.
37. PAYNE, B. R., 1970. Water balance of Lake Chala and its relation to groundwater from tritium and stable isotope data. *J. Hydrol.*, **2**, 47–58.
38. FONTES, J.-C., GONFIANTINI, R. and ROCHE, M. A., 1970. Deuterium et oxygène-18 dans les eaux de lac Tchad. *Isotope Hydrology*, IAEA, Vienna, pp. 387–402.
39. GONFIANTINI, R., TOGLIATTI, V. and TONGIORGI, E., 1962. Il repporto $^{18}O/^{16}O$ nell-acqua del lago di Bracciano e nelle falde a sud-est del lago. *Notiziario CNEN*, **8**, 6, 39–45.
40. MORGANTE, S., MOSETTI, F. and TONGIORGI, E., 1966. Moderne indagini idrologiche nella zona di Gorizia. *Boll. Geofis. Teor. Appl.*, **8**, 14–38.
41. CONRAD, G. and FONTES, J.-C., 1970. Hydrologie isotopique du Sahara Nord-Occidental. *Isotope Hydrology*, IAEA, Vienna, pp. 405–19.
42. CONRAD, G. and FONTES, J.-C., 1972. Circulations, aires et périodes de recharge dans les nappes aquifères du Nord-ouest Sahariens: données.
43. HANSHAW, B. B., BACK, W. and RUBIN, M., 1965. Radiocarbon determinations for estimating groundwater flow velocities in Central Florida. *Science*, **148**, 494–5.
44. PEARSON, F. J. Jr, and WHITE, D. E., 1967. Carbon-14 ages and flow rates of

water in Carrizo Sand, Atacosa County, Texas. *Water Res. Res.*, **3**, 251–62.
45. DAVIS, G. H., LEE, C. K., BRADLEY, E. and PAYNE, B. R., 1970. Geohydrologic interpretations of a volcanic island from environmental isotopes. *Water Res. Res.*, **6**, 99–109.
46. FONTES, J.-C., LETOLLE, R., OLIVE, P. and BLAVOUX, B., 1967. Oxygène-18 et tritium dans le bassin d'Evian. *Isotopes in Hydrology*, IAEA, Vienna, pp. 401–15.
47. AIREY, P. L., BENTLEY, H., CALF, G. E., DAVIS, S. N., ELMORE, D., GOVE, H., HABERMEHL, M. A., PHILLIPS, F., SMITH, J. and TORGERSON, T., 1983. Isotope hydrology of the Great Artesian Basin, Australia. *Intl Conf. on Groundwater and Man, Sydney, Australia*, pp. 1–11.
48. KAUFMAN, W. J. and ORLOB, G. T., 1956. Measuring groundwater movement with radioactive and chemical tracers. *J. Amer. Water Works Assn*, **48**, 559–72.
49. KNUTSSON, G. and FORSBERG, H. G., 1967. Laboratory evaluation of ^{51}Cr–EDTA as a tracer for groundwater flow. *Isotopes in Hydrology*, IAEA, Vienna, pp. 629–52.
50. GUIZERIX, J., GRANDCLEMENT, G., GAILLARD, B. and RUPY, P., 1963. Appareil pour la mesure de vitesses relatives des eaux souterraines par la methode de dilution ponctuelle. *Radioisotopes in Hydrology*, IAEA, Vienna, pp. 25–35.
51. KAUFMAN, M. I. and TODD, D. K., 1962. Application of tritium tracer to canal seepage measurements. *Tritium in the Physical and Biological Sciences*, Vol. 1, IAEA, Vienna, pp. 83–94.
52. HALEVY, E., MOSER, H., ZELLHOFER, O. and ZUBER, A., 1967. Borehole dilution techniques: a critical review. *Isotopes in Hydrology*, IAEA, Vienna, pp. 531–64.
53. BOROWCZYK, M., MAIRHOFER, J. and ZUBER, A., 1967. Single-well pulse technique. *Isotopes in Hydrology*, IAEA, Vienna, pp. 507–19.
54. HALEVY, E. and NIR, A., 1962. The determination of aquifer parameters with the aid of radioactive tracers. *J. Geophys. Res.*, **61**, 2403–9.
55. LENDA, A. and ZUBER, A., 1970. Tracer dispersion in groundwater experiments. *Isotope Hydrology*, IAEA, Vienna, pp. 619–41.
56. HAZZAA, I. B., 1970. Single-well technique using ^{32}P for determining direction and velocity of groundwater flow. *Isotope Hydrology*, IAEA, Vienna, pp. 713–24.

CHAPTER 13

Geophysical Investigations of Groundwater — I. Surface Methods

Groundwater is not visible at the surface of the Earth, but it can be examined therefrom, although the results are usually incomplete. However, surface methods are extensively employed because they are cheaper than subsurface ones (which will be discussed in Chapter 14). Surface methods include conventional geological approaches such as field reconnaissance of the ground truth with especial reference to hydrologic data obtainable from streamflow, springs, boreholes and wells, groundwater recharge and discharge together with varying associated water levels, water quality and pollution, etc., all very useful observations which require no particularly expensive equipment and can be regarded as quite economical. In recent years, the development of remote sensing has made a great contribution generally, but its applicability to water resources is still progressing.

13.1. ABOVE-SURFACE METHODS

Aerial photographs are valuable; black and white ones may be examined stereoscopically in order to distinguish rock and soil types, identify patterns of fractures which can be related to porosity and permeability, detect springs and marshes which indicate relatively shallow depths to groundwater, and facilitate classification of a region into areas of good, fair and poor groundwater prospects. Geobotanical surveying is also feasible and recognition of the various types of vegetation may be valuable. For instance, phreatophytes transpire water from shallow water tables of unconfined aquifers and define depths to groundwater; halophytes tolerate soluble salts and, when associated with white efflorescence on the ground, demonstrate that shallow brackish or saline groundwater occurs; finally, xerophytes are desert plants requiring minimal water and indicate considerable depth to the water table of an unconfined aquifer.

In some situations, colour photographs may be preferred because the tonal difference is more easily distinguishable than with black and white panchromatic photography. Also, false colour imagery can be utilized. Photography is able to provide an instantaneous synoptic picture of an entire region visible through the camera lens. Owing to image geometry disturbance caused by the central perspective in stereoscopic vision, it is extremely time-consuming to transfer photographic data to a base map, which is often inaccurate. This can be overcome through orthophotography, which is the optical conversion of the central projection of ordinary photographs to a parallel projection at uniform scale without relief displacement; thereafter, the transfer of orthophoto data to a base map is easy and fast. Production and application of orthophotographs has been discussed by J. Visser and his associates in 1980.[1] The technique was introduced in 1964 and, in 1976, the OR1 camera appeared with a digital steering, automatic orthophoto printer. In 1980, another camera, the Orthocomp Z2, was introduced by Zeiss Oberkochen. It was an expensive instrument, costing at that time, US$250 000.

The smallest scale for useful photomaps seems to be about 1:10 000 except in arid terrains and the US Geological Survey orthophotomaps at a scale of 1:24 000 may be considered as constituting provisional substitutes for maps. Stereo-orthophoto pairs can be produced as well. Photomaps are becoming ever more generally utilized, and, in city areas, a scale of 1:1000 is available. Normally they are in black and white, but they can be in colour too. The US Geological Survey has reported bad experiences with colour, however, indicating that when high altitudes (i.e. 12 000 m or more) are involved for 1:24 000 orthophotoquad production, poor contrast occurs due to atmospheric scattering of the shorter wavelengths. US Geological Survey 1:24 000 orthophotoquads are made at high altitudes; in 1978, US$4·5 million and 125 man-years were spent in producing 4400 orthophotos. Orthophotography involves resolution expressed as the smallest object recognizable on the image by its salient characteristics, this depending upon the scale of the photographs, the type of film employed and the sort of lens used.

Satellite imagery is available in data provided through manned space excursions in the Gemini, Apollo, Skylab and Soyuz programmes, but

the coverage of the planet is both haphazard and small-scale.
An interesting development in photography was described by L. S. Cluff and G. E. Brogan in 1974, relating to sun angle giving patterns of linear shadows which can reveal morphological lineaments.[2] A limitation is that only those lineaments which strike approximately between north-west and south-west can be observed.

Multispectral scanning is a recent advance of great interest. Multispectral optical–mechanical scanners are used in recording the electromagnetic radiant energy which is emitted or reflected by the surface of the Earth in several precisely defined bands of the electromagnetic spectrum between 0·3 and 0·03 μm. Data on radiation intensity are integrated over a certain region of the terrestrial surface, the area of which is determined by the 'instantaneous field of view' (IFOV), the opening angle of the scanner following the flying altitude. For scanners in aircraft, the individual picture element is termed a pixel: it covers 3 m × 3 m of the Earth's surface at a flight altitude of 2000 m with a scale of 1:15 000 fully enlarged. With LANDSAT imagery, the scanners provide a pixel of 60 m × 80 m using a much smaller angle. The equivalent scale is 1:3 500 000, but extremely detailed data on the radiation intensity are provided. These are stored on tape after being corrected electronically for geometrical distortions arising from the actual scanning operation. The data can later be converted to imagery strips showing the radiation intensity of one spectral band in black and white; alternatively, two or three radiation ranges may be used. These are then superimposed to make a 'colour composite'. Multispectral scanning (MSS) gives information on the spectral signatures of areas. This differs from the results of aerial photography, the latter being a delineation-detection tool whilst MSS imagery is an areal sensing technique. However, MSS can recognize vegetation, soil and rock types by these spectral signatures and, hydrologically, it is valuable in that it easily distinguishes between land and water which have very different spectral signatures. Thus, delineation of drainage systems is a simple matter.

Radar is another useful tool. Side-looking airborne radar, SLAR, and synthetic aperture airborne radar (SAAR) are important remote sensing techniques which depend upon the emission of pulses of radio from an aircraft, the return signals being recorded with an antenna. The range in the electromagnetic spectrum from 8 mm to several metres can be used, but the usual range is 0·8–15 cm. The approaches can be employed both at night and during the day and, with longer wavelengths, e.g. with SAAR, heavy rain showers can be penetrated. The intensity of the return signals is transformed into grey tones which, when corrected for geometrical distortion, can be used to form radar image strips of the areas on both sides of the aircraft. As there is a very wide angle involved, the methods cannot usually be applied to hilly terrain, as J. F. M. Mekel pointed out in 1972.[3] However, recent high-altitude SAAR flights suggest that they may be applicable to some hilly areas. In hydrology, large-surface water bodies provide specular reflections and are apparent on radar images as a result of their strong contrast with the surrounding soils and rocks. Some small drainage lines may be visible also, but only if they are related to the geomorphology, when those parallel with rather than perpendicular to the line of flight are better defined in the radar image. Slope changes may be visible in radar images.

During November 1981, the space shuttle *Columbia* conducted a radar experiment of great interest, this despite a reduction in the length of the mission from 124 to 54 h because of a faulty fuel cell. The pertinent results were described by J. F. McCauley and his associates in 1982.[4] A 50 km wide swath across the Sahara desert was made. Normally, radar waves penetrate only a few centimetres into the surface of the Earth, but, in the hyper-arid conditions prevailing in Egypt and the Sudan, they pierced to depths of 5 or 6 m, reflecting from bedrock. The signals obtained were processed through image-enhancing computer techniques and the results analyzed by the US Geological Survey, Flagstaff, Arizona, the University of Arizona, the Jet Propulsion Laboratory of the California Institute of Technology in Pasadena, California, and the Egyptian Geological Survey and Mining Authority. A buried topography was revealed, one without trace at the surface. The images showed stream channels, broad flood plains and river valleys of which some were as wide as the Nile. The trend of the ancient rivers is south and west, i.e. the opposite of existing movements, and they may have joined up into one large basin of interior drainage resembling the present-day Caspian Sea. Clearly, the region was once sufficiently wet to support plant, animal and human life. It is known that rainy episodes, the pluvials, occurred during the Pleistocene and these are believed to have taken place about 200 000, 60 000 and 10 000 years ago. Tools, supposedly of *Homo erectus*, have been located in the area as well, in fact at sites along the banks of the concealed river beds now revealed in the radar-derived maps. From these results, it appears likely that the same type of radar scanning can be applied to other arid regions for the detection of waterways not too far below the surface of the ground. Potential sites would be those where groundwater intersects the surface.

Another technique available is thermography, which is the accumulation of data on the superficial temperatures of the Earth. These are obtained by making a record of radiation in the 3–5 m and 8–13 m infrared using optical–mechanical scanners, as with MSS, or making a record in the microwave range (8 mm–20 cm) using a passive microwave radiometer with a scanning antenna, as in radar. From the data derived during a scanning exercise, a thermal image can be derived which shows temperature differences by various tones of grey or colours. Factors which may interfere include air temperature, terrain, inclination direction, wind velocity, atmospheric moisture content and soil moisture content. In cases where there is only a minimal interference the technique may be useful, for example in detecting freshwater springs in the sea, sources of geothermal energy or heat emission losses from subterranean pipe systems.

In terms of expense, photography is the cheapest of the techniques mentioned above. In 1979 for example, for areas exceeding 1000 km^2 on scales between 1:20 000 and 1:10 000, the price was US$5–10 per km^2. MSS imagery is dearer, but radar imagery is very expensive indeed.

13.2. ON-SURFACE METHODS

On the terrestrial surface, geophysical exploration may be used. This may be defined as the scientific measurement of the physical properties of the terrestrial crust in order to investigate geological structures or mineral deposits; it developed greatly following the discovery of oil by geophysical methods in 1926. Application to groundwater investigations was much slower because oil is commercially more valuable than water, but it has increased in recent years until, today, almost all organizations involved with groundwater utilize geophysical methods.

The most important of the physical properties measured are electrical resistivity and seismic refraction. The electrical resistivity of a rock restricts the amount of current passing through it when an electrical potential is applied and it is defined as the resistance in ohms between opposite faces of a unit cube of the material concerned. If this resistance be denoted by R and the material in question has a cross-sectional area of A and a length of L, then the following expression may be given:

$$\rho = \frac{RA}{L}$$

where ρ is the resistivity, the units of which are ohm m^2/m (sometimes written as ohm-m).

The great variability in the resistivities of the various types of rock is caused by many factors, such as density, porosity, pore size and shape, water content and quality together with temperature. There are no fixed limits. Igneous and metamorphic rocks yield values ranging from 10^2 to 10^8 ohm-m and, for sediments, some typical figures are 10^0–10^1 ohm-m for clay, 10^2–10^3 ohm-m for sand (and sandstone) and a similar range for porous limestone, dense limestone having a range extending from 10^3 to beyond 10^6 ohm-m. In porous formations, resistivity is more affected by water content and quality than by the resistivity of the actual rock material. In the case of aquifers which are composed of unconsolidated materials, resistivity decreases with the degree of saturation and the salinity of the groundwater. Clays have a lower resistivity than permeable alluvial aquifers because they have minerals within them which conduct electrical current through their matrix.

In practice, resistivities are determined from apparent resistivities calculated from measurements of current and potential differences between two potential electrodes which result from an applied current through two other electrodes beyond them, but in line; see Fig. 13.1.

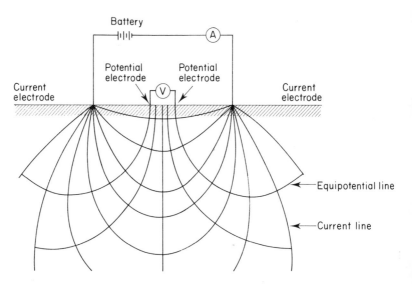

Fig. 13.1. Electrical circuit for determination of resistivity and the electrical field for an isotropic subsurface situation. A, ammeter; V, voltmeter.

If the resistivity is uniform in the subsurface zone under the electrodes, an orthogonal network of circular arcs is formed by the current and equipotential lines. The measured potential difference is a weighted value over a subsurface zone controlled by the shape of the network. Hence, the measured current and potential differences give an apparent resistivity over an unspecified depth. If the spacing between the electrodes is increased, then a deeper penetration of the electrical field will take place and of course a different apparent resistivity is the result. Hence, apparent resistivities alter as electrode spacings are increased, but not in proportion. Changes of resistivity at great depth exert only a slight influence on the apparent resistivity in comparison with those at shallow depths, so the method is not very effective in determination of actual resistivities below a few hundred metres.

The electrodes are metal stakes which are driven into the ground; sometimes the potential electrodes comprise porous cups filled with a saturated copper sulphate solution which inhibit any electrical fields from arising around them. Various spatial arrangements exist. In the Wenner arrangement (Fig. 13.2.) the potential electrodes are located at

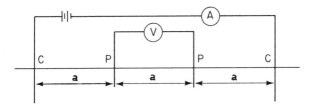

FIG. 13.2. The Wenner array. V, voltmeter; A, ammeter; C, current electrode; P, potential electrode.

the points spaced at one-third and two-thirds of the distance between the current electrodes. The apparent resistivity is given by the ratio of voltage to current times a factor of spacing and the following expression is valid:

$$\rho_a = 2\pi a \frac{V}{I}$$

where a is the distance between adjacent electrodes, V is the voltage difference between the potential electrodes and I is the applied current. In the Schlumberger arrangement, Fig. 13.3, the potential electrodes are

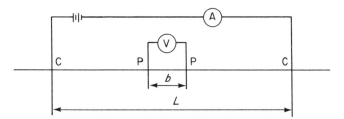

Fig. 13.3. The Schlumberger array. Symbols as in Fig. 13.2.

closer together and the apparent resistivity is given by:

$$\rho_a = \pi \frac{(L/2)^2 - (b/2)^2}{b} \frac{V}{I}$$

where L and b are the current and potential electrode spacings respectively. Whilst theoretically $L > b$, in practice satisfactory results can be obtained so long as $L \geq 5b$.

In the case where apparent resistivity is plotted versus electrode spacing (a in the Wenner array and $L/2$ in the Schlumberger array) for various spacings at one location, a smooth curve can be drawn through the points. Its interpretation is difficult and the solution is obtainable in two sections. Firstly, the approach envisages various layers of real resistivities at their actual depths. Secondly, the actual resistivities are interpreted in terms of the subsurface geological and groundwater conditions. For the first stage, computed resistivity/spacing curves for two-, three- and four-layer cases are useful; such curves, together with explanations of matching techniques, have been published for the Wenner configuration by R. G. Van Nostrand and K. L. Cook in 1966[5] and for the Schlumberger configuration by E. Orellana and H. M. Mooney in the same year.[6] The second stage depends upon supplemental information. If actual resistivity variations with depth are compared with data from an adjacent logged test borehole, a correlation can be established with subsurface geological and groundwater conditions. Such information can be used in interpretation of resistivity measurements in surrounding areas. Resistivity investigations can cover vertical variations at selected sites by variation of the electrode spacings and they may be utilized to produce horizontal profiles of apparent resistivity or apparent resistivity maps of a region by using a constant electrode spacing.

It is important to note that any factors which disturb the electrical field near electrodes can invalidate the resistivity measurements. Such factors include anisotropic geology, buried pipelines and cables, together with wire fences at the surface.

This method is extremely widely used. Areal resistivity changes may define aquifer limits whilst vertical variations (or soundings) may indicate aquifers and water tables, salinities, impermeable formations and the depth to bedrock.

Among recent new applications of the method is its employment in detecting polluted areas. Results fit best in groundwater terms where a conductive pollutant such as dissolved salt moves in a shallow zone with rather uniform geological conditions; see for instance a study of fresh water lens configuration on the Cayman Islands by S. F. Bugg and J. W. Lloyd in 1976.[7]

The seismic refraction method entails initiation of a small shock at the surface of the Earth by appropriate means, e.g. a small explosive charge, thereafter measuring the time required for the consequent shock wave to travel known distances. Seismic waves obey the same laws of propagation as light waves, so they can be reflected or refracted at any interface where a change in velocity takes place. Seismic reflection is suitable for assessing geological structures thousands of metres down, but the seismic refraction technique gives data up to about 100 m deep, so it is excellent in groundwater studies. The actual travel time of a seismic wave is dependent upon the nature of the media traversed, velocities being highest in compact, consolidated and crystalline rocks such as granite and lowest in loose, unconsolidated materials. Typical velocities in unsaturated and saturated materials were given by the American Society of Civil Engineers in 1972[8] and listed in Table 13.1. These values enable the nature of subterranean materials to be identified.

In coarse alluvium, the seismic velocity increases sharply from unsaturated to saturated zones; this facilitates the mapping of the depths to the water table. Such changes in seismic velocities are related to changes in the elastic properties of the formations. Obviously, the greater the contrast in these properties, the more clearly the formations and their boundaries can be discerned. It may be noted that in sedimentary rocks texture and geological history are more significant than mineral composition. Porosity tends to decrease wave velocity, but water content increases it.

In consolidated formations, where there is a uniform distribution of small pores, e.g. in a sandstone, velocity and porosity are related thus:

$$\frac{1}{v} = \frac{n}{v_L} + \frac{1-n}{v_s}$$

TABLE 13.1
APPROXIMATE VELOCITIES OF SEISMIC WAVES IN VARIOUS MATERIALS[8]

Velocity (m/s)	Medium
(a) Unsaturated materials:	
100–300	Topsoil and organic material
300–500	Loose sand
500–1000	Silt
500–1300	Gravel
1000–1800	Clay and till
1200–3500	Sandstone, graywacke and conglomerate
1800–5000	Shale, tillite and argillite
2500–5000	Limestone and dolomite
1300–2500	Weathered and fractured rocks in general
2500–6000	Metamorphosed sedimentary rocks, acid volcanic, igneous
4500–over 6000	Basic igneous, volcanic rocks and high grade metamorphics.
(b) Saturated materials	
1400–1700	Topsoil, organic materials, loose sand, silt, gravel, clay and till
1400–2500	Sandstone, graywacke and conglomerate
1700–5000	Shale, tillite, argillite
2000–5000	Limestone and dolomite
1700–2800	Compacted clay and till
1500–3500	Weathered and fractured rocks of all types

where v is the measured velocity, v_L is the velocity in the liquid saturating the rock, v_s is the velocity in the solid matrix of the rock and n is the porosity of the rock. In water, under typical groundwater conditions, the seismic velocity is roughly 1460 m/s.

When a seismic shock is initiated at the surface of the ground, the resultant shock wave travels outwards at a velocity which depends upon the medium through which it passes. If an unconsolidated, but isotropic and homogeneous, unconfined aquifer is involved, the seismic shock wave reaches the water table and then travels along the interface. On the ground surface, the first wave to be received must emanate either directly from the initial shock or by a refracted path. By measuring the times of arrival of this shock wave at various distances from the point of its origin, a graph of time versus distance can be constructed. If v_1 is the velocity above the water table of an unconfined aquifer and v_2 the velocity below it, then:

$$H = \frac{s}{2}\sqrt{\frac{v_2 - v_1}{v_2 + v_1}}$$

where H is the depth to the water table and s is the distance from the point of origin to the point of intersection of the shock waves v_1 and v_2. This can be considered as a two-layered case and multilayered problems may be approached similarly, frequently with the assistance of nomographs.

In the field, it is not difficult to apply the seismic refraction method because of the availability of compact and efficient equipment. A small dynamite charge is placed in a hole made by auguring to a depth of about 1 m, then the hole is back-filled. Detectors termed seisometers or geophones are laid out along a line from the point of detonation at distances of 3–15 m apart. These are the recipients of the shock wave and they convert the tremor into electrical impulses; an electrical circuit connects the detectors to an amplifier and a recording oscillograph which automatically records both the moment of detonation and also the various initial arrivals of the ensuing shock wave. Using such equipment, it is possible to obtain satisfactory results from depths of 60–100 m and, sometimes even to 300 m. In the case of very shallow depths, say 20 m or less, instead of dynamite, a blow from a sledgehammer on the ground surface is capable of producing a suitable shock wave.

Interpreting information so obtained requires the assumption that the layers involved are homogeneous and bounded by interfacial planes. In the event that no distinct boundary of this type exists, there will be a transition zone: on the time–distance graph there will be a curve rather than a sharp break in slope as was the case with shock waves v_1 and v_2. The great advantage of the method is that it obviates the necessity for drilling, but its major disadvantage is that it is restricted to largish areas, distances of several hundred metres being necessary for the line of detectors. Also, local interference such as airports, roads, construction sites and so on can be detrimental.

Since specialized technical personnel are a requisite for its successful application, the method has not been used as much as it might have been in groundwater studies. However, it has been applied to the mapping of cross-sections of alluvial valleys, from which the variations in thickness of unconfined aquifers can be determined.

Some other surface methods must be mentioned. One is the gravity method, in which density differences on the surface of the Earth are measured, and possibly provide information on geological structure. If the geology is appropriate, e.g. a buried valley, an aquifer configuration is detectable by gravity measurements. Usually, however, the differences in water content in subsurface beds do not produce measurable differences in density at the surface and in this case the method is not useful.

The magnetic fields of the Earth can be mapped by magnetic methods. Again, there is little relevance to groundwater studies because magnetic contrasts are not normally associated with the occurrence of aquifers, although some rare exceptions are reported in ref. 9. However, the methods have been extensively used in oil and mineral exploration. Using an accuracy of ± 1 gamma, fluxgate and proton magnetometers may be employed on surveys where magnetic anomalies of between 5 and 1000 gammas are interesting. The fluxgate magnetometer is valuable in areas of very high magnetic gradients and, when utilized with digital sampling at eight times per second, this instrument has an advantage over proton magnetometers. The latter are widely used, however, and in optimum circumstances can provide a noise envelope of ± 5 gammas. In some types of exploration, a higher-sensitivity optical pumping magnetometer is applied and, in flat magnetic areas with good control of all survey variables, it is possible to produce magnetic contour maps with a contour interval of 0.25 gamma, although it is expensive to do so. Every type of magnetometer can be combined with both analogue and digital recording, the latter being more advantageous in that it can filter out high-frequency noises.

From other magnetic measurements, information relevant to groundwater studies may be obtained; for instance dykes which constitute aquifer boundaries can be identified as reported by A. A. Zohdy and his associates in 1974.[10]

Finally, dowsing may be mentioned. This is sometimes referred to as water divining or water witching and it involves using a forked twig, usually of hazel. This is held in the hands and is said to be attracted downwards, this movement of the free end indicating subterranean water. There is a huge body of literature on this subject which has been accumulated over at least the last 400 years. Indeed, dowsing has great psychological interest because of the publicity it has received over such a long time. D. K. Todd in 1980 referred to an American survey indicating that there are, on average, 181 water witches per million population in that country.[11] Almost unbelievably, during a drought in 1977 in San Francisco, a local newspaper reported that a suburban flowing well had been located by the use of a bent coat hanger. The detection of magnetic fields arising from groundwater and correlation of these with water dowsing has been discussed by D. G. Chadwick and L. Jensen in 1971.[9] The author was the recipient of an impassioned, pro-dowsing letter from a reader of the first edition of this book, accompanied by a paper by this reader, one more addition to the existing voluminous literature. His main objection was to the author's 1980 statement that: "there is absolutely no scientific evidence whatsoever that it has the slightest value in the search for ground water. It is not surprising that the US Geological Survey advises against employing the 'technique'". No reason is seen to change this view.

REFERENCES

1. VISSER, J., BOUW, T., GRABMAIER, K., GRAHAM, R., HOWARD, E., KUNJI, B., LORENZ, R. and VAN ZUYLEN, L., 1980. Orthophotos: production and application. *ITC Journal*, **4**, Sp. Issue: *Proc. Intl Inst. for Aerial Survey and Earth Science. Post-Cong. Seminar, Hamburg, following XIVth Cong. Intl Soc. for Photogrammetry*, pp. 638–59.
2. CLUFF, L. S. and BROGAN, G. E., 1974. Investigation and evaluation of fault activity in the U.S.A. *Proc. 2nd Intl Cong., Intl Soc. Eng. Geol.*, 1, Theme II, Sao Paolo, Brazil.
3. MEKEL, J. F. M., 1972. The geological interpretation of radar images. In: *ITC Textbook of Photo-Interpretation*, ITC Enscheide, The Netherlands, Chapter VIII.
4. MCCAULEY, J. F., SCHABER, G. G., BREED, C. S., GROLIER, M. J., HAYNES, C. V., ISSAWI, B., ELACHI, C. and BLOM, R., 1982. Subsurface valleys and geoarcheology of the Eastern Sahara revealed by shuttle radar. *Science*, **218**, 1004–20.
5. VAN NOSTRAND, R. G. and COOK, K. L., 1966. *Interpretation of Resistivity Data*. US Geol. Surv. Prof. Paper, 499, 310 pp.
6. ORELLANA, E. and MOONEY, H. M., 1966 *Master Tables and Curves for Vertical Electrical Sounding over Layered Structures*. Interciencia, Madrid, 150 pp.
7. BUGG, S. F. and LLOYD, J. W., 1976. A study of fresh water lens configuration on the Cayman Islands using resistivity methods. *Quart. J. Engrng Geol*, **9**, 291–302.
8. AMER. SOC. CIVIL ENGINEERS, 1972. *Ground Water Management*, Manual Engrng Practice 40, NY, 216 pp.
9. CHADWICK, D. G. and JENSEN, L., 1971. *The Detection of Magnetic Fields Caused by Groundwater and the Correlation of Such Fields with Water Dowsing*. Utah Water Research Lab., Logan, 57 pp.
10. ZOHDY, A. A. et al., 1974. *Application of Surface Geophysics to Groundwater Investigations*. US Geol. Surv. Techniques of Water Resources Investigations, Chapter D1 Bk 2, 116 pp.
11. TODD, D. K., 1980. *Groundwater Hydrology*. J. Wiley, New York, 535 pp., p. 426.

CHAPTER 14

Geophysical Investigations of Groundwater — II. Subsurface Methods

14.1. GEOLOGICAL METHODS

Drilling small-diameter boreholes enables much valuable information about groundwater conditions to be gathered and, if necessary, these holes can be enlarged later and converted into pumping wells. Such test holes are useful also in measuring water levels and conducting pumping tests. In unconsolidated formations, cable tool and rotary methods are employed in the drilling work; these have been described in Section 6.2. Whilst cable tool methods are slower, they provide more accurate samples from the bailer. In cases where only shallow holes are required, augering or jetting, also described earlier, may be utilized. The result of the drilling activities is a geological log, which is constructed from the sample cuttings collected at frequent intervals while the drilling is going on. A log of this type can provide information on the nature of the geological materials encountered, the thickness of each bed in terms of depth and the total depth reached.

In practice, it is often very difficult to prepare an adequate geological log, because the cuttings obtained are usually small and of course they are mixed with lubricant mud. In rotary drillings, silt and clay may be masked by the drilling mud itself. Where satisfactory samples are collected, they must be stored systematically so that, at a later stage, not only can the geological log described be constructed, but it also is feasible to make correlations with adjacent boreholes and derive grain size analyses. A valuable supplement to test drilling is the drilling-time log which, as the term implies, is an accurate record of the actual times taken, in minutes and seconds, to penetrate each unit depth of the hole.

The approach is used mostly in connection with hydraulic rotary drilling, but it can be employed with other kinds of drilling as well.

The depth-to-water is a vital parameter to measure because it facilitates definition of groundwater flow direction, chronological variations in water level and the effects of pumping tests. To measure the depth-to-water, a steel tape may be lowered into a well, its end being chalked so that the length of submersion is at once apparent. If repeated measurements are required, or depths exceeding 50 m, an electrical water-level sounder may be used; it comprises a battery, a voltmeter, a calibrated two-wire cable with reel and an electrode. On contacting the water, the electrode completes a circuit and the associated voltmeter is deflected. The depth may be read directly from graduations along the cable. The air-line technique is also employed: a small-diameter tube is inserted into the annular space between a pump column and the casing. Usually, the tube is attached to the pump column so they are installed at the same time. The tube extends below the surface of the water and is connected to a small air compressor with a pressure gauge. Air is pumped into the tube until a maximum pressure is noted and this, converted into depth of water, indicates the gap between the lower end of the tube and the surface of the actual water. The air-line technique is not as accurate as the others mentioned, but it may be utilized in those water wells where splash and turbulence preclude them.

D. M. Stewart in 1970 devised an interesting rock technique.[1] He determined empirically the time needed for a 1·55 cm glass marble, a standard ball bearing or air rifle shot to fall to the surface of the water together with the time required for the sound of the splash to reach the surface of the ground. A stopwatch is used and the depth-to-water can be read directly from Table 14.1 In the case where the depth-to-water exceeds 57 m, the sphere will attain a constant terminal velocity and the following expression is applicable:

$$d = 27 \cdot 3t - 47 \cdot 6$$

where d is the depth-to-water (m) and t is the time interval. Pebbles cannot be substituted for spheres in this technique, because they have irregular shapes.

Sometimes it is necessary to record short-term fluctuations in water level, e.g. near other, intermittently operated wells, and, in this case, an automatic water level recorder can be utilized. This comprises a float with a counterweight, a rotating chart on drum and a recording pen driven across the chart.

TABLE 14.1
DEPTH-TO-WATER FROM SPLASH TIMING

Time (s)	Distance (m)	Time (s)	Distance (m)	Time (s)	Distance (m)
0·0	0·0	1·4	9·6	2·8	33·3
0·1	0·0	1·5	11·0	2·9	35·4
0·2	0·2	1·6	12·4	3·0	37·6
0·3	0·4	1·7	13·9	3·1	39·8
0·4	0·8	1·8	15·3	3·2	42·1
0·5	1·2	1·9	16·8	3·3	44·4
0·6	1·8	2·0	18·4	3·4	46·9
0·7	2·4	2·1	20·0	3·5	49·2
0·8	3·1	2·2	21·7	3·6	51·6
0·9	4·0	2·3	23·5	3·7	54·1
1·0	4·9	2·4	25·4	3·8	56·7
1·1	5·9	2·5	27·3	3·9	59·1
1·2	7·1	2·6	29·3	4·0	61·6
1·3	8·3	2·7	31·3	4·1	64·3

14.2. GEOPHYSICAL METHODS

Geophysical logging involves lowering appropriate sensing probes into boreholes and thereby recording physical parameters which can be interpreted to shed light on the characteristics of aquifers and formations, including the quality and amount together with the movement of groundwater and the physical structure of the borehole involved. In 1971, W. S. Keys and L. M. MacCrary summarized logging applications to groundwater hydrology; their presentation is given in Table 14.2.[2]

Geophysical logs are much more used in petroleum exploration than in groundwater work, mainly because there is more money available in the oil industry. Also, most water wells are shallow and have small diameters; as they supply domestic water, such logging is not required. Where larger-diameter and deeper wells are concerned, i.e. for irrigation, municipal or injection purposes, geophysical logging may provide data from which improved well construction and performance can result.

Apart from expense, another drawback to wider use is the fact that experienced personnel are rare and, as the techniques develop and become increasingly complex, this problem is likely to become greater.

TABLE 14.2
LOGGING APPLICATIONS TO GROUNDWATER HYDROLOGY

Data required	Feasible logging techniques
Lithology, stratigraphic correlation between aquifers and associated rocks	Resistivity, sonic or caliper logs made in open holes; radiation logs made in open or cased holes.
Total porosity or bulk density	Calibrated sonic logs in open holes; calibrated neutron or gamma–gamma logs in open or cased holes.
Effective porosity or true resistivity	Calibrated long–normal resistivity logs.
Clay or shale content	Natural gamma logs.
Permeability	Under certain conditions, long–normal resistivity logs.
Secondary permeability, i.e. fractures and solution openings	Caliper, sonic or television logs.
Specific yield of unconfined aquifers	Calibrated neutron logs.
Grain size	Possible relationship to formation factor derived from resistivity logs.
Location of water levels or saturated zones	Resistivity, temperature or fluid conductivity logs, neutron or gamma–gamma logs in open or cased holes.
Moisture content	Calibrated neutron logs.
Infiltration	Time-interval neutron logs.
Dispersion, dilution and movement of waste	Fluid conductivity or temperature logs; natural gamma logs for some radioactive wastes.
Source and movement of water in a well	Fluid velocity or temperature logs.
Chemical and physical characteristics of water, including salinity, temperature, density and viscosity	Calibrated fluid conductivity or temperature logs; resistivity logs.
Construction of existing wells, diameter, position of casing, perforations, screens	Gamma–gamma, caliper, casing or television logs.
Guide to screen setting	All logs giving data on the lithology, water-bearing characteristics and correlation and thickness of aquifers.
Cementing	Caliper, temperature or gamma–gamma logs; acoustic logs for cement bond.
Casing corrosion	In certain conditions, caliper, casing or television logs.
Casing leaks and/or plugged screen	Fluid velocity logs.

14.2.1. Resistivity Logging

It is possible to lower current and potential electrodes into an uncased well in order to measure the resistivities of the surrounding rocks, thus obtaining a trace of variations with depth which is termed a resistivity or electrical log. Such a log is influenced by fluid in a well, by the diameter of the well, by the nature of the beds penetrated by the well and by the groundwater. In order to minimize the effects of drilling fluid and well diameter, several electrodes are utilized (often four, of which two emit current and two make potential measurements), the curves recorded being either normal or lateral according to the arrangement of the electrodes. Figure 14.1 illustrates the two systems: the short-electrode spacing is the more effective for locating the boundaries of formations which have different resistivities.

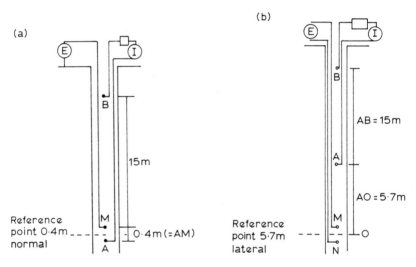

FIG. 14.1. Electrode arrangements and standardized distances for resistivity logs: (a) short normal; (b) lateral.

The electrical log obtained from a well comprises vertical traverses recording short and long normal, lateral and spontaneous potential curves; the last of these is discussed below in Section 14.4. Accurately to interpret resistivity logs is a daunting task, but resistivity curves are valuable in indicating the lithologies of the rock strata which the well penetrates and also in permitting fresh water in them to be distinguished from saline water.[3] Resistivity logs may be employed in determining the specific resistivities of strata. The resistivity of an unconsolidated aquifer was considered in Section 13.2, where it was mentioned that this parameter decreases in value with the degree of saturation and the salinity of the relevant groundwater. Thus, the extent of saturation is crucial and relates both to porosity and packing. Temperature is also significant. Generally speaking, shale, clay and saline water sand have low specific resistivity, fresh water sand moderate to high specific resistivity and cemented sandstones and non-porous limestones high specific resistivity. As might be expected, casings have very low resistivities indeed.

Groundwater resistivity depends upon ionic concentration and mobility of the salt solution, the mobility being related both to the molecular weight and to the electrical charge. Consequently, differences exist between various solutes.

As groundwater gets warmer, ionic mobility rises and the water becomes less viscous. Thus there is an inverse relationship between resistivity and temperature, which is expressed as a correction factor. If resistivity at the measurement temperature is multiplied by the correction factor for that temperature, the resistivity for the standard temperature of 25°C is obtained. Obviously, the correction factor for 25°C is unity; for 10°C it is 0·5 and for 50°C it is 1·5.

Electrical logs are commonly used to ascertain the correct position in which to site a well screen. The log gives the basis for selection of the proper length and also for setting it against the optimum formations. A. N. Turcan, Jr. in 1966 extended the utility of electrical logs in Louisiana aquifers in the USA,[4] where resistivity logging permitted an estimation of groundwater quality to be made. Initially a field-formation factor, F, for an aquifer is determined from previous data by:

$$F = \frac{\rho_0}{\rho_w}$$

where ρ_0 is the resistivity of the saturated aquifer and ρ_w is the resistivity of the groundwater contained in it, this being the reciprocal of its specific conductance (see Section 9.2), in the relationship $\rho_w = 10^4/E_c$, where ρ_w is in ohm-m and E_c is the specific conductance in microsiemens (equivalent to micromhos/cm), $\mu S/cm$. Specific conductance is related to the chloride or total dissolved solids content of the aquifer in a known manner and, with E_c and F available, ρ_0 can be read from the long–normal resistivity curve in an aquifer enabling ρ_w and the salinity of the groundwater to be calculated. This approach gives the best results in uniform clastic aquifers, e.g. sand and sandstone, comprising inter-granular pores which are saturated with water.

Resistivity logs can also assist in the recognition of wells which intersect both fresh water and saline water zones. Circulation in a well of this type, if there is no pumping activity, depends upon the relative hydrostatic heads, the respective water densities, the locations and thicknesses of the aquifers, the physical construction of the well and its condition.

14.2.2. Spontaneous Potential Logging

The spontaneous potential logging technique measures the natural electrical potentials, sometimes termed the self-potentials, SP, occurring below the surface of the Earth. These measurements are normally in millivolts and are made with a recording potentiometer which is connected to two like electrodes. One electrode is lowered into an uncased well and the other is connected to the surface of the ground. The potentials are produced mainly by electrochemical cells formed by the electrical conductivity differences of drilling mud and groundwater where boundaries of permeable formations intersect a borehole. Sometimes the electrokinetic effects of fluids which move through permeable beds also create spontaneous potentials. Hence, potential logs indicate permeable zones and can also aid in determining the lengths of casings and in estimating the total dissolved solids contents of groundwaters. Potential logs are not very useful if there are no sharply defined boundaries in permeable zones and, in additon, they may be interfered with by industrial activities which can set up spurious earth currents.

Potential values cover a spectrum from zero to hundreds of millivolts. They are conventionally read as positive and negative deviations from an arbitrary baseline normally associated with an impermeable, very thick formation. The sign of the potential is dependent upon the ratio of the salinity (or resistivity) of the drilling mud to the formation water. Spontaneous potentials which result from electrochemical potentials can be expressed by:

$$SP = (64 \cdot 3 + 0 \cdot 239 T) \log \frac{\rho_f}{\rho_w}$$

where ρ_f is the drilling fluid resistivity in ohm-m, ρ_w is the groundwater resistivity in ohm-m and T is the borehole temperature (°C). Hence, for measured SP, ρ_f and T values, the resistivity and therefore the salinity of groundwater can be determined.[2] However, this equation applies only to highly saline groundwater, the predominant salt being sodium chloride. Potential and resistivity logs are usually recorded together and one can supplement the other.

14.2.3. Radioactive (Nuclear) Logging

This approach involves measuring fundamental particles emitted from unstable radioactive isotopes. Appropriate logs which apply to groundwater include natural gamma, gamma–gamma and neutron; see the Technical Reports Series 126 written by the Working Group on Nuclear Techniques in Hydrology and issued by the International Atomic Energy Agency in 1971.[5]

14.2.3.1. Natural Gamma Logging

All rocks emit natural gamma radiation and this, recorded, constitutes a natural gamma log; the source of the radiation is unstable isotopes of the elements potassium, uranium and thorium and the method depends upon the recognition of their decay products (a typical log is shown in Fig. 14.3). Energies vary and quantitative estimates are usually related to equivalent radium, RaEq, i.e. the quantity of radium which would emit a similar dosage of radiation. V. I. Feronsky has given interesting data[6] (see Tables 14.3 and 14.4).

Significant in groundwater studies is the capacity of the method to identify lithologies, especially those such as clayey or shale-bearing sediments which are characterized by high gamma intensities. Actually,

TABLE 14.3
RADIOACTIVITY EMITTED FROM SELECTED SEDIMENTARY ROCKS

Type	RaEq (10^{-12} g/g of rock)
Anhydrite	0·5
Brown coal	1·0
Rock salt	2·0
Dolomite	0·5–10
Limestone	0·5–12
Sandstone	1·0–15
Clayey sandstone	2–20
Clayey limestone	2–20
Carbonaceous claystone and shale	3–25
Claystone and shale	4–10
Potassium salt	10–45
Deep sea clay	10–60

TABLE 14.4
RADIOELEMENTAL CONTENT OF SELECTED SEDIMENTARY ROCKS

Type	Ra (10^{-12} g/g of rock)	U (10^{-6} g/g of rock)	Th (10^{-6} g/g of rock)
Sandstone	up to 1·5	up to 4·0	—
Quartzite	0·5	1·6	—
Clay	1·3	4·3	13·0
Claystone and shale	1·1	3·0	—
Limestone	0·5	1·5	0·5
Dolomite	0·1	0·3	—

most of the gamma radiation detected originates within 15–30 cm of the wall of the borehole. Consequently, gamma logs made before and after a well is developed can show zones from which clays have been removed. The natural relationship between a lithological type and its radioactive (gamma) content offers a means of detecting individual zones within an aquifer and, perhaps, its stratification.

Gamma measurements are influenced by a number of factors including the dimensions of the borehole concerned, the fluid content, the casing and the associated gravel pack.

14.2.3.2. Gamma–gamma Logging

In this method, a sonde is employed which contains both a source (gamma emitter) and a detector which measures scattered gamma radiation as the sonde is moved along a borehole (Fig. 14.2). The source usually comprises cobalt-60 or caesium-137 and it is shielded from a sodium iodide detector. The quantity of gamma radiation received by the latter is a function of the density of the surrounding fluid and rock; thus density can be assessed and, in addition, saturated materials can be distinguished from unsaturated ones. Porosity can be determined by:

$$n = \frac{\rho_G - \rho_B}{\rho_G - \rho_F}$$

where n is the porosity, ρ_G is the grain density obtained from cuttings or cores, ρ_B is the bulk density obtained from a calibrated log and ρ_F is the fluid density. Within the same geological formation, it ought to be possible to derive specific yield from the difference in bulk density measured above and below the water table. Gamma–gamma logs may

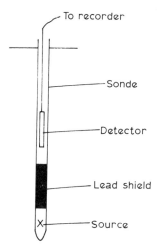

FIG. 14.2. Schematic diagram of a gamma–gamma sonde.

assist also in the placing of casing, collars, grout and zones of enlargement of the hole. Readings may be affected by such factors as fluid and borehole conditions.

14.2.3.3. Neutron Logging

The probe for moisture measurement consists of a radioisotope neutron source which emits fast neutrons and a detector of thermal or epithermal neutrons. The emitted fast neutrons are slowed down, thermalized and diffused in the surrounding geological material. The process is governed by elastic collisions with hydrogen ions and ends when the epithermal neutrons reach low energies around 1 eV. It is influenced by the hydrogen ion content of the geological materials and their bulk density. The former parameter is taken to represent the moisture content. Care must be exercised because some elements may occur which possess high cross-sections for the adsorption of thermal neutrons. In this case, the incidence of thermal neutrons is reduced.

In the source, an appropriate target such as beryllium is bombarded with alpha particles in order to produce neutrons thus:

$$(^9Be, {}^4He)\ ({}^{12}C, {}^1n)$$

Some appropriate alpha particle emitters are listed in Table 14.5.

TABLE 14.5
ALPHA PARTICLE EMITTERS

Isotope	Half-life (years)	Gamma output
^{226}Ra	1,620	High
^{227}Ac	22	Low
^{210}Po	138 days	Very low
^{239}Pu	2.4×10^4	Very low
^{241}Am	458	Very low

The yield of the neutron sources is of the order of 2×10^6–10^7 n/s Ci. The energy spectrum resulting from all α–n reactions in the source is continuous from 0 to 10 MeV. Usually, sources are cylindrical in shape with diameters between 10 and 20 mm, the activity being of the order of a few millicuries.

The detectors include gas proportional and scintillation counters. In a detector, the thermal neutrons interact with lithium or boron to cause emission of an alpha particle which is registered by the gas or scintillation counter. Lithium-loaded glass may be utilized as a scintillator and boron trifluoride (enriched in ^{10}B) gas may be used to fill proportional counters.

As well as operating in subsurface environments, neutron moisture probes may be employed on the surface of the ground. In the first case, they have slightly larger diameters than those indicated above and are inserted into aluminium access tubes pressed into the selected borehole. Surface probes work in a 2π geometry and neutron reflectors are placed above them so as to increase the neutron flux.

The lateral penetration of neutron logs is of the order of 0·2–0·6 m. This defines a 'sphere of importance', which is largest where the moisture content is lowest and vice versa.

One of the problems is calibration of the probe. This can be done in the laboratory in a number of 'infinite' models which are uniformly filled with a soil of known moisture content for a constant dry bulk density. Field calibration is possible by collecting and comparing many field measurements by the neutron probe with measurements made by other means on the same soil. Usually, a suitable calibration curve is supplied by the manufacturer. Accuracies to 0.01 g H_2O/cm^3 are obtainable.

14.2.3.4. Neutron–Gamma Logging
This records the gamma radiation emitted by nuclei in the formation

when they capture thermal neutrons emitted by a fast neutron source in the sonde (Fig. 14.3). Rough calculations indicate that neutron-capture gamma yield in a homogeneous medium depends upon the moderating properties of that medium. The technique is particularly suitable for detecting the boundaries between fresh and saline waters as a result of the fact that, if chlorine is present, high neutron-capture gamma yield is noted. An adverse factor is the presence of elements such as boron, cadmium, iron and potassium which decrease the apparent moisture content due to neutron-capture effects.

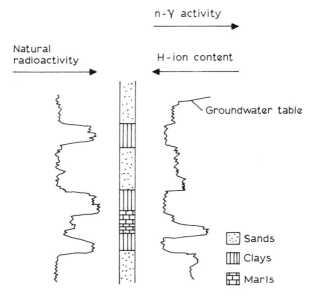

FIG. 14.3. Borehole logging using natural gamma logging (left) and neutron–gamma logging (right).

Groundwater is present in the zone of saturation, which it reaches through the overlying unsaturated zone. Shallow occurrences of subterranean water can be investigated using this approach. Also, soil density is determined by a transmission-type density gauge based on the principle of attenuation of gamma radiation by matter, thus:

$$I = I_0 \exp(-\mu \rho z)$$

where I is the transmitted intensity of gamma radiation, I_0 the primary intensity, μ the mass adsorption coefficient of the soil, ρ its density and z the distance between the detector and source of radiation.

Radiologging is important in determination of aquifer stratification from deep wells. Normally, there are three components, namely the logging sondes themselves, a surface unit for controlling them and recording data from them, and a cable on a winch which transmits information to the surface and also powers sondes and sources. Properly effected operations provide information regarding the profiles of formations penetrated by a borehole. These can be carried out at great depth and give high accuracy.

14.2.4. Temperature Logging

A recording resistance thermometer can be utilized in order to obtain a vertical measurement traverse of groundwater temperature which can be invaluable in analyzing the subsurface conditions.

As mentioned earlier (Section 8.1), temperatures increase with depth in accordance with the geothermal gradient of approximately 1°C per 30 m, but any variations on this can provide information about circulation or geological conditions in the well. Very cold conditions imply the presence of gas and high temperatures imply the occurrence of geothermal waters.

14.2.5. Caliper Logging

This gives a record of the average diameter of a borehole: the requisite tool may have four extensible arms with an electrical resistor motivated by them. The technique is to insert such a device with the arms folded into a well until the bottom is reached. The arms are then released by detonation of a small charge and the average hole diameter logged as a continuous graph by the recording of resistance changes while the caliper is being raised.

14.2.6. Fluid-conductivity Logging

This is a continuous record of the conductivity of fluid in a borehole. A probe measures the AC-voltage drop across two adjacent electrodes and relates it to the resistivity of the fluid between them. Fluid resistivity is assessed in ohm-m and its reciprocal, conductivity, is measured in $\mu S/cm$. The fluid conductivity log must be carefully distinguished from a resistivity log which measures rock and fluid conditions outside a borehole. Usually, temperature logging is effected in association with fluid-conductivity logging so that the values obtained from the latter can be correlated with a standard temperature. The main uses of this method are to identify saline water zones and to provide a means of extrapolating water sample data from a well. Also, it gives information on fluid flow within a well.

14.2.7. Fluid-velocity Logging

This entails measuring fluid movement within a borehole; the data so obtained show strata which contribute water, inter-stratal flow, casing leaks and hydraulic differences between aquifers penetrated by a well. There are several types of borehole flowmeter, all being compact and sensitive to water movement and direction. Vertical flow measurements utilizing suitable radiotracers can be made and have been discussed in Section 4.1

14.2.8. Television Logging

This involves lowering into a well a television camera of the wide-angle type; usually under 7 cm in diameter, equipped with lighting; thus a continuous visual inspection of the well is provided which can be recorded on videotape at the surface. A number of phenomena can be examined, including changes in subterranean geological beds, the state of the casing and screen, and so on. A photorecord of these is termed a photolog.

14.2.9. Casing Logging

The casing collar locator is useful in recording the locations of casing collars, perforations and screens.[2] The instrument comprises a magnet which is enclosed with a wire coil. Voltage fluctuations caused by changes in the mass of metal cutting the flux lines from the magnet are recorded to constitute the log.

14.2.10. Acoustic Logging

This is a sonic technique involving the measurement of the velocity of sound through a rock in which there is an uncased, fluid-filled hole. This velocity is governed by the fluid within pore spaces and also by the actual rock matrix itself, but, from the former, it is apparent that the more porous a rock is, the closer the measured velocity of sound is to that of the fluid. The main uses are in determination of the depth and thickness of porous zones, the estimation of porosity and the identification of zones of fracturing.

14.2.11. Environmental Isotope Logging

Discrete water bodies can be identified by employment of the environmental isotope logging technique and also by the stratification of aquifers. This stratification may arise because of aquicludes such as clays, or because of variation in density and the age of subterranean waters resulting from inadequate mixing.

In phreatic aquifers, a tritium gradient may be observed which reflects high input of this radioisotope in recent years, so that higher concentrations are found to occur in shallow levels near the water table. An example of the application of this approach is described by G. H. Davis in 1968,[7] with reference to results obtained from the Vienna Basin in Austria. This is a flat valley which, to the south, encompasses $1600 \, km^2$ and includes the city of Vienna as well as about one-third of the population of Austria. Geologically, it is a graben, a down-faulted block, bounded to the west by the limestones of the eastern Alps, to the east by crystalline rocks of the Alpine central zone and to the north by a structural ridge which brings impermeable Tertiary sediments to the surface to act as a barrier to groundwater flow. Bordering faults preclude subterranean flow. Thus, practically all inflow and outflow is by means of superficial streams. The main aquifer is a $35 \, km \times 4 \, km$ (length × width) Pleistocene body of gravel which varies in depth from 100 to 150 m, this acting as a pipeline receiving discharge from an alluvial fan at its upper end and discharging to streams in the lower portion. The isotopic results reveal the following data in regard to the area.

(a) All groundwater is from a common source, probably snowmelt in the Alps (shown by oxygen-18 stable isotope contents).
(b) The large, non-thermal springs of the Eastern Alps emit waters which comprise a mixture of current year recharge with an older baseflow component (shown by tritium analyses).
(c) The thermal springs which emerge along bordering faults represent a triple blend of current year recharge, recharge from the post-1965 period and antique water with a ^{14}C content of 30% modern (shown by tritium and ^{14}C analyses).
(d) The main aquifer has an upper, unconfined zone of approximately 20 m depth overlying a lower zone with uniform tritium content which indicates good mixing (shown by tritium analyses).
(e) The estimated transit time is about 4 m/day (this agrees well with the figure obtained using conventional methods).

Test wells in the Vienna Basin demonstrate marked tritium stratification. Those at Mitterndorf and Haschendorf were drilled by the percussion method and water samples were taken from the open-bottomed and cased wells as the drilling was being effected. Sufficient water was pumped to account for the volumes of water in the casings prior to sampling, in order to avoid any possible contamination from surface water which contains relatively high tritium contents. Records from the two boreholes indicate a rapid decrease in tritium content down to a certain depth, below which quite uniform low tritium values occur. It is interesting that this change corresponds to a lithological boundary. The results were interpreted as indicating a sharp separation between an overlying zone with incomplete mixing and an underlying zone of good mixing. It is believed also that there is lateral flow between the two wells.

Another isotope application is that in 1982 of M. A. Geyh and G. Michel, who studied the isotopic differentiation of groundwater of different hydrogeological origin.[8] Variations in ^{14}C, tritium and $\delta(^{13}C)$ values can differentiate young groundwater which was recharged in limestone from that recharged in sandstone areas. This is ascribed to the process of isotopic maturation of groundwater taking place at different rates in different geological environments.

On the topic of mixing, A. Issar in 1983 referred to the source of water from the thermo-mineral springs of Lake Kinneret in Israel.[9] He stated that chemical, ^{18}O and deuterium composition of these springs is similar to saline water from deep wells dug when exploring for oil in the Coastal Plain of that country, but different from the waters near the Dead Sea. It is believed that an ancient brine filling deep, non-flushed aquifers is driven from the south towards the Rift Valley by a piston action and is mixed with paleowater and contemporaneous meteoric water before emerging as the thermo-mineral springs in question.

The Sinai–Negev paleowater has been discussed by R. Nativ and his associates in 1983 and by A. Issar in 1985.[10,11] The Sinai–Negev peninsula is regarded as a miniature plate and contains artesian springs and wells separated by hundreds of kilometres which attest to the existence of an extension of the Nubia sandstone aquifer below them. At 'Ayun Musa, radiocarbon dating established the age of the spring water as about 30 000 years, and water from artesian wells in the Nubia sandstone near the Dead Sea is stated to have the same age. The aquifer is believed to contain 200×10^9 m^3 of water, of which about 35% is under the Negev. At present, approximately 23×10^5 m^3 annually of the Nubia sandstone water is utilized for industrial and agricultural purposes in Israel. One result could be to convert the Arava rift valley into a densely populated, agricultural area.

REFERENCES

1. STEWART, D. M., 1970. The rock and bong techniques of measuring water levels in wells *Ground Water*, **8**, 6, 14–18.
2. KEYS, W. S. and MACCRARY, L. M., 1971. *Application of Borehole Geophysics to Water-Resource Investigations.* US Geol. Surv. Techniques of Water-Resources Inv., Bk 2, Chapter E1, 126 pp.
3. KEYS, W. S. and MACCRARY, L. M., 1973. *Location and Characteristics of the Interface between Brine and Fresh Water from Geophysical Logs of Boreholes in the Upper Brazus River Basin, Texas.* US Geol. Surv. Prof. Paper 809-B, 23 pp.
4. TURCAN, A. N., Jr, 1966. *Calculations of Water Quality from Electrical-Logs Theory and Practice.* Water Resources Pamphlet 19. Louisiana Geol. Surv. Baton Rouge, 23 pp.
5. WORKING GROUP ON NUCLEAR TECHNIQUES IN HYDROLOGY, 1971. *Nuclear Well Logging in Hydrology.* Tech. Repts Ser. No. 126, IAEA, Vienna, 90 pp.
6. FERONSKY, V. I., 1968. Sratification of aquifers. In: *Guidebook on Nuclear Techniques in Hydrology*, Tech. Repts Ser. No. 91, IAEA, Vienna, pp 156–64.
7. DAVIS, G. H., 1968. Saturated zone. In: *Guidebook on Nuclear Techniques in Hydrology*, Tech. Repts Ser. No. 91, IAEA, Vienna, pp. 125–6.
8. GEYH, M. A. and MICHEL, G., 1982. Isotopical differentiation of groundwater of different hydrogeological origin. *J. Hydrol.*, **59**, 161–71.
9. ISSAR, A., 1983. On the source of water from the thermomineral springs of Lake Kinneret (Israel). *J. Hydrol.*, **60**, 175–83.
10. NATIV, R., ISSAR, A. and RUTLEDGE, J., 1983. Chemical composition of rainwater and floodwaters in the Negev desert. *Israel. J. Hydrol.*, **62**, 1/4, 201–23.
11. ISSAR, A., 1985, Fossil water under the Sinai–Negev peninsula. *Sci. Amer.*, **253**, 1, 82–9.

CHAPTER 15

Groundwater in Construction

15.1. GROUNDWATER AS A HAZARD

Groundwater may constitute a hazard in construction, for example in tunnelling, where its occurrence may necessitate extensive and expensive remedial work. This is especially manifest in soft rock tunnels which have to be drilled in unconsolidated materials of relatively low tensile and shear strength. Combating water influx in such circumstances may be a continuous activity and require the sinking of well points as a means of lowering the water table to an acceptable level at which it does not interfere with the labour. Well points comprise perforated, screened pipes usually of about 3·8 cm diameter and a metre or more in length. One of these can be attached to a riser pipe of the same diameter, the entire unit being inserted into the ground with or without jetting. Individual riser pipes are fixed to a header pipe or manifold which leads to a pump. Lines of such well points are arranged in accordance with the physical characteristics of the water table surface of the relevant unconfined aquifer and operate, basically, as de-watering equipment. In normal construction work, well points can be separated by 1 or 2 m, centre to centre, and attainable suction lifts may range from 4·5 to 7·5 m or more. Much depends upon the efficiency of the system, which in turn depends to some extent upon the altitude of the operation. Where excavation work is very deep, well points are normally arranged on a series of descending steps from 4·5 to 6 m high; this is termed a multiple-stage set-up.

The effects of groundwater influx into a soft rock tunnel are apparent from the excavation of Box Tunnel on the main line of the Great Western Railway between London and Bristol, England, commenced in 1836 under the direction of Isambard Kingdom Brunel and planned to be 3·2 km long. The water problem delayed its completion until June

1841; a lining of specially made bricks to the number of about 30 million had to be installed in the clay, blue marl and oolitic beds, by candlelight only (1 tonne of candles per week), and of course this added greatly to the costs.

Another instance in the UK is the Kilsby tunnel, which Robert Stephenson had to drain temporarily as a means of lowering the water table.

Occasionally, such lowering of the water table may have deleterious effects. This was the case in the Brooklyn area of Long Island, where there is a very high consumptive usage of water. Here, sewer construction and street paving work had promoted a drop in the level of the water table to as much as 10·7 m below sea level which led to seawater intrusion and saline contamination in a number of wells. Continuous monitoring of the situation from its inception in the 1930s accompanied rising anxiety, culminating in 1947 in cessation of practically all groundwater pumping for public supply. Thereafter, Brooklyn and its environs were supplied from surface water sources available through the New York Board of Water Supply. Groundwater was still pumped, but only for air-conditioning; after use, it was returned underground. Consequently, the water table gradually rose again; this caused flooding of basements and required costly remedial action. One way of avoiding deliberate lowering of the groundwater in areas where it might be adverse is to resort to freezing as an alternative.

15.2. BUILDING FOUNDATIONS

Groundwater can have significant effects in building foundations. For example, some of the very old piles of timber under Winchester Cathedral, in England, have decayed as a consequence of varying groundwater levels throughout the history of the building. Another case is the Pepoli Museum at Trapani in Italy which was constructed in calcareous tuff masonry and showed traces of stress at many points. The building was observed to be out-of-plumb and there was a bulging of all four sides of the porch and along the west end. According to F. Lizzi in 1982, the basic cause was subsoil instability movements arising from its nature and accompanied by variations in the water level due to exploitation of groundwater by nearby wells.[1] Root piles were used in remedial work. These are really friction piles which respond rapidly to any

movement, however small, of the sturcture being underpinned.
If rotting of piles under buildings is not attended to in time, expensive remedial work is mandatory. This happened with Strasbourg Cathedral, which was commenced in 1439 with stone footings supported on timber piles. A new drainage system was installed in the eighteenth century; this interfered with the local groundwater conditions and damaged the piles. Therefore, a large-scale underpinning operation had to effected in the 1920s.[2] A similar problem arose around 1929 with the Boston Public Library, Massachusetts, under which almost half of the piles had rotted. Approximately 40% of the building had to be underpinned.[3]

An interesting situation was encountered at Milwaukee, Wisconsin, where there is normally a rather high-level water table and the central part of the city is underlain by variable soils. Close to the waterfront is the head office of the Northwestern Mutual Life Insurance Company, erected in 1912. In 1930, it was proposed to add an extension which would complete an entire city block belonging to the company. This was founded on timber piles grouped under mass concrete footings. To ensure a satisfactory result, the architectural design provided every concrete footing with a 10 cm pile capped at the level of the sub-basement floor. Groundwater levels have been recorded in these every subsequent month. Where a drop in level occurs, water is introduced until the usual level is restored. Good results have been achieved, no doubt due to careful planning.[4]

Various grouts may be utilized for underpinning. For example, the lignochrome TDM was applied at Great Cumberland Place (Bilton Towers) in London, England. In this case, underground garages were planned for an erection including 14 floors, so deep excavations were necessary. Some of the adjacent buildings were over two centuries old and had to be protected. The relevant strata included the Taplow terraces, sands of various types associated with saturated gravels, these being underlain by the London Clay. After grout injection, practically uniform results in sandy gravels and adequate results of lower consistency in laminated sands were obtained.

Another example of unusually deep excavation work is that of the New York World Trade Center in Lower Manhattan, New York City. This cost US$600 million and consists of two buildings, the tallest in the world, supported by concrete piers which lie directly on the Manhattan schist, a rock with a high bearing capacity but often showing slippage along joint planes. The actual level of the schist is about 21 m below street level and therefore a six-storey basement had to be constructed.

From maps dating back to 1783, it was found that at that time the Hudson River shoreline ran two blocks inland of its present location. Many wharves and other buildings had been built on it; these were covered later by miscellaneous fill and lay on mixed beds of organic silt and sand, all which are water-bearing. As a result, groundwater occurs near the existing ground surface. In the prevailing circumstances of streets and buildings together with subterranean services, it was not possible to lower the water table around the selected site. It was decided to enclose this area using a concrete wall placed by means of a slurry trench, which had to be very carefully excavated and was infilled subsequently with bentonite and water. Concrete was placed in the trench from the bottom upwards and a tremie pipe was employed (tremie injection involves insertion of grout material through open-ended pipes embedded in a mound of the relevant material, a technique often utilized in cavernous rocks for constructing grout curtains). Steel grills were also placed in the trench so that they would become encased in the concrete and reinforce the structure. As the concrete rose, it displaced bentonite and reinforced the sides of the trench, thus making it safer. After excavation to total depth was executed, the walls were anchored by inclined steel tiebacks drilled and grouted into the underlying basement schist. All excavated material was transported to a nearby site in the Hudson River which was enclosed by a sheet pile retaining structure; after this dumping, the city had acquired a new land area which is worth at least US$100 million!

In some cases, building foundations have to be constructed in regions where there are sulphates in groundwater: an example is the St Helier Hospital at Carshalton, Surrey, England, built in 1938 with a capacity of 750 beds and comprising four multi-storey ward blocks with central services in a main block and subways connecting the various parts. Concrete foundations were placed on brown clay overlying the London Clay and, in the spring of 1959, some foundation concrete was exposed during maintenance work. It was seen to have deteriorated and a consequent study demonstrated that most of the foundation was deteriorating progressively. A high sulphate content was found in substratal water; although the beds concerned had quite adequate bearing capacity, damage had arisen from this other cause. Chemical attack had weakened the high-alumina cement content of the concrete, which had been affected also by a warm, damp atmosphere. A major underpinning operation had to be undertaken which necessitated the provision of new foundations for the entire building complex.

15.3. TUNNELS

One of the most important findings arising from tunnel construction is that actual lengths are much less significant than the characteristics of the rocks to be excavated. The hydrogeology is a key factor because groundwater is a serious potential hazard. An influx of groundwater into a tunnel can increase the construction costs by at least 20% because it will necessitate installation of waterproof lining, and also treatment of the drainage behind the lining. Such water may be hot enough to scald and even to kill. For instance, in the Tecolote tunnel, USA, in January 1951, drillers were almost drowned and suffered extensive burns after a methane explosion at the work face followed a hot-water influx. Another example is the Simplon tunnel, which was commenced in 1898 and completed in 1906 to constitute what is still the longest continuous railway tunnel in the world (over 19 km). At one stage, the work was halted by a water influx of almost 40 m^3/min. Water present at the level of the tunnel can be drained temporarily, but the usual approach is to install barriers and increase the air pressure in the working area, thus holding back groundwater during tunnelling. However, such an operation involves risk, including the possibility of personnel suffering 'bends'.

An interesting case arose during the construction of the Victoria underground line in London. At one particular locality, compressed air was in use in Woolwich and Reading Bed sands, which contain water and underlie the London Clay. This diverted 'fossil' air, entrapped in the voids of these sands, into a nearby tunnel which was being constructed simultaneously; as this second tunnel was oxygen-deficient, fatal injuries could have resulted—but, fortunately, they did not. Similar phenomena are recorded from Seattle, Washington and Melbourne, Australia.

When large-scale subterranean structures, such as power stations, are located in large caverns, it is necessary to drain the lining because it cannot resist groundwater pressure. On the other hand, external water pressure is sometimes appropriate in water-pressure tunnels. In sites below river valleys, drainage is not feasible because of very high inflows.

It is usually necessary to estimate the volume of inflows of water at various points specified along the proposed axis of a tunnel to be constructed. Their nature must also be assessed, i.e. are they likely to be dispersed influxes or concentrated flows? The work is even more vital with tunnels which are downwardly driven.

Water influx badly affected the railway tunnel between Bologna and

Florence in Italy. This tunnel was driven through the Apennines, and was excavated from both portals as well as from an inclined adit midway along the route. There was such a high influx of water into the downwardly driven section that the pumping station had to be enlarged more than once. In total, there were 37 pumps with a total discharge of 1200 litre/s before completion.

Inflow of water into tunnels from aquifers can take place through several inlets, of which the following are significant.

(a) Fractures resulting from the tectonic disruption of rocks (including joints and faults).
(b) Porous and permeable rocks, e.g. sandstone and conglomerate. Some crystalline rocks, e.g. granite, may become permeable if weathered, and some volcanic rocks are highly permeable, including vesicular basalts in which water occurs in the vesicles which were originally filled with gas.
(c) Karst openings, which can be very wide, in karstified limestones. Some of the limestone was karstic in the Harmanec tunnel, Slovakia, where, in January 1938, the face of the advanced head became flooded by a concentrated influx of about 400 litre/s.

Of course, it is extremely difficult to predict how much water will flow into a tunnel, but a reasonable estimate can be made if the area of the catchment is known accurately and also if details of the annual precipitation are available. Where a tunnel is constructed to convey water (i.e. a water-pressure tunnel), the groundwater pressure must be assessed since knowledge of it is essential for determining the appropriate thickness of the lining.

15.4. SETTLEMENT

Settlement occurred in Long Island, which is approximately 192 km long with a maximum width of 35 km. It constitutes a coastal plain which borders New York State and Connecticut and is divided into four counties, of which two (the smallest at the western end) constitute boroughs of the city of New York. Geologically, there is bedrock approaching the surface on the northern side, but this lies about 600 m below the surface on the southern side; overlying it is a set of beds which are mostly water-bearing with some aquicludes such as the Raritan and Gardiner clays. The superficial deposits, which were laid down in the

Pleistocene, are pervious so that water may seep down through them into lower sands and gravels. When the settlement commenced, groundwater conditions were in equilibrium. The infiltration of precipitation maintained the water table at a high level and the balance was preserved because of the movement of water through springs and also in small water courses on the island. There was probably some seepage from the sea as well. Under the sea, of course, the groundwater is saline. Consequently, there must be an interface under the island which separates underlying saline water from overlying fresh water. This may be about 300 m down. At first, the inhabitants obtained an excellent water supply from springs and rather shallow wells, using cesspools for waste disposal. These cesspools were allowed to drain back into water-bearing beds, which caused problems of pollution. Water was drawn from the Jameco gravel (Pleistocene) and the Magothy formation (Cretaceous) and, although pumping increased, the equilibrium was not disturbed seriously because most of the water returned to the water-bearing beds after use. However, as Brooklyn developed, servicing there entailed loss of much waste water to the sea or to Long Island Sound, so it was not being re-cycled. This caused a decline in the water table, which was corrected by restriction in groundwater usage and accompanied by flooding when the water table again rose after the resumption of recycling (see Section 15.1). The groundwater situation under Long Island shows all phases of the development at the western end discussed previously, but at the eastern end the natural conditions still exist.[4] This end is much less developed. The island is believed to constitute one of the largest usage regions for any one well-defined subterranean aquifer anywhere in the world. Construction activities on the island are involved closely with groundwater problems; these may be categorized as, from the east westwards, settlement of individual dwellings, usage of public wells, urban zone sewerage and seawater intrusion. Long Island is a very interesting location for groundwater studies and plans have been made for future conservation, including artifical recharge wells (some now operating, and additional ones), reclamation of water from sewage and even desalination of seawater.

Another instance of settlement in the USA occurred in the area including Norfolk around Hampton Roads in Chesapeake Bay, Virginia, where there are some very unusual traffic problems. A sunken tube tunnel, the Second Elizabeth River Tunnel, was opened in September 1962 at a cost of US$23 million. It is 1·275 km long from portal to portal, and comprises 12 precast concrete tubes which were floated into place.

The short approach section to the portals were constructed within cofferdams and the first stage of test boring took place in 1957, drillings being effected at 152 m intervals. This spacing pattern was considered satisfactory because of the rather uniform geological conditions prevailing locally. Borings were spaced at 61 m intervals on the shore, however, because of the greater geological complexity involved and also because meanders occurred in the relevant area. This spacing proved to be inadequate because the geology turned out to be so difficult that a second stage of exploration was necessary together with a rigorous testing of undisturbed soil samples derived from the site. Extra borings were made as well. A soil profile along the route of the tunnel showed a basal marl overlaid by clay and silt, the tunnel itself lying mostly in the clay. There was also a zone of organic silt running for about 53 m along the centre line of the tunnel. This zone was so limited that it had not been detected by test borings (they had missed it). It was very important, nevertheless, because of its high degree of incompetency; if subjected to load, it would have settled at least 15 cm over the long term. To avoid this, the zone was loaded with a fill surcharge 1·83 m high, left *in situ* for 15 months. Calculations showed that, during this time, excess load would promote settlement approximately the same as that anticipated under normal load. Actually, the settlement exceeded 15 cm slightly, after which the construction work proceeded quite satisfactorily. No subsequent settlement has been observed

15.5. DAMS

These are large-scale hydraulic structures, in fact the largest engineering accomplishments of Man, and they constitute potentially the most dangerous of all human constructions, as E. Grunner in 1963 indicated.[5] Their erection is intimately associated with the geological characteristics of a site, which includes not only the dam but also its impounded reservoir. There is a variety of dam types including embankment, concrete (arch and dome or multiple arch and dome), concrete gravity and gravity arch, concrete slab and buttress, and combined. All of these have foundations to which rock and soil mechanics theories are applicable.

The influence of groundwater on dam foundations is considerable. A case reported by K. E. Robinson in 1977 may be cited.[6] This is a tailings dam located on a very loose, saturated, sandy silt soil at Kimberley, British Columbia, the investigation and design for which were completed in 1974. It was intended to replace an existing embankment of marginal stability and allow for the extension of a waste pond to a total height of 30 m or so. The design was such that it permitted retention of iron tailings to the north in a first phase and gypsum tailings to the south later (after recovery of the iron tailings). Especially significant was the existence of a loose saturated layer of iron tailings which, in the design stage, was envisaged as forming the foundation for about 240 m of the dam length. Iron and silica tailings waste had been stored in an area to the north of the dam site for over half a century. South of the site, recent gypsum pond construction is expected to result in gypsum tailings contacting the southern edge of the dam. Over most of the length, gravel or glacial till constitutes the foundation. Several embankment failures occurred in the old iron tailings disposal area where embankments are about 0·8 km north of the new dam. The worst was a shear failure in the retention dyke which, in 1984, caused liquefaction of the contained tailings, approximately a million tons of which flowed down the valley towards the new dam site of which a part is located on up to 6 m of loose saturated tailings. It was found that gypsum tailings were accumulating to the south much faster than iron tailings.

The planned embankment had to satisfy several requirements, including adequate storage for iron tailings, containment of gypsum tailings and allowance for the future removal of the former. Stability under static and dynamic loading conditions had to be attained as well. The site is flat and grassy with occasional peaty or boggy ground and two groundwater zones in the foundation materials. Immediately above or at the surface of the bedrock, there is an artesian aquifer with a piezometric level exceeding 3 m above the level of the ground. Groundwater within the iron tailings is within 0·3–0·6 m of the surface. Glacial till provides a rather impervious barrier between the aquifers. There exists a very high water table associated with very loose tailings from which a 'quick' condition can develop easily if anyone so much as walks over them. During probe tests, the tailings did actually liquefy underfoot when a man stood too long in one spot. Previous liquefaction failures occurred as a consequence of dyke displacements when the phreatic level rose to critical height in late winter and the safety factor fell below unity. Resultant high pore pressures induced over-stressing in the tailings and the resultant displacements were sufficient to liquefy saturated sandy silt. Vibration-induced liquefaction of impounded tailings was considered and various concentrations of saturated materials at low relative density modelled at different positions relative to the dam face. Various positions of the phreatic surface within the tailings with regard to toe drains and the embankment were simulated in a wedge analysis, and foundation liquefaction was considered as well. The dam design had to be such that long-term stability conditions could be achieved and, in fact, stability was obtained during construction.

As a result of pore pressures built up during initial filling over the tailings foundations, it was thought advisable to monitor activities using piezometers and settlement plates. Normally, the piezometers were installed to depths equal to 75% of the thickness of tailings and daily measurements were made. As quick response was anticipated from the tailings, simple standpipe piezometers were used, comprising 1·3 cm PVC-coated conduits with slots cut in the lower 0·3 m, the slotted parts protected by a 200-mesh brass screen surrounding the pipe. In addition, the interior was sand-filled to a depth of almost 0·5 m. Settlement plates were located over the piezometer pipes. There was a response in the piezometers as the initial lift of fill approached to within a few metres of them and pre-pressure build-up was enough to cause overflow in the PVC tubes. Initial settlements were quite high.

Tailings dams can be dangerous. The earthdam near Stava, just over 100 km northeast of Trento, Italy, which collapsed on 19 July, 1985, killing hundreds of people and destroying three tourist hotels, ten homes and a number of holiday chalets along the 5 km long Fiemme Valley, may have been of this type since it impounded reservoirs owned by a fluorite mining company. The area is not too far from Longarone where 2000 were killed in the Vaiont dam disaster in October, 1963.

Groundwater may have to be drained entirely away from soil in some operations, e.g. the reclamation of marshes or slope stabilization. The problem is complicated by the fact that replenishment may take place by infiltration or by subsurface flow into the zone of saturation.

Dams and other large-scale hydraulic structures are influenced greatly by water; beds show a decreasing bearing capacity as their water content increases. Shales are a special case because they possess an extremely low modulus of elasticity, 2·07 minimal to 68·31 maximal, in $kN/m^2 (\times 10^6)$ and demonstrate plastic behaviour so that, under great load, they can be squeezed out from beneath the foundations of buildings. Shales slake on contacting water while drying, although there is one type which resists weathering and does not slake: it is stronger and has a higher bearing capacity.

In addition, groundwater may exert adverse pressures affecting foundations. Λ. W. Bishop and his associates in 1963 discussed the development of uplift pressures downstream of a grouted cutoff during the impounding of the Selset reservoir behind an earth dam constructed in a valley containing boulder clay.[7] At this site, a concrete-filled cutoff trench extended by a cement grout curtain failed to prevent uplift pressure and relief wells had to be installed. The reservoir was located in a valley which is blanketed by the boulder clay; artesian conditions existed underneath prior to construction, the bedrock being Carboniferous limestone. The area concerned is in England north of the Cotherstone syncline, so that rocks dip south at about $10°$ and consist of alternating shales, sandstones and thin limestones overlying massive limestone. The latter outcrops within the area of the reservoir and no attempt was made to extend the cutoff into it. The shales are laminated and have fine fissures. The sandstones have bulk permeabilities assessed as of the order of 10^{-3} to $^{-4}$ cm/s and the bulk permeabilities of the limestone are larger by at least an order of magnitude. The boulder clay has a very low permeability, the measured coefficient of this parameter *in situ* being in the range 10^{-7}–10^{-8} cm/s. The groundwater flow is of course restricted to rocks underlying the boulder clay and produces several springs, but there are rather few of these so that artesian pressures built up over the general area. It is clear that the groundwater conditions are an important factor in determining the cutoff problem at any dam which requires extensive hydrogeological investigation in the relevant region.

Uplift problems may occur with concrete dams also, as happened with the famous Boulder Dam in the USA. In that case, the extent of uplift diminished after reconstruction of the grout curtain. Just occasionally, for instance in a case in Czechoslovakia, water pressure tests actually initiate upheaval of bedrock surfaces as described by J. Verfel in 1957.[8]

Water may have significant effects in loess, an aeolian and mainly siliceous soil of glacial or desert origin and a very well graded material. Loess consolidates rapidly when wetted and, if supporting an earth dam, can cause this structure to undergo settlement. This phenomenon is termed hydroconsolidation and arises from one of two causes. In the USA, loess is associated with the Pleistocene glaciations and possesses clay films around its constituent silt-sized particles. Added water lubricates these and promotes sliding between them.[9] Elsewhere, settlement may result from a combination of this lubrication effect and also a removal by the water of calcium carbonate cement from the loess.[10] Loess is extremely porous and also has very high permeability, which is greater vertically than horizontally because of the presence of vertical tubing originating from plant roots. Settlement problems of this type are resolved by using appropriate grout, applied for example by pumping a loess–bentonite slurry into boreholes drilled into the foundation material; in Nebraska, USA, local materials were employed in this technique and effected an increase in the weight of loess by 20% after treatment.[11]

15.6. EXCAVATION BENEATH ROTTERDAM

This is a very interesting case described by Robert F. Legget.[4] Rotterdam is a beautiful city in the Netherlands with a tasteful blend of antiquity in its older streets and modernity in its elegant centre which was rebuilt after the Second World War during which it had been devastated by low-flying German bombers. There is a river, the Nieuwe Maas, which handles a huge volume of traffic as well as the largest port in the world dealing with ocean-going ships and Rhine River vessels. There are bridges over the Nieuwe Maas, but they are unable to cope with the great numbers of passengers travelling to and from central Rotterdam. Hence, between 1938 and 1942, a vehicular tunnel was constructed under it. Later, another tunnel was driven to form part of a rapid-transit urban subway system.

The geology consists of a Pleistocene sand and gravel deposit 16–18 m below local datum (mean sea level) transported by two rivers, the Rhine and the Maas. It is water-bearing, sometimes extending to a depth of 40 m in the east of the city. It has been utilized as a foundation for many city structures, which transmit their loads through piles. Overlying the sand and gravel is an alluvial clay about 1 m thick surmounted by a peat layer which, in turn, is succeeded by a marine clay roughly 2 m thick. Manmade fill concludes the sequence to the surface of the ground, on which roads, etc., have to be founded. Settlement is a constant problem and must be compensated for by infilling and rebuilding. The general level of Rotterdam is below sea level, so everywhere groundwater is near the surface and, in fact, is never more than 2 m below the ground. However, the level is quite constant and therefore there is practically no deterioration in the many timber piles used. Of course, deep excavations seriously interfere with the water table; this is the main problem encountered in the construction of the subway system mentioned above. All rapid-transit lines on the north side of the Nieuwe Maas in the city centre had to be constructed in a tunnel emerging into an elevated line south of the river. The initial 7·6 km of tunnel had 3·1 km in the tunnel and 4·5 km on a pre-stressed reinforced concrete superstructure. The bottom of the tunnel was 11 m below ground; construction was hazardous due to the propinquity of many piles and also due to the high water table. It was calculated that excavation of open trenches using cut-and-cover methods would require a large-scale pumping operation for every section of trench over at least five years, even with controlled pumping and efficient well points. To avoid this, it was decided to build inner sections of the subway tunnel as pre-cast concrete tubes and float them into position in the centre of the city. This was achieved and this part of the project was completed by 1967. The first section of the subway opened on 9 February, 1968.

15.7. PERMAFROST

This is permanently frozen ground below 0°C; ice need not be present. Where it is not, dry permafrost exists, but this does normally occur.

The upper limit of perennially frozen ground, the permafrost table, is almost impermeable. Continuous permafrost comprises material with an annual temperature of $-5°C$ at 10 m depth. The permafrost table is usually 0·5 m or so below ground, but, in granular materials, it may be deeper. Discontinuous permafrost is much thinner and may be interrupted in places by thawed areas; it has a lower permafrost table and the mean annual temperature at 10 or 15 m depth is between $-5°C$ and $-1·5°C$.

Above the permafrost table, there is an active layer subjected to alternate freezing and thawing, a zone where freezing may take place from the bottom upwards as well as from the top downwards and one with the maximum fluctuation of temperature. It is termed the frost zone, its irregular lower surface constituting the frost table. If the active layer and frost zone leave a residual thawed ground zone between the frost and permafrost tables, this so-called 'talik' may act as a viscous liquid and flow.

The importance of this subject is that permafrost or frozen ground underlies about one-fifth of the entire land surface of the Earth, mostly in the northern hemisphere, where there is much more land. Drilling has been effected in such areas and geophysical methods such as electrical resistivity and seismic techniques have been applied to the problem of determining the thicknesses of the various zones.[12] Thermal profiles have been constructed using glass thermometers, thermocouple resistance thermometers and thermistors.[13]

Temperatures show their widest fluctuation at the surface and in the active zone in response to daily and seasonal changes in atmospheric values. This fluctuation decreases with depth until a level of zero annual amplitude is reached, below which temperature increases with depth. Permafrost reaches a maximum thickness of 1500 m in Siberia, but in Europe and Asia its average is 300–460 m and, in North America, the range is from 245 to 365 m.[14]

The planetary distribution of frozen ground reflects a thermal equilibrium arising from the relations between the climatic regimen, heat flow from the interior of the Earth, the properties of soils and rocks and the presence of water. Climate is probably the most important factor, and optimum conditions for permafrost development are cold, long winters with little snow followed by short, dry and rather cool summers.[15] Attempts have been made to predict the depth of freezing and thawing using climatic indices. One of these is the freezing index, based on a cumulative total of degree-days, being the area under the curve between the maximum and minimum points on a cumulative curve constructed from appropriate temperature data plotted against time in a freezing season. There is a satisfactory correlation between the freezing index and the penetration of frost into the ground when dry unit weight and moisture contents of the soils are taken into account. Although the existing thermal equilibrium can be correlated with prevailing conditions, there is no doubt that this equilibrium has been in existence over thousands of years with its general pattern probably being modified by secular changes in climate and other controls. The significant point is that construction promotes degradation or aggradation, involving usually irreversible changes and initiating a new thermal equilibrium in the ground. In regions of permafrost, the groundwater is in the ice phase; this ground ice is important, taking a variety of forms as listed below.

(a) Open-cavity ice forms by sublimation of ice crystals directly from atmospheric water vapour.
(b) Single-vein ice forms when water enters and freezes in an open crack penetrating permafrost.
(c) Ice wedges form through the thermally induced cracking of frozen ground.
(d) Tension-crack ice grows in cracks resulting from mechanical rupture of the ground.
(e) Closed-cavity ice forms by vapour diffusion into enclosed cavities in permafrost.

(f) Segregated ice grows in lenses in material older than itself (epigenetic ice) or develops when the permafrost table slowly rises (aggradational ice).
(g) Intrusive ice forms by the intrusion of water under pressure which, on freezing, promotes ground heave.
(h) Pingo ice forms when intrusive ice domes the overlying ground surface.
(i) Pore ice forms and holds soil grains together.

The processes and control of frost action and solifluction necessitate the investigation of frost heave in association with segregated ice in supersaturated materials. Ground ice may cause vertical displacement of the ground surface, which may disrupt engineering structures by many centimetres and require expensive remedial action. Also, vertical particulate sorting may take place with upward migration of large stones from mixed sediments to above ground, where they create stone circles or nets. Ground heave may produce hummocks (palsas or mima mounds) which can attain heights of 10 m. Frost heaving can be reduced by chemically treating soils, e.g. by injecting calcium chloride to reduce the freezing temperatures of water in soils and minimize loss of strength, also by grouting with polymers or resins and, of course, by proper structural design measures such as increasing the strength of pavements by thickening them. Solifluction is a phenomenon combining flow and slip which involves the surface layer on a frozen substratum and takes place in the Arctic and high mountains. Thaw in such regions occurs to depths of only 50 cm or so during the short summers and meltwater accumulates because the ground is impenetrable, remaining frozen. The overlying, waterlogged bed later flows downhill under gravity as a dense sludge. Solifluction was much more common in Pleistocene ice stages than it is now, but still may constitute an engineering hazard by disrupting linear objects such as cables and pipes inserted through the mobile layer. Mostly, it produces quite characteristic landforms such as terraces, which may be delimited by stone banks covered by vegetation, occurring on slopes exceeding 5°.

Permafrost can be ignored where adequate drainage conditions exist and ground materials are not susceptible to severe frost action and solifluction. Sometimes, where permafrost is thin, sporadic or discontinuous and the thawed ground has adequate bearing capacity, frozen ground may be removed by thawing the ice by means of steam, by excavation or by otherwise treating the ground. Material not susceptible to frost action may be placed on the surface; this is an approach termed the active method, as opposed to another, the passive method, in which the permafrost may be preserved by insulating the surface using gravel blankets or vegetation mats.

In siting buildings, optimum conditions include thin active layers, bedrock near the surface, thawed materials with adequate bearing capacities, good drainage, absence of frost-susceptible materials, and a stable site removed from potential water-seepage and icing. An important factor is the adfreezing strength of frozen ground, i.e. the resistance to the force necessary to pull the frozen ground apart from objects such as foundations to which it is frozen. This strength is usually higher in sands than in clays. It is significant in emplacement of piles, which must be frozen into the ground to a depth of at least twice the thickness of the active zone with adfreezing avoided in the active zone by insulation or lubrication in that zone. Upward pile movement is prevented by a combination of the downwardly acting force in the permafrost zone together with the superincumbent load. Other techniques include trenching round buildings so as to eliminate possible horizontal stresses, elevating the floor by a metre or more so as to permit circulation of air, insulating the floor to restrict temperature rise underneath it, and providing skirting insulation between the ground and the floor with air vents which can be closed in summer and opened in winter in order to allow free air movement. Occasionally, jacks are employed so as to adjust the level of a building if there is any movement. Pads of concrete, wood or gravel may be sited on natural vegetation to procure an insulating foundation for small temporary buildings. Similar problems may arise in road or airport runway construction, where the passive technique requires the creation of adequate natural or artificial insulating layers whereas the active technique necessitates the replacement of frost-susceptible surface material with surface material which is not frost-susceptible.

REFERENCES

1. Lizzi, F., 1982. *The Static Restoration of Monuments.* Sagep, Genoa.
2. Anon, 1923. Modern engineering to save medieval tower. *Engrng New-Record*, **91**, 505.
3. Anon, 1965. Foundation reinstatement without interference to services. *The Engineer, London*, **220**, 791.
4. Legget, R. F., 1973. *Cities and Geology.* McGraw-Hill, New York, p. 233.

5. GRUNER, E., 1963. *Dam Disasters*. Inst. Civ. Engrs, London.
6. ROBINSON, K. E., 1977. Tailings dam constructed on very loose saturated sandy sill. *J. Can. Geotech. Soc.*, **14**, 3, 399–407.
7. BISHOP, A. W., KENNARD, M. F. and VAUGHAN, P. R., 1963. The development of uplift pressures downstream of a grouted cutoff during the impoundment of the Selset reservoir. In: *Grouts and Drilling Muds in Engineering Practice*. Butterworths, London, pp. 98–105.
8. VERFEL, J., 1957. *Provadeni Vodnich Tlakovych a Injekenich Zkousek*. Brno.
9. KRYNINE, D. and JUDD, W. R., 1957. *Principles of Engineering Geology and Geotechnics*. McGraw-Hill, New York.
10. SCHIEDIG, A., 1934. *Der Loess und Seine Technische Eigenschaften*. T. Steinkopf, Leipzig.
11. JOHNSON, G. E., 1953. Stabilization of soil by silt injection method. *Proc. Amer. Soc. Civil Engrs, J. Soil Mechs and Fdns Divn*, **79**, 323, 1–18.
12. BARNES, D. F., 1966. Geophysical methods of delineating permafrost. *Proc. Permafrost Intl Conf.*, NAS/NRC Publn 1287, pp. 349–55.
13. HANSEN, B. L., 1966. Instrument for temperature measurements in permafrost. *Proc. Permafrost Intl Conf.*, NAS/NRC Publn 1287, pp. 356–8.
14. BLACK, R. F., 1954. Permafrost — a review. *Bull. Geol. Soc. Amer.*, **65**, 839–55.
15. MULLER, S. W., 1947. *Permafrost or Permanently Frozen Ground and Related Engineering Problems*. Edwards, Ann Arbor, Michigan.

CHAPTER 16

Groundwater Modelling

16.1. MODELS FOR GROUNDWATER CHARACTERISTICS

These have been devised and utilized especially to study, under natural and artificial boundary conditions, the flow and distribution of groundwater, a water resource which, being subterranean, cannot be observed directly and therefore must be analyzed by simulation. The applications of groundwater models of various types have been discussed by T. A. Prickett in 1979.[1] They can be classified as porous media models, non-electrical analogue models, electrical analogue models based upon the similarity between Darcy's law and Ohm's law, and digital computer models capable of providing numerical solutions of aquifer flow equations as well as improving the efficiency of management of large-scale resources of groundwater.

16.2. POROUS MEDIA MODELS

These are hardware models and entail the physical construction in the laboratory of appropriate scale representations of the aquifer with its boundaries included, but with the spatial distribution and value of its permeability suitably modified.

Sand models have been used for many years; indeed, the first one dates from 1898, when P. Forscheimer made it in order to study flow in a well in Graz, Austria. They have been made in many different sizes and shapes, often rectangular, in corresponding watertight boxes. Unconfined aquifers can be modelled with the water table acting as the upper boundary. Confined aquifers are replicated by using an impermeable cover which permits application of pressure. Since it is difficult to see the water table, both this and, in the confined aquifer model, the piezometric surface are identified from small piezometer tubes inserted into the models. These are kept small so that there will be minimal interference with the patterns of flow. Chemical dyes, e.g. potassium dichromate, can be employed in order to reveal them visually. It has been noted that coarse sand which has been placed below water in small amounts and consistently compressed to remove included air gives a uniform permeability inside the model. With layers of sand of different grain size, it is perfectly possible to obtain anisotropic conditions and various permeabilities.

It is noteworthy that the capillary rise which takes place in such models is far larger than that which actually occurs in a real field situation, but this effect is not too important if confined aquifers are simulated. On the other hand, in the case of an unconfined aquifer, it is more significant and some correction may have to be applied. In any case, the extent of the capillary rise can be reduced by utilizing a coarse-grained medium with high porosity in association with a more viscous liquid. Velocities and flow rates are obtainable by the application of Darcy's law (Section 4.1); the flow rate will be given by dividing the volume of the fluid by time, this latter deriving from considerations of specific yield and model dimensions, the product of which is divided by the measured hydraulic conductivity.

It is possible also to make a transparent porous media model, by matching the refractive index of the fluid with that of the medium, e.g. employing crushed glass in association with a suitable mineral oil, both being contained in a glass vessel. In the case of such a transparent model, it is feasible to use chemical dyes and visually to observe the flow of the fluid, in particular its dispersion.

16.3. NON-ELECTRICAL ANALOGUE MODELS

A natural analogue is an analogous natural system which is both simpler and easier to observe than the original system. Representative and experimental basins provide natural analogues of this type and the final report of the International Hydrological Decade inter-governmental meeting of experts issued by UNESCO in 1964 may be consulted for further details.[2]

Non-electrical analogue models include the following.

16.3.1. Viscous Fluid Models

Since the nineteenth century, it has been known that the flow of groundwater may be simulated using two closely spaced parallel plates, the movement of a viscous fluid between them paralleling that of groundwater flow in two dimensions. In the case of laminar flow, the corresponding flow lines comprise a two-dimensional potential flow field and, using the Navier–Stokes equations of motion, for vertical plates and steady flow, and taking into account the Dupuit assumptions, the mean flow velocity may be obtained from:

$$v_m = -\frac{b^2 \rho_m g}{12 \mu_m} \frac{dh}{dx} \qquad (16.1)$$

where b is the distance between the plates, ρ_m is the fluid density, μ_m is the viscosity, g is the acceleration due to gravity and dh/dx is the hydraulic gradient. By analogy with Darcy's law, the hydraulic conductivity is given by:

$$K_m = \frac{b^2 \rho_m g}{12 \mu_m} \qquad (16.2)$$

indicating that the plate separation and fluid can be so chosen as to correspond with any permeability required. In order to maintain laminar flow, the plates are usually kept about 1 mm apart.

J. Bear in 1960 gave appropriate scales of viscous analogy models for the study of groundwater.[2]

The models are made from two sheets of glass or plastic mounted a fixed distance from each other with end-reservoirs to control the flow of fluid, which may be oil or glycerin, sometimes with added chemical dyes. These dyes define the free surface for unconfined flows, and point sources of them placed along inflow boundaries demonstrate flow lines. If the separation of the plates is small enough, water may be utilized because it also will then undergo laminar flow.

Vertical viscous fluid models have been constructed in order to represent vertical cross-sections through an aquifer, facilitating analyses of seepage and drainage, the intrusion of seawater, artificial recharge and other phenomena. R. G. Thomas in 1973 gave an interesting account of such groundwater models and M. A. Collins and his associates in 1972 made a Hele-Shaw model of the Long Island aquifer system.[3,4] The Hele-Shaw model is another name for the viscous fluid model; another synonym is the parallel plate model. H. S. Hele-Shaw made experiments on the nature of surface resistance in pipes and on ships in 1897 from which this model developed.[5] A great advantage of this type of model is that not only steady-, but also non-steady-state free surface and flow directions of a viscous fluid can be studied by its application. Since the rate of flow varies with the cube of the width of the model, it is clear why the plates must be no further apart than 1 mm if laminar flow is to be achieved. In addition, in order to maintain constant fluid properties, the temperature must be controlled. Lastly, it may be desirable to introduce corrections for capillary effects.

16.3.2 Membrane Models

In 1952, V. E. Hansen indicated that complicated well problems can be solved by using the membrane analogy.[6] A thin stretched rubber membrane has small slopes on its surface which are expressible in polar coordinates, thus:

$$\frac{d^2z}{dr^2} + \frac{1}{r}\frac{dz}{dr} = -\frac{W_m}{T_m} \qquad (16.3)$$

where dz is the deflection at a radial distance dr from a central deflecting point, W_m is the weight of the membrane per unit area and T_m is a uniform membrane tension. The Laplace equation in cylindrical coordinates is given by:

$$\frac{\partial^2 h}{\partial r^2} + \frac{1}{r}\frac{\partial h}{r} + \frac{1}{r^2}\frac{\partial^2 h}{\partial \theta^2} + \frac{\partial^2 h}{\partial z^2} = 0 \qquad (16.4)$$

in which h is the hydraulic head, r is the radial distance, θ is the angular coordinate and z is the vertical coordinate. In the case of steady, axially symmetrical flow in an incompressible fluid (steady radial flow to a well in a homogeneous, isotropic aquifer), then:

$$\frac{d^2h}{dr^2} + \frac{1}{r}\frac{dh}{dr} = 0 \qquad (16.5)$$

The analogy between this and eqn (16.3) is close when the weight of the membrane is very small or when it is placed vertically. Equation (16.5) is a reasonable approximation to well flow in the case of an unconfined aquifer where there is only a small drawdown.

In actual studies, the membrane is clamped under uniform tension over a circular boundary positioned horizontally. A central probe deflects the membrane and represents a pumping well, the deflection being analogous to drawdown. Deflections can be measured by micrometer or other means and the approach can be adapted for unconfined well flow or for multiple well systems as well as for complex aquifer boundary conditions.[6,7]

16.3.3. Moiré Pattern Models
Although these are mainly useful for demonstrations, they are also valuable for displaying equipotentials or streamlines of two-dimensional groundwater flow situations. A Moiré pattern is an optical phenomenon which takes place when two sets of grid lines are superimposed; it is formed by the loci of points where the two grids intersect. R. A. Freeze in 1970 utilized such patterns.[8]

16.3.4. Blotting Paper Models
R. J. Sevenhuysen in 1970 employed this device to simulate groundwater flow. The technique is to hang blotting paper, cut to a suitable shape, in a vertical position and so that it is contacted by water introduced through other saturated sheets at selected boundary edges.[9] If an appropriate pattern of ink dots is utilized, the flow pattern is demonstrated.

16.3.5. Thermal Models
This is a very interesting application of the fact that the heat flow in a uniform body of material accords with the Laplace equation and, therefore, moves as a potential flow system in a manner similar to that of groundwater. The analogous features are listed in Table 16.1.

TABLE 16.1
ANALOGIES BETWEEN AQUIFERS AND THERMAL MODELS

Aquifer parameter	Thermal model analogy
Hydraulic conductivity	Thermal conductivity
Flow rate of water	Flow rate of heat
Head	Temperature
Storage coefficient	Model thickness × density × specific heat

The thermal model was utilized in order to investigate the theory of transient flow to a partially penetrating well by I. Javandel and P. A. Witherspoon in 1967. Non-steady flow was involved and a thick steel slab was taken to represent the aquifer, this being placed between styrofoam insulation which represented confining layers.[10] There was a heater in the steel slab and the temperatures at various distances were measured by thermocouples.

The major disadvantage of this model is that it is rather difficult to design and supply with appropriate instrumentation.

16.4. ELECTRICAL ANALOGUE MODELS

These are hardware models based upon the analogy between Ohm's law and Darcy's law. The flow of an electrical current is expressed by the former thus:

$$I = -\sigma \frac{dE}{dx} \quad (16.6)$$

where I is the electrical current per unit area through a material possessing a specific conductivity σ, dE/dx being the voltage gradient. This satisfies the Laplace equation and is comparable with the following form of the Darcy equation:

$$v = -K \frac{dh}{dx} \quad (16.7)$$

from which it may be seen that the velocity, v, is analogous to the electrical current, I, the hydraulic conductivity is analogous to the specific conductivity and the head corresponds to the voltage.

There are two approaches: either a continuous system or a discontinuous system may be employed.

16.4.1. Continuous Systems

In these, the properties of an aquifer are simulated using an electrical conductive medium which is spatially continuous; the possible liquids include copper sulphate, or a conductive solid such as carbon can be employed.

The conductive liquid technique involves filling an insulated tank with a suitable electrolyte. If dilute $CuSO_4$ is used, copper electrodes are inserted so as to create equipotential surfaces and lines of constant potential drop are traceable by a probe which is connected to an oscilloscope, a voltmeter circuit and a pantograph. Of course, the boundaries of the tank are so scaled as to represent the boundaries of the aquifer. The equipotential and flow lines are mappable by reversing the conducting and non-conducting boundary surfaces.

Such models usually demonstrate two-dimensional steady-state conditions, but they can be modified in order to examine multiple aquifers and even to study three-dimensional cases. Unfortunately, there is no analogous substitute for gravity and hence, no water table is produced; however, this may be located on the basis that the decrease in head is proportional to the decrease in electrical potential along the surface. The method has been utilized by, among others, B. E. Debrine in 1970[11] and D. K. Todd and J. Bear in 1961.[12]

Conductive solid models may utilize pressed carbon to represent radial flow to a well, or a saline gelatin mixture for three-dimensional flow to a partially penetrating well. Models may be designed so as to represent desired aquifer boundaries and, usually, they are applied to the study of steady flow in confined aquifers; see for instance V. M. Shestakov in 1968.[13]

16.4.2. Discontinuous Systems

The two main types in this group are discussed below.

Firstly, there is the resistance–capacitance network. Such networks are valuable in investigating confined aquifers under non-steady, two-dimensional flow conditions and they permit such water reservoirs to be modelled together with associated effects from pumping wells. The aquifer is represented by individual electrical elements forming a small-scale model of it, suitable electrical voltage and current sources being connected to individual junctions (nodes) of the network so as to create sources, external boundaries and so on for the water-bearing body. Voltage-measuring instruments assess the distribution of voltage through the network. After a scale conversion, the head distribution through the aquifer may be obtained.

The analogy between the two-dimensional groundwater flow in a homogeneous isotropic confined aquifer and the flow of electrical current in a resistance–capacitance network is apparent from a comparison of the finite-difference form of the non-steady, two-dimensional equation which describes the former, i.e.:

$$T(h_2 - h_3 - h_4 - h_5 - 4h_1) = a^2 S \frac{\partial h_1}{\partial t} \qquad (16.8)$$

where T is the local aquifer transmissivity in m^2/day, the various heads are represented by h with subscripts referring to locations of nodes, a is the length of each side of one (square) unit of a finite-difference grid which is superimposed over the aquifer, S is the local aquifer storage coefficient and t is time, and the relationship from Kirchhoff's current law, which is given for node number 1 by:

$$\frac{1}{R}(V_2 - V_3 - V_4 - V_5 - 4V_1) = C\frac{\partial V_1}{\partial t} . \qquad (16.9)$$

where the V terms, with subscripts, refer to the voltages at the corresponding (numbered) nodes.

There are additional analogies as well, e.g. amount of water is analogous to electrical charge, head loss corresponds to drop of voltage, etc., so it is possible to state that an electrical pulse generator is analogous to a water pump and the oscilloscope acts as a water-level recorder.

Such resistance–capacitance network analogues are applicable to a number of aquifers, not just one, because the number of nodes can be increased to cover them. Three-dimensional cases can also be examined and it is possible to include non-homogeneous conditions. The only real problems arise in considering non-linear conditions of varying transmissivity in unconfined aquifers and two-fluid flow situations.

The second type of discontinuous system comprises resistance networks. These resemble the resistance–capacitance networks; models consisting of an electronic analyzer coupled to an analogue model, but this model comprises a set of resistors only;[14] capacitors are absent and so no storage elements are necessary, the analogue being applicable only to steady-state conditions.

16.5. DIGITAL COMPUTER MODELS

Digital computers have become ever more widely distributed and used in recent years. They may be applied to the mathematical modelling of aquifers and, as their capabilities increase and the techniques of programming improve, more and more types of groundwater situation can be simulated; see for instance T. A. Prickett and C. G. Lonnquist in 1973.[15] There are several approaches, some of which will be discussed below.

The finite-difference method is based upon division of an aquifer into a grid followed by analysis of the flows which are associated within a single zone. The flow equation derives from the hydrologic (continuity) equation discussed earlier (Section 1.1)

$$I - O = \pm \Delta S \qquad (16.10)$$

where I is the total infiltration, O is the total outflow and ΔS is the change in storage. For a sufficiently restricted part of an aquifer, it is feasible to re-state this in the following form:

$$\sum \text{subsurface flows} + \text{net flow to or from the surface} = \Delta S \qquad (16.11)$$

Involving Darcy's law for the equation of motion provides:

$$\frac{\partial}{\partial x}\left(T\frac{\partial h}{\partial x}\right) + \frac{\partial}{\partial y}\left(T\frac{\partial h}{\partial y}\right) - Q = S\frac{\partial h}{\partial t} \qquad (16.12)$$

where T and S are the transmissivity and storage coefficient of the aquifer, Q is the net external inflow, h is the head and t is time. This equation can be expressed in finite-difference form, thus:

$$\sum_i \frac{W_{iB} T_{iB}}{L_{iB}} (h_i^{j+1} - h_B^{j+1}) - A_B Q_B^{j+1} = \frac{A_B S_B}{\Delta t}(h_B^{j+1} - h_B^j) \qquad (16.13)$$

in which W, T and L are the zonal boundary width, transmissivity and length of flow path respectively, A is the area of a single zone, the superscript j denotes points along a time coordinate with Δt being one step and the subscripts i and B refer to a contiguous zone and the zone in question respectively. Q is a quantity representing the algebraic sum of the extraction flows (i.e. pumpage) and replenishment flows (i.e. precipitation, irrigation excess, imported water, stream percolation and artificial recharge). Knowing the zonal configurations which define the values of W, L and A, together with estimates from hydrogeological information for S, T and Q, the time variations of h over the aquifer can be computed by solving the system of simultaneous equations. Past records of the levels of the relevant groundwater aid in verifying the model: S, T and perhaps Q are adjusted as necessary, so that agreement is attained between the calculated water-level responses and the actual data from the area. In this way the model is calibrated, and thereafter it can be utilized to examine the dynamic behaviour of a basin so as to design future conditions of operation. Probably the first digital computer model using this finite-difference method was the one developed by the California Department of Water Resources as a means of studying the dynamic behaviour of the Los Angeles coastal plain groundwater basin which covers about 1240 km^2. Subsequently, other means have been devised for solving the simultaneous equations.[16-18] Many groundwater topics have been covered using this method, including well flow, unsaturated flow, dispersion, saltwater intrusion, land subsidence, mass transport and management.[19-27] The finite-element method involves solution of a differential equation for the flow of groundwater using variational calculus; see for instance G. F. Pinder and W. G. Gray in 1977.[28] The equation for two-dimensional, non-steady groundwater flow in a non-homogeneous aquifer is given by:

$$\frac{\partial}{\partial x} K_x b \frac{\partial h}{\partial x} + \frac{\partial}{\partial y} K_y b \frac{\partial h}{\partial y} + Q_s = S \frac{\partial h}{\partial t} \qquad (16.14)$$

where K_x and K_y are hydraulic conductivities in the coordinate directions, L is the head, b is the thickness of the aquifer and Q_s is a source (sink) function. The solution is equivalent to solving h so that the variational function is minimized:

$$F = \int\int \left[\frac{K_x}{2}\left(\frac{\partial h}{\partial x}\right)^2 + \frac{K_y}{2}\left(\frac{\partial h}{\partial y}\right)^2 + \left(S\frac{\partial h}{\partial t} - Q_s\right)h \right] dx\, dy \qquad (16.15)$$

In deriving a numerical solution of eqn (16.15), it is necessary to divide the aquifer into finite elements of which the shapes are arbitrary, often being triangular or quadrilateral. The elements can be non-uniform and disorderly, but should be smallest where the flow is concentrated, e.g. adjacent to a well. Parameters K_x, K_y, S and Q_s are kept constant for a given element, but they may vary from one element to another. In order to minimize eqn (16.15), the differential $\partial F/\partial h$ is evaluated for each node and equated to zero. The resulting set of simultaneous equations are then quite easily soluble by a digital computer.[28]

The selection of either the finite-difference or the finite-element method described above depends upon the following variables: the complexity of the flow system; the time required for the solution by the computer; the problems of stability and truncation error; together with the overall applicability of computer programs.

Some other computer-based models include the hybrid type consisting of a combination with a resistance network analogue. It was developed to reduce the time required for finite-difference solutions in the computer, which provides input data such as aquifer properties and boundaries which are expressed electrically by a digital–analogue converter and connected with the resistance network through a distributor.

The hybrid type is particularly suited to solving problems of non-steady flows in unconfined aquifers.

The mathematical approaches outlined above provide possible ways of improving efficiency in groundwater management, and they can act also in the decision-making process. The numerical modelling of groundwater is a relatively recent development which really started in the mid-1960s, when digital computers of adequate capacity became available; see for instance C. A. Appel and J. D. Bredehoeft in 1976.[29] Nevertheless, there are still deficiencies in application, perhaps resulting from the lack of contact between water managers and modellers, inadequate input data and inaccessibility of existing models.[30]

More development is necessary to cover additional matters such as the flow of immiscible fluids, considerations of stochasticity and pollutant transport. Some interesting work has been done recently. For instance, groundwater residence times and recharge rates were studied in 1984 by M. E. Campana and E. S. Simpson, using a discrete-state compartment model and ^{14}C data.[31] Earlier, E. J. Reardon and P. Fritz in 1978 gave details of an appropriate isotope sub-routine written for WATERF, a Fortran IV version of a water analysis treatment program WATERQ aimed at modelling both ^{13}C and ^{14}C.[32] Tritium data were handled by linear programming by W. E. Bardsley in 1984 for the conservative estimation of groundwater volumes.[33] In that same year, a groundwater model for simulating the rise of water table under irrigated conditions was published by R. Singh and his associates.[34] A review and an indicative model of Holocene depletion and active recharge of the Kalahari groundwaters in South Africa was proposed by J. J. de Vries in 1984.[35] A. Vandenberg in 1982 offered an alternative conceptual model of groundwater flow.[36] Estimation of groundwater flow parameters and quality modelling of the Equus beds aquifer in Kansas, USA, were presented in 1984 by M. A. Sophocleous.[37]

Also in 1984, P. W. Kasameyer and his associates described the development and application of a hydrothermal model for the Salton Sea geothermal field in southern California.[38] Here, a simple lateral flow model explained many of the relevant features. Earthquake swarms, a magnetic anomaly and aspects of the gravity anomaly are all indirect indicators that the ultimate heat source is igneous activity. Heat is believed to be transferred from this area of intrusion by lateral spreading of hot water in a reservoir under an impermeable cap rock. A two-dimensional analytical model encompassing this transport mechanism matched the general characteristics of the thermal anomaly and, indeed was utilized to estimate the age of the existing thermal system. This was calculated by minimizing the variance between the observed surface heat flow data and the model, and values ranged from 3000 to 20 000 years.

A very recent and extremely interesting piece of work was published in

1985 by F. W. Schwartz and A. S. Crowe.[39] It is known that coal strip mining can influence groundwater levels in the western plains of Canada and the USA and may have deleterious effects on the environment. However, almost nothing is known about the size, extent and timing of such changes. These investigators developed a finite-element model and showed that, using it, it is possible to solve for hydraulic head distribution within a two-dimensional region and also for the position of the water table and seepage faces in a mined area as a function of time. Sensitivity analyses conducted with this model demonstrated that strip mining does not simply influence water levels during mining and the early post-mining period, but is perhaps most significant during the later re-establishment of a steady-state flow system following mining. These authors believed that the magnitude and extent of changes in groundwater levels are controlled by at least the following four important hydrogeological parameters:

(a) the presence of local areas of discharge or of sources of recharge;
(b) the nature of the post-mining landscape;
(c) changes in the rate of recharge through the spoil;
(d) changes in the hydraulic conductivity of the spoil vis-à-vis the undisturbed geological materials.

The actual timing of the water level changes depends upon features of the flow system, e.g. how efficiently water can move to the low in hydraulic potential caused by mining and also the storativity of the mine spoil. All the simulations indicated that long times are necessary for the flow system to re-establish steady-state conditions following mining activities. The results imply that the maximum impact of mining at sites may not be apparent until many years after it has ceased.

REFERENCES

1. PRICKETT, T. A., 1979. Ground-water computer models — state of the art. *Ground Water*, **17**, 167–73.
2. BEAR, J., 1960. Scales of viscous analogy models for ground water studies. *J. Hydraulics Divn, Am. Soc. Civ. Engrs*, **86**, HY2, 11–23.
3. THOMAS, R. G., 1973. *Groundwater Models*. Irrig. and Drainage Paper 21, FAO, United Nations, Rome, 192 pp.
4. COLLINS, M. A. et al., 1972. Hele-Shaw model of Long Island aquifer system. *J. Hydraulics Divn, Am. Soc. Civ. Engrs*, **98**, HY9, 1701–14.
5. HELE-SHAW, H. S., 1897. Experiments on the nature of the surface resistance in pipes and on ships. *Trans. Inst. Naval Architects*, **39**, 145–56.

6. HANSEN, V. E., 1952. Complicated well problems solved by the membrane analogy. *Trans. Amer. Geophys. Union*, **33**, 912–6.
7. ZEE, C. H. et al., 1957. Flow into a well by electric and membrane analogy. *Trans. Amer. Soc. Civ. Engrs*, **122**, 1088–112.
8. FREEZE, R. A., 1970. Moiré pattern techniques in groundwater hydrology. *Water Res. Res.*, **6**, 634–41.
9. SEVENHUYSEN, R. J., 1970. Blotting paper models simulating groundwater flow. *J. Hydrol.*, **10**, 276–81.
10. JAVANDEL, I. and WITHERSPOON, P. A., 1967. Use of thermal model to investigate the theory of transient flow to a partially penetrating well. *Water Res. Res.*, **3**, 591–7.
11. DEBRINE, B. E., 1970. Electrolytic model study for collector wells under river beds. *Water Res. Res.*, **6**, 971–8.
12. TODD, D. K. and BEAR, J., 1961. Seepage through layered anisotropic porous media. *J. Hydraulics Divn, Amer. Soc. Civ. Engrs*, **87**, HY3, 31–57.
13. SHESTAKOV, V. M., 1968. *On the Technique for Solving Hydrological Problems using Solid and Network Models*. Intl Assn Sci. Hydrol., Publn 77, pp. 353–69.
14. BATURIC-RUBCIC, J., 1969. The study of non-linear flow through porous media by means of electrical models. *J. Hydrol. Res.*, **7**, 31–65.
15. PRICKETT, T. A. and LONNQUIST, C. G., 1973. *Aquifer Simulation Model for Use on Disk Supported Small Computer Systems*. Circ. 114, Illinois State Water Survey, 21 pp.
16. TODD, D. K., 1980. *Groundwater Hydrology*. J. Wiley, New York, 535 pp, p. 400.
17. TYSON, H. H., Jr and WEBER, E. M., 1964. Groundwater management for the nation's future — computer simulation of groundwater basins. *J. Hydraulics Divn, Amer. Soc. Civ. Engrs*, **90**, HY4, 59–77.
18. TRESCOTT, P. C. et al., 1976. *Finite-difference Model for Aquifer Simulation in Two Dimensions with Results of Numerical Experiments*. US Geol. Surv. Techniques of Water-Resources Investigations, Bk 7, Chapter C1, 116 pp.
19. BRUTSAERT, W. F. et al., 1971. Computer analysis of free surface well flow. *J. Irrig. Drain. Divn, Amer. Soc. Civ. Engrs*, **97**, IR3, 405–20.
20. GREEN, D. W. et al., 1970. Numerical modeling of unsaturated groundwater flow and comparison of the model to a field experiment. *Water Res. Res.*, **6**, 862–74.
21. SHAMIR, U. Y. and HARLEMAN, D. R. F., 1967. Numerical solutions for dispersion in porous mediums. *Water Res. Res.*, **3**, 557–81.
22. SHAMIR, U. Y. and DAGAN, G., 1971. Motion of the seawater interface in coastal aquifers: a numerical solution. *Water Res. Res.*, **7**, 644–57.
23. GAMBOLATI, G. and FREEZE, R. A., 1973; 1974. Mathematical simulation of the subsidence of Venice. *Water Res. Res.*, **9**, 721–33; **10**, 563–77.
24. KONIKOW, L. F., 1977. *Modeling Chloride Movement in the Alluvial Aquifer at the Rocky Mountain Arsenal, Colorado*. US Geol. Surv. Water Supply Paper 2044, 43 pp.
25. CALIFORNIA DEPARTMENT OF WATER RESOURCES, 1974. *Mathematical Modeling of Water Quality for Water Resources Management.* 2 Vols. The Resources Agency, Sacramento, 304 pp.

26. MADDEUS, W. O. and AARONSON, M. A., 1972. A regional groundwater resource management model. *Water Res. Res.*, **8**, 231–37.
27. O'NEIL, P. G., 1972. A mathematical programming model for planning a regional water resources system. *J. Inst. Water Engrs*, **26**, 1, 47–61.
28. PINDER, G. F. and GRAY, W. G., 1977. *Finite Element Simulation in Surface and Subsurface Hydrology.* Academic Press, New York, 295 pp.
29. APPEL, C. A. and BREDEHOEFT, J. D., 1976. *Status of Ground-water Modeling in the US Geological Survey.* US Geol. Surv. Circ. 737, 9 pp.
30. BACHMAT, Y., et al., 1978. *Utilization of Numerical Groundwater Models for Water Resources Management.* Rept EPA-600/8–78-012. U.S. Environmental Protection Agency, Ada, Oklahoma, 177 pp.
31. CAMPANA, M. E. and SIMPSON, E. S., 1984. Groundwater residence times and recharge rates using a discrete-state compartment model and ^{14}C data. *J. Hydrol.*, **72**, 171–86.
32. REARDON, E. J. and FRITZ, P., 1978. Computer modelling of groundwater ^{13}C and ^{14}C isotope compositions. *J. Hydrol.*, **36**, 201–24.
33. BARDSLEY, W. E., 1984. Conservative estimation of groundwater volumes: application of linear programming to tritium data. *J. Hydrol.*, **67**, 183–93.
34. SINGH, R., SONDHI, S. K., SINGH, J. and KUMAR, R., 1984. A groundwater model for simulating the rise of water table under irrigated conditions. *J. Hydrol.*, **71**, 165–79.
35. DE VRIES, J. J., 1984. Holocene depletion and active recharge of the Kalahari groundwaters — a review and an indicative model. *J. Hydrol.*, **70**, 385–6.
36. VANDENBERG, A., 1982. An alternative conceptual model of groundwater flow. *J. Hydrol.*, **57**, 187–201.
37. SOPHOCLEOUS, M. A., 1984. Groundwater-flow parameter estimation and quality modeling of the Equus beds aquifer in Kansas, USA. *J. Hydrol.*, **69**, 197–222.
38. KASAMEYER, P. W., YOUNKER, L. W. and HANSON, J. M., 1984. Development and application of a hydrothermal model for the Salton Sea geothermal field, California. *Bull. Geol. Soc. Amer.*, **95**, 1242–52.
39. SCHWARTZ, F. W. and CROWE, A. S., 1985. Simulation of changes in groundwater levels associated with strip mining. *Bull. Geol. Soc. Amer.*, **96**, 253–62.

CHAPTER 17

Groundwater Management

17.1. GROUNDWATER BASINS

In order to develop groundwater resources most efficiently, whole groundwater basins have to be considered: the principal factor involved in planning is to realize that these are subterranean water reservoirs of which the use by any one person affects the supply of water to all others, so that the main objective of management should be to exploit and operate them in a manner beneficial to the entire community. Thus not only geological and hydrological, but also sociological, considerations are involved. However, the generally agreed aim of the exercise is to secure the maximum amount of water economically and ensure that it has optimum quality as well.

17.2. EXTRACTION OF GROUNDWATER

This process is analogous to the extraction of other minerals such as oil or gas, but in the case of water the resource is renewable. To ensure that this is true in practice, it is necessary to create a balance between water which is recharged to the basin from surface sources and water which is pumped by wells from inside the basin. The development of such a basin will begin with a few wells, but, as time passes and the population increases, more and more will be drilled and the rate of extraction rises. The end result may be that the quantity of water being pumped out exceeds the natural recharge capability and, if pumping continues without change, very deleterious consequences may follow including depletion of the groundwater resource. If the inflow and outflow from a basin can be regulated, a groundwater aquifer can function optimally and indefinitely in the same manner as a surface reservoir of water. It is useful to consider the pros and cons of subsurface and surface reservoirs; they are summarized in Table 17.1, which is derived from data published by the US Bureau of Reclamation (Department of the Interior).[1]

TABLE 17.1
RELATIVE MERITS OF SURFACE AND SUBSURFACE RESERVOIRS[1]

Surface reservoirs	Subsurface reservoirs
Disadvantages	*Advantages*
Few new sites free (in the USA).	Many large-capacity sites available.
High evaporative loss, even where a humid climate prevails.	Practically no evaporative loss.
Need large areas of land.	Need very small areas of land.
May fail catastrophically.	Practically no danger of failure.
Varying water temperature.	Water temperature uniform.
Easily polluted.	Usually high biological purity, although pollution can occur.
Easily contaminated by radioactive materials.	Not rapidly contaminated by radioactive fallout.
Water must be conveyed.	Act as conveyance systems, thus obviating the need for pipes or canals.
Advantages	*Disadvantages*
Water may be available by gravity flow.	Water must be pumped.
Multiple usage.	Only uses are storage and conveyance.
Usually low mineralization of water.	Water may be highly mineralized.
Maximum flood control value.	Minor flood control value.
Large flows.	Limited flow at any one point.
Head available.	Head usually not available.
Relatively easy to investigate and manage.	Expensive and difficult to investigate and manage.
Recharge dependent on the annual precipitation.	Recharge dependent on surplus surface flows.
No treatment of recharge water is necessary.	Recharge water may need expensive treatment.
Low maintenance costs.	Maintenance of recharge areas or wells must be continuous and is expensive.

17.3. THE EQUATION OF HYDROLOGIC EQUILIBRIUM

This was discussed in the Sections 1.1 (eqn (1.1)) and 16.5 (eqn (16.10)). In the case of a particular groundwater basin, it may be written thus: (surface inflow + subsurface inflow + precipitation + imported water + decrease in surface storage + decrease in groundwater storage) = (surface outflow + subsurface outflow + consumptive use + exported water + increase in surface storage + increase in groundwater storage).

As mentioned earlier, it is possible to eliminate some items from this equation on grounds of negligibility, or because they do not influence the solution in a meaningful manner. This has the result that an exact balance will never be achieved, but this is acceptable if the degree of the unbalance does not exceed the limits of the accuracy of the data. If it does, then further investigations are necessary.

17.4. THE INVESTIGATION OF GROUNDWATER BASINS

The fundamental problem is to determine the safe yield, which should be known before the groundwater resources are developed. This is not often possible because of financial constraints; investigations usually follow some critical situation involving danger to the water supply. In the UK, management studies are undertaken by the various water Authorities which superseded the old River Boards and in the USA local government agencies carry out this work. There are four stages in such studies which produce data applicable to the equation of hydrologic equilibrium. Incidentally, any system of units can be used, provided that they are consistent, in this equation which refers to any size of hydrological structure such as an aquifer or a river valley. The relevant water year with which it is associated is not the calendar year, but a 12-month period extending from 1 October to 30 September.

The necessary stages of study recommended by the American Society of Civil Engineers are as follows:[2]

(i) preliminary survey;
(ii) reconnaissance;
(iii) feasibility;
(iv) project.

To accumulate the necessary information, the geology must be evaluated, starting with adequate topographic data embodied in contour maps, aerial photographs, benchmarks related to mean sea level and an associated network for levelling. These data are invaluable in detection of wells, assessment of levels of groundwater, and so on.

Additional geological information in the form of surface and subsurface mapping and drilling records, including well logs and the results of geophysical surveying, is required. Part of the drilling programme will entail pumping tests on wells, from which both storage coefficients and transmissivities may be obtained. From subterranean data acquired, it may be feasible to derive reasonably accurate pictures of the main aquifers and their extents and conditions (areas of confinement and unconfinement). Additionally, structural features such as faults, joints, etc., which can have a great influence on groundwater may be located.

Hydrological data are basic to the solving of the equation of hydrologic equilibrium and the following are requisites. Firstly, the surface inflow and outflow, imported and exported water are measured by standard methods, as is precipitation, using automatic and non-automatic recording rain gauges. The consumptive use involves all that water from the surface and below ground which is removed to the atmosphere through the processes of evaporation, transpiration and evapotranspiration. Finding out the quantities of water involved parallels the construction of a water balance in an area and requires a soils and land use survey, for which aerial photographs may be valuable. Satellite imagery, often false coloured, is applicable also on regional scales. In the equations, methods based on available heat are satisfactory, such as the Thornthwaite method discussed earlier (Section 2.1). The Blaney–Criddle method is also applicable. This has the form:

$$U = K \sum_{1}^{m} pt = KF$$

where U is evapotranspiration and F summarizes factors other than K, which is a coefficient varying with vegetation, month and locality; m is the number of inches per month, t is the mean monthly temperature, and p is the received solar input. p can be computed from appropriate sunshine tables. Other evapotranspiration equations include the Penman method (Section 2.1). In addition, there are the Hedke method,

$$U = kH$$

where H is the available heat for the growing season, summed, and k is a coefficient (0·0004 in the Cache La Poudre Valley in Colorado, USA), and the Lowry–Johnson method,

$$U = 0.000\,156H - 0.08$$

H being available heat which is assumed to have a linear relationship with the evapotranspiration, etc. Locally, there should be adequate evaporation records available that can also be utilized. Major difficulties arise in urbanized industrial areas, where the sewage outflow must be assessed.

Surface storage changes are obtained directly from changes in water levels in reservoirs and lakes within a region. Soil moisture changes can be monitored by a neutron probe or other suitable device. Changes in the groundwater storage are computed from changes in groundwater levels in wells, but all data relating to pumping tests, artificial recharge, etc., are also relevant. The specific yields of unconfined aquifers are determinable by testing samples in the laboratory, and storage coefficients can be obtained from pumping tests of wells in an aquifer. Probably the most difficult terms of the equation of hydrologic equilibrium to evaluate accurately are the subsurface inflow and outflow. In fact, one of these variables, or the difference between them may be obtained as the single unknown in the equation. In the case where the two cancel each other out, the equation is solved much more easily. An especially problematic situation arises where water may flow underground from one basin to another, but the direction of flow can be ascertained from water table or piezometric gradients and computed from Darcy's law if groundwater slopes and transmissivities are known.

17.5. SAFE YIELDS

Safe yields are the quantities of water perennially available from a groundwater basin; they are limited solely by the adverse side-effects which may arise as a result of pumping and basin operation.

The various concepts of basin yield are discussed below.

17.5.1. Mining Yield

Where the rate of withdrawal of groundwater exceeds the recharge, a mining yield exists and the aquifer undergoes depletion. This situation is found in a great many American groundwater basins, such as the High Plains of eastern New Mexico and western Texas where there is low rainfall and high evaporation. The aquifers have been isolated from an abundant water supply by the downcutting of river valleys. This did not matter until the present century, at the turn of which it was realized that water could be obtained from these aquifers so that an ideal climatic regimen for farming could be created. Lack of planning resulted in a rich farming community arising on a perched water supply which, after more than half a century, decreased in many places by more than 40%.[3] The water table declined almost 30 m and, therefore, there has been a great increase in the costs of pumping. One estimate is that the situation will not return to normal for at least 4000 years. It was proposed that artificial recharge should be used, but this is probably too expensive to justify in terms of the yields from farming. Consequently, the farmers of the region seem to have a rather bleak future in store for them. Since this appears to be an unwelcome development, it is surprising that groundwater mining elsewhere has been justified on the grounds that the stored groundwater is of no use until it is exploited; see for instance R. T. Sasman and R. J. Schicht in 1978.[4]

17.5.2. Perennial Yield

This defines the rate at which water can be withdrawn perennially from a groundwater basin under specified conditions of operation and without producing an adverse result such as progressive depletion of the water resource, land subsidence or degradation of the quality of the groundwater. Extraction in excess of perennial yield may be described as overdraft and is considered to be a hazard for the future welfare of the associated communities.

Deferred perennial yield comprises two pumping rates, the initial one being larger so as to exceed the perennial yield, thus reducing the level of the groundwater. This is an organized overdraft situation and provides storage water at low cost with the additional benefit that the lower water table is less liable to lose water to the atmosphere by evaporation.

Maximum perennial yield is a term applied to the maximum amount of groundwater which is available perennially. It is correlated with the natural and artificial recharge to the groundwater basin. Perennial yield varies with different patterns of recharge, and overdraft must be avoided if groundwater is to be treated as a renewable natural resource and only a certain amount of it is to be withdrawn annually from a groundwater basin. Clearly, recharge in this case refers to water which reaches the zone of saturation and becomes available for extraction. Other factors which may affect perennial yield include the relationship between the quantity of water actually supplied to a groundwater basin and its storage volume.

Water quality may be affected in cases of overdraft in coastal aquifers which may induce seawater intrusion; elsewhere, adjacent polluted water may be drawn into a pumped aquifer or lowered levels of groundwater may produce pumping of deep-lying connate brines. Of course, if the quality controls are relaxed, the perennial yield may be increased.

Overdraft is the largest potential groundwater problem in the USA. There is no solution other than rational management and reduction of overdrafts to perennial yields in the affected groundwater basins.

Perennial yields vary not only through considerations of recharge and overdraft as mentioned above, but also with time. Then, too they may vary with the groundwater level within a basin so that, if it drops, the subsurface inflow will rise and the subsurface outflow will fall, recharge from losing streams will increase and discharge from gaining streams will decrease. Evapotranspirative losses will be reduced. The converse is true where the groundwater level rises.

It is interesting that the perennial yield in an unconfined basin which is supplied by an adequate source of recharge can be increased not only by increased pumping, but also by rearranging the pattern of such pumping. This may be effected by concentrating the wells near the source of recharge so that there will be an augmented inflow with the advantage that an enhanced supply can be obtained without the necessity of increasing the pumping lifts.

In the case of a confined aquifer having a recharge area rather removed from the area of pumping, the rate of flow of water through the aquifer governs the perennial yield. With large aquifers of this type, the pumping of water from storage can proceed for years without establishing any equilibrium with the basin recharge. The slope of the piezometric surface increases, but the aquifer permeability is seldom sufficient to maintain a compensating flow in the basin.[5]

Another cause of variation in perennial yield is vegetation, which affects infiltration and percolation. Finally, the urbanization of a region is usually accompanied by the installation of sewage systems (*usually*, because this is not always so; for example Tehran in Iran, a city of about five million people, had no sewage system in 1978 when the author was there) and it is always accompanied by greater surface interception and runoff.

Changes in water usage affects the rate of pumping: if irrigation is replaced by industrialization, there may be greater pumping lifts and also greater perennial yield.

17.6. SALINITY

The maintenance of stability in a groundwater basin necessitates a constant salinity, i.e. the value of this parameter should not increase with time to a stage where it is deleterious and pollutes the water supply. A dynamic balance of total salt entering and leaving a groundwater basin is the ideal, so that in the longer-term basin:

$$\sum_n (CQ)_i = 0$$

where $(CQ)_i$ is the product of salt concentration C and the discharge Q of the ith flow component to or from the groundwater basin. In nature, such a situation is hardly ever obtained; this is a result of the fact that dissolved solids are added to water during most of its uses and thus recharge to the groundwater. Salt can derive from the solution of the geological materials of the aquifer, from rainfall, from surface water inflow, from subsurface inflow, from connate brines and, where saline intrusion takes place, from seawater. In addition, evapotranspiration removes water, thus increasing the residual salt concentration, which may also be raised through both domestic and industrial water usage as well as by the employment of fertilizers and other chemicals in agricultural regions. On the other hand, salt is taken away from the groundwater basin as a result of natural outflow or drainage, or by pumping. Wherever irrigation is practised in arid and semi-arid areas, the salt content becomes an urgent matter; see for instance O. J. Helweg in 1977.[6] The end consequences may be as bad as the total destruction of the relevant land for agricultural activities. The major solution is to lower the water table, reduce the salt content by leaching and construct a drainage system capable of removing the water from the basin concerned. However, sometimes groundwater of good quality occurs in adverse salt situations and sometimes, even the reverse can be true. This is a result of the fact that good mixing is rare in groundwaters so that good- and bad-quality groundwaters may comprise horizontal layers within the groundwater basin or they may be separated vertically. Although salt may be a longer-term danger, therefore, it may not pose an immediate threat at all.

17.7. CONJUNCTIVE USAGE

This is a procedure which involves coordinating the operation of both surface water and groundwater resources. It is based upon the assumption that surface reservoirs impound streamflow which is thereafter transferred to groundwater storage. The purposes of the impounded water and the groundwater are rather different. The former is intended to supply water for various uses on an annual basis whereas the latter is stored with the aim of catering for cyclic storage requirements, i.e. providing a reserve for years in which there is a lower-than-usual precipitation.

When excess precipitation occurs, the surface resources may be utilized to the maximum extent possible and some may also be introduced below ground by artificial recharge which ultimately may add substantially to the groundwater reserve. On the other hand, when drought prevails, the below-normal surface resources can be supplemented by pumping out more groundwater. Careful planning is necessary in order to ensure that there is sufficient space to store the water which is recharged and to ensure also that there is always water in storage which can be withdrawn by pumping when it is required. Then, too, the mandatory physical facilities must be available for conjunctive use of water: these include those related to distribution as well as those relevant to pumping and artificial recharge. A management study for conjunctive usage entails acquiring data on both surface and groundwater resources together with a good knowledge of the geology in the region, also the purposes for which water supply is necessary and how wastewater is to be handled. Inevitably, this approach incorporates investigating system dynamics, making a mathematical model and verifying this so that a viable simulation of coordinated operations can be made available. Such a basin model provides a means of determining probable responses of the real basin to variations in some of the parameters, e.g. natural and artificial recharge and pumping. As a result, it becomes feasible to devise optimal operational procedures.

At this point it is germane to compare the pros and cons for the conjunctive usage of surface water and groundwater: see Table 17.2.[7]

L. C. Fowler in 1964 listed the basic necessities which will ensure that an optimum management plan for water resources utilization exists, as follows.[8]

(a) Surface and subterranean storage capacities must be integrated in

TABLE 17.2
CONJUNCTIVE USAGE OF SURFACE AND GROUNDWATER[7]

Advantages	Disadvantages
1. Greater conservation of water and smaller surface storage with smaller surface distribution system.	1. Less hydroelectric power.
2. Smaller drainage system, reduced canal lining, greater flood control.	2. Greater power consumption. canal lining, greater flood control.
3. Easy integration with existing development, smaller evapotranspiration losses.	3. Decreased efficiency of pumping, increased water salination, greater operational complexity.
4. Greater control over outflow, improvement of power load and pumping plant usage.	4. Requires artificial recharge which is expensive; therefore, more difficult cost allocation.
5. Less danger from dam failure, reduction in weed seed distribution, better timing of water distribution.	5. Danger of land subsidence.

order to obtain the most economic use of local storage resources and the optimal quantity of water conservation.

(b) The surface distribution system must be integrated with the groundwater basin transmission characteristics so as to provide a minimally expensive distribution system.

(c) An operating agency must be available with sufficient power to manage the surface water resources, the groundwater recharge sites, the surface water distribution facilities and groundwater extractions.

Some instances of groundwater management are given below.

17.8. CASE HISTORIES

17.8.1. High Plains, New Mexico/Texas, USA

The earlier reference (Section 17.5.1) to this groundwater basin will be amplified here. The aquifer in question is the Ogallala Formation which is believed to have contained water resources of about $250 \times 10^9 \, m^3$ in 1958. Excessive pumping increased the irrigated acreage by 362% from 1948 to 1958 and the number of wells rose from 8356 to 45 522. By 1958, about 50×10^9 m^3 of water had been withdrawn and the pumping rate was 9×10^9 m^3 annually, this being approximately one hundred times the rate of recharge. It is a classic case of water mining and led to drastic control measures in some places; for example in New Mexico groundwater extraction was severely restricted on a township basis and it was believed that a firm water supply may be obtainable thereby for the next few years or so (up to the beginning of the 21st century).[9]

17.8.2. Los Angeles Coastal Plain, California, USA

This groundwater basin covers 1240 km^2 and provides approximately half of the water supply for the city of Los Angeles; in recent years, the basin has been over-exploited with consequent decline in groundwater levels accompanied by seawater intrusion. The California Department of Water Resources in 1968 made a detailed management study so as to devise a plan of operation capable of meeting the increasing water demand and conserving the maximum quantity of locally available water as well as reducing the effects of overdraft.[10] The Los Angeles coastal plain is actually served by a network of water sources, including local surface water and groundwater, reclaimed wastewater and imported water obtained from the Owens, Colorado and Feather Rivers. These vast increases necessitated evaluation of the dynamic response of the basin to both recharging and pumping, in order that the maximum usage of the subterranean reservoir could be obtained through an appropriate operating plan which took into account factors such as rates of withdrawal of water and artificial recharge, methods for controlling seawater intrusion, and so on.

17.8.3. Indus River Valley, Pakistan

This is the largest single irrigated area on Earth, covering about 9×10^6 ha, i.e. approximately the total irrigated acreage in the USA. The Indus River, which supplies the water involved, has an average annual discharge ten times greater than that of the Colorado River and twice that of the Nile.

The development of this truly vast irrigated area is associated with the introduction of canal irrigation before 1900 in the days of the British Raj. There has been a considerable leakage from the canals which has brought the water table very close to the ground surface over considerable parts of the area. The result has been the creation of over 2.5×10^6 ha of reduced fertility consequent upon salinity and waterlogging. Because of inadequate drainage and a minimum application of drainage water, coupled with a very high rate of evaporation, the salts continued accumulating, thus increasing the affected area. The solution must embrace the reduction or elimination of salinity and waterlogging and the increasing of agricultural production.

Remedial actions include the drilling of a network of deep, high-capacity wells about 1·6 km apart and reaching depths of 100 m, from which water is pumped and introduced into existing local canals for use in irrigation.[11] The wells in question aid in lowering the water table, providing extra irrigation water for agriculture and supplying water for the leaching of the saline soils. Leaching does not actually remove the salt, but it does disperse it through the aquifer, the salinity of which will in consequence slowly rise, in future, a part of the groundwater may be pumped off to waste in order to attain a salt balance. As a result of these measures, agricultural productivity in the area is rising.

17.8.4. Libya

A vast irrigation network is in process of construction in Libya: the area encompassed would, if superimposed over Europe, extend from Britain to Switzerland. The objective is to irrigate the Mediterranean coast of the country and the work is being done by a private construction company, the Dong Ah construction conglomerate of South Korea.

In the middle 1970s, a huge aquifer was discovered by geologists working in the southern desert; it was estimated to contain a quantity of water equivalent to the flow of the Nile over two centuries. A plan resulted which envisages pumping 476 000 m^3 of water daily and transporting it north to irrigate the rich, desiccated soils along the Gulf of Sidra. Around Kufra, some agricultural development has commenced and wheat, barley and alfalfa are being grown. The first phase of the project foresees expenditure of US\$3·3 × 10^9 in order to convert over 1·19 × 10^6 km^2 of land between Benghazi to the east and Sirte to the west into lush farmland. A second phase may extend this to Tobruk on the east and to Tripoli on the west to include almost all of the coastline region of Libya. In the first phase, a total pipeline length of about 1880 km will be utilized, including, two parallel lines and two coastal branches. Huge quantities of concrete will be required — some have called its association with the distribution network a 'concrete Nile'.[12] At present, construction is proceeding on the Tripoli–Benghazi section and on the parallel lines from the coastal town of Brega south to the Sarir Well Field together with the single line to the Tazerbo Well Field. Projected construction will involve extending this latter line along a separate branch to Kufra and running a line from Tripoli south to the Fezzan Well Field.

17.8.5. USSR

A tremendous project which will have a correspondingly great effect on groundwater over a very extensive area is that of switching Siberian river waters from the Arctic Ocean to the dry lands in the south. The Minister of Land Reclamation and Water Resources (N. Vasiliev) has asserted that there will be only local environmental changes and no overall global ones. This is certainly desirable.

The plan is to divert 6% of the River Ob water about 2400 km to the south by means of a new canal so as to provide irrigation water for new farming areas in Kazakhstan and compensate for over-exploitation of the water resources of this region. The local Aral Sea, supplied by only two rivers, both of these being very heavily employed in irrigation, has declined in level by about 10·5 m during the last 30 years, so the improvement which the project offers is critical in alleviating bad harvests caused by water shortages and is held by the USSR to justify the proposed expenditure of about US$67 $\times 10^9$. Opposition by conservationists has produced a more restricted project than that originally envisaged: this was to reverse the flows of both the Ob and the Yenisei Rivers completely. In consequence, now only the Ob is involved and a mere 6% of its flow is to be affected, at least initially, although the Minister implied that a further extension may take place later. He indicated that 75% of river systems flow north, but the agriculture is mainly in the south, where 1 ha (2·471 acres) of irrigated land is said to produce five times the crops of the same area of non-irrigated land. On the agricultural matter, it is certainly true that at present the USSR has just over $18·6 \times 10^6$ ha of irrigated land and about $14·2 \times 10^6$ ha of drained land, together constituting a mere 12% of the arable area of the country. However, this produces over a third of the crops in the USSR. The plan envisages a doubling of this area over the next 15 years.

17.8.6. Federal Republic of Germany

The exploitation of brown coal resources in the lower Rhine River valley in North Rhine Westfalia state has had a huge, but controlled, influence on the groundwater of the affected area which lies west of the city of Cologne (Fig. 17.1). The total area involved exceeds 2500 km^2 and industrial development began in the middle of the last century, but is concentrated now in a zone including the north-west–south-east strip

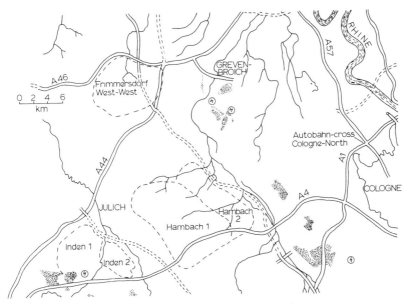

Fig. 17.1. Rhineland brown coal region.

from Grevenbroich to Brühl and the region between Eschweiler to the west (near Aachen), Düren to the east and Jülich to the north, a triangular region. The brown coal is obtained by open-cast mining extending to a total depth of well over 450 m in the Hambach locality. This has necessitated considerable de-watering and lowering of the groundwater table in the upper of two aquifers, to the extent that the quantities shown in Table 17.3 are estimated to have been removed by canalization to the local stream since 1955.

The situation in the affected region between Grevenbroich to the south-east and Mönchen-Gladbach to the north-west (Fig. 17.2) includes

TABLE 17.3
WATER VOLUMES REMOVED FROM OPEN-CAST MINES SINCE 1955

Site of mine	Volume of water (m^3)
Garsdorf	$(17.2–18.2) \times 10^9$
Bergheim	up to 27.1×10^9
Hambach	$(38.5–44.5) \times 10^9$

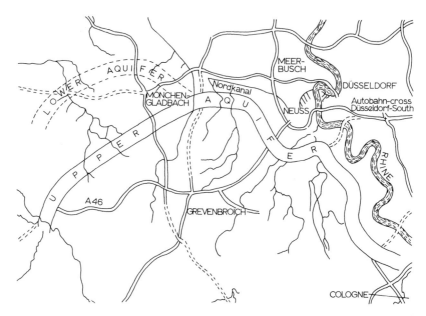

Fig. 17.2. Situation in affected area in 1979.

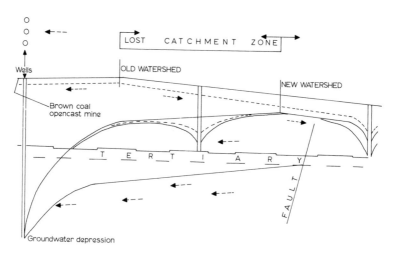

Fig. 17.3. Diagrammatic representation of the watershed movement in the Grevenbroich area.

the effects both in the upper and lower aquifers mentioned above. Some of these are illustrated in Fig. 17.3, a schematic diagram demonstrating the lost catchment and the shifting of the watershed eastwards. Also, a cone of depression resulting from one of three wells is partially developed and paralleled in the other two wells. Consequently, there is a considerable lowering of the groundwater table in the upper aquifer. Relevant groundwater contours for the upper aquifer are shown in Fig. 17.4 and

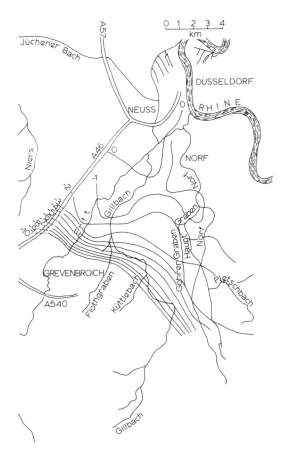

FIG. 17.4. Groundwater differences for the upper aquifer in the Grevenbroich area, 1953–1977.

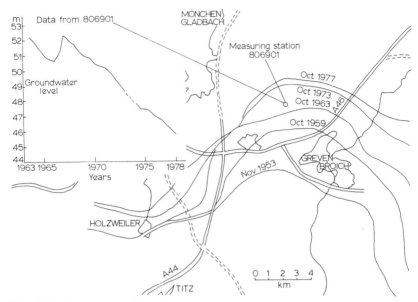

FIG. 17.5. Movement of the watershed near Grevenbroich and (inset) decline in groundwater level at one site (Measurement station 806901).

relate to the years 1953–77, demonstrating a decline of 10 m. The northeasterly movement of the watershed is shown in Fig. 17.5 and the decline in water level in the groundwater in metres above mean sea level at the measuring station number 806901 is also given.

It is interesting to consider the effects of the lowering of the groundwater levels on groundwater and surface water, as well as on vegetation and the animal world, soils and climate in the relevant region, der Regierungsbezirk Düsseldorf (the Düsseldorf Government District). The effects include not only those upon the groundwater table, but also those influencing the piezometric surface of the lower, confined aquifer.

17.8.6.1. The Upper, Unconfined Aquifer
The lowering of this aquifer varies over the affected area; the data given in Table 17.4 are relevant. From these figures, it may be inferred that there has been continuing lowering in the past few years, this demonstrating a continuing tendency towards lowering. The adverse effects which had to be controlled included extra costs in pumping, land

TABLE 17.4
LOWERING OF UPPER, UNCONFINED AQUIFERS

Locality	Lowering of aquifer (m)
Grevenbroich	20–40
Jüchen	~10
Glehn	~2·5
Büttgen	~2·0
Neukirchen	~2·0
Hoeningen	4–7

subsidence and the rendering of the surface environment unsuitable for a number of sensitive plants and animals. As a result of the latter, not only the numbers of individual plants and animals diminish, but also the numbers of their species. The ecological consequences include increased erosion and a process of aridification. In addition, the contribution made by groundwater to surface flow declined.

17.8.6.2. The Lower, Confined Aquifer
If water is withdrawn from this lower aquifer, the piezometric level will decline. Figure 17.2 shows places where this effect occurs and indeed affects the upper aquifer as well.

17.8.6.3. Watershed Movement
This is associated with the lowering of the groundwater level in the area of pumping. The groundwater availability is also reduced, not only in the immediate area, but also wherever withdrawal occurs up to the shifted watershed. Since the groundwater supply is less for a specific aquifer width, in order to obtain the same amount of water as before a greater width must be involved. Also, deeper and wider wells may become necessary.

17.8.6.4. Surface Streams
In some surface streams the water flow has been changed; in the Norf and its tributaries, flow has greatly diminished, and indeed many have become desiccated. In the Niers, the water loss is indicated by the data given in Table 17.5. Thus, at the Wickrathberg benchmark, there is a loss of 340 litre/s. Associated with this is the drying-up of the source of the Niers and its tributaries, now far advanced. Above Wickrathberg, there is

TABLE 17.5
WATER LOSS IN THE NIERS CATCHMENT AREA

	Subterranean area of possible withdrawal (km^2)	Available water in (10^6 m^3/a)^a	Water deficit (litre/s)
1953–54	58·4	16·17	520
1977	20·1	5·56	180
Loss	38·3	10·61	340

^{a} 1a = 10^4 m^2

no groundwater contribution and, during dry periods in Mönchen-Gladbach, stagnant water pools become eutrophic and also smell rather badly. The Niers, in fact, is virtually dead during these periods.

17.8.6.5. Remedies

The most important measure is probably to increase observation. An important effort by the private companies involved extensive re-afforestation; for instance at Hambach an artificial hill complete with small lakes, nature walks, deer, etc. has been created. However, the water balance remains affected through the pumping activity so that, at present, the groundwater availability is reduced by about 43×10^6 m^3/a.

The organizations involved realize that the remedial measures proposed must be carried out as rapidly as possible so as to repair the damage to the natural environment. The project is the largest open-cast coal mining operation in the world, involving removal of hundreds of metres of overburden and associated transportation of it, and coal, by a conveyor-belt system over 14 km in length. This system has been so constructed that noise emission by the moving belts (of which there are two parallel lines) is kept to an absolute minimum, another measure to reduce interference with nature. Re-siting of villages has been accomplished on a large scale, without undue hardship to the local inhabitants affected. But the fact remains that adverse groundwater changes have occurred and the surface water has also been affected adversely in parts of the region.[13] It must be remembered that, as F. W. Schwartz and A. S. Crowe indicated in 1985, this type of mining can have such effects in many places; they referred specifically to the western plains region of Canada and the USA and developed a finite-element model (see Section 16.5).[14]

17.8.7. North Africa and the Arabian Peninsula

The enormous area of desert in North Africa extends into the Arabian Peninsula and is underlain by some of the largest artesian basins in the world, covering the colossal extent of more than $5 \times 10^6 \text{ km}^2$. In these basins, the water-bearing beds are mostly sandstones which range in age from the Paleozoic to the Mesozoic and attain thicknesses from 100 to 3000 m. This comprises the famous Intercontinental Calaire to which reference was made earlier (Section 12.2), the so-called Nubia Sandstone also being involved, all constituting deposits laid down under epicontinental conditions according to A. A. Shata in 1983.[15] Regional basins function as primary water sources used for urban and industrial purposes; they are liable to deterioration due to defective regional water supply management. Thus, there has been successive lowering of the water levels and also of rates of production for wells, phenomena connected with the augmentation of desertification. The affected area of desert lies over $10 \times 10^6 \text{ km}^2$ of the Sahara (North Africa) and the Arabian peninsula, i.e. between 10° N and 30° N latitude, including more than 100 million inhabitants. Groundwater is critical in almost all of the countries in this region, where it is needed in all the water uses such as domestic, industrial and agricultural. There is typical arid zone hydrology.

Information on the sandstone aquifers is comprehensive; in North Africa up to 10° N latitude, the strata are almost horizontal and mostly undisturbed. The basement is composed of igneous–metamorphic Precambrian rocks and is usually not deeper than 300–500 m. North of 20° N latitude, there is folding and faulting on a large scale with uplifts climaxing centrally and associated downthrusts into deep, sedimentary basins. The aquifer depths in these basins range up to 2000 m and the total aquifer thickness may exceed 500 m. The contained water is 'fossil' with a radiocarbon age of 20 000–40 000 years. D. J. Burdon in 1975 described the possible mechanisms which alone or in combination can maintain the flow of the fossil aquifers under conditions of no recharge.[16] Since the Second World War, there has been very extensive exploitation of this fossil groundwater, causing the water levels to fall. In Egypt, this had serious consequences; in the New Valley project area, at least half of the 350 deep wells drilled have ceased to flow and pumping is required. In the Arabian peninsula, there is a shield (the Nubian–Arabian shield extending into Africa too) of igneous–metamorphic rock outcropping along the Red Sea and overlaid with a sedimentary cover which ranges in age from Cambrian to Holocene, the included sandstones being water-bearing if the conditions are suitable. The water is present both under unconfined and artesian conditions. Total dissolved solids content is under 4000 ppm in the west towards the outcrop, but rises to 21 000 ppm down the hydraulic gradient in the north-east direction.

The management problems arise from a number of causes. Firstly, there is a lack of data; problems also stem from the unreliability of some of the information which is actually available. This applies particularly to the sandstone aquifers, for which far too few pumping tests have been made. There are so many horizontal and vertical variations that a great many tests are necessary in order for statistically accurate information to be derived. In addition, confusion has arisen because of various different names being given to the same geological formation, uncertainty regarding fossil identifications and a ridiculously small number of hydrologic observations from which sweeping generalizations have been derived. For example, U. Thorweihe and his associates in 1984 calculated a groundwater influx in the Bir Tarfawi area of southern Egypt which they themselves stated 'has a big margin of error; it could exceed one order'.[17] Such rashness is particularly unwise in this region where critical hydrogeological parameters such as permeability, regional transmissivity and aquifer geometry are hardly known at all at present. As M. Heinl and R. Holländer stated in the same year, 'the distance between different groundwater observation points frequently exceeds 100 km'.[18]

The impact of human activities is often adverse. Examples are seawater injection into oil fields, and uncontrolled drilling and pumping of water wells, for example at Kharga-Dakla and at Kufra in Egypt and Libya respectively. Appropriate legislation is required, and also enforcement procedures. It is to be hoped that further studies on a much larger scale will be able to cope with these problems in the future.

17.9. WATER BALANCES

The management of water is easier to effect efficiently if the water budget or water balance of a region is known or can be calculated to a reasonable degree of accuracy. In fact, its determination regionally and on a planetary scale is one of the most important, if not the most important, objective of hydrologic studies. It constituted an aspect of the International Hydrological Decade (1965 to 1975) initiated by UNESCO and other specialized agencies of the United Nations Organization and in which they and 60 countries participated.

Water balances in small experimental basins and surface lakes of fresh water and salt water have been assessed, sometimes by the use of the isotopic approach as exemplified by the investigations described below.

17.9.1. Modrý Dul
This is a small experimental basin in northern Czechoslovakia examined by T. Dinçer and his associates in 1968.[19] The equation used was:

$$rc_s + (1-r)c_g = c_t$$

where r and $(1-r)$ are the relative contributions of surface water and groundwater to the streamflow there, c being concentrations of tritium in (c_s) surface runoff, (c_g) groundwater and (c_t) streamflow. Only a few samples were necessary. A relationship was found between total discharge and subsurface flow, the latter being separable from direct snowmelt water contributing to the stream channel.

17.9.2. Lake Chala
B. R. Payne investigated Lake Chala, and his conclusions appeared in 1970.[20] The lake is volcanic in origin, has an area of 4·2 km², a maximum depth of 100 m, a volume of 3×10^8 m³, no surface inflow or outflow and was originally thought to feed the Njoro Kubwa and other springs on the west bank of the Lumi River to the south; the Lumi River discharges into Lake Jipe through the Lumi delta swamp and, thereafter, into the Ruvu River in Tanzania. The lake examined is sited in extrusive volcanic rock overlying basement igneous and metamorphic rock at Mount Kilimanjaro. Tritiated water was introduced into the lake, found to be homogeneous and monitored over five years. One result was to find an annual subsurface inflow and outflow and, using stable isotope analyses, a contribution to the springs of only 6% of their discharge by Lake Chala.

17.9.3. Turkey
T. Dinçer in 1968 showed how natural concentrations of stable isotopes of hydrogen and oxygen (D, ^{18}O) can be used to determine the water balance of lakes in a sub-humid climatic regimen in the southwest of Turkey.[21] He later extended the work to salt lakes.

17.9.4. Other Cases
F. Begemann and W. F. Libby in 1957 were the first to utilize environmental isotopes in water balance assessments, one being developed

for the Mississippi River valley.[22] R. M. Brown in 1970 used variations in tritium content in river waters as a function of the tritium content of precipitation and the storage of the river in the Ottawa River basin.[23] Earlier, E. Eriksson also investigated this river, introducing the concept of transit time distribution.[24] An environmental isotope balance for Lake Tiberias in Israel was developed by J. R. Gat in 1970.[25]

17.9.5. Global Water Balance

This has been studied by a number of investigators. In 1962, M. I. Budyko used the equation:

$$P + r = E \pm \Delta W$$

where P is the precipitation, E is the evaporation, r is the runoff and ΔW represents the water exchange between oceans.[26] Results are displayed in Tables 17.6 and 7.7.

TABLE 17.6
WATER BALANCE OF THE OCEANS FROM M. I. BUDYKO[26]

Ocean	Precipitation (cm/y)	Runoff from adjacent land areas (cm/y)	Evaporation (cm/y)	Water exchange with other oceans (cm/y)
Atlantic	78	20	104	−6
Arctic	24	23	12	35
Pacific	121	6	114	13
Indian	101	7	138	−30

TABLE 17.7
WATER BALANCE OF THE CONTINENTS[26]

Continent	Precipitation (cm/y)	Evaporation (cm/y)	Runoff (cm/y)
Africa	67	51	16
Asia	61	39	22
Australia	47	41	6
Europe	60	36	24
North America	67	40	27
South America	135	86	49

The inference is that the precipitation and evaporation for the Earth are of the order of 100 cm/y, which would give the figures shown in Table 17.8 describing global water balance. They represent necessarily a rather crude picture, but there can be no doubt that the global water balance is dependent upon the circulation of water in the atmosphere and oceans which, in turn, is closely connected with the global energy budget.

TABLE 17.8
GLOBAL WATER BALANCE

Units	Precipitation (cm/y)	Evaporation (cm/y)	Runoff (cm/y)
Oceans	112	125	−13
Continents	72	41	31
Earth	100	100	0

REFERENCES

1. BUREAU OF RECLAMATION, 1977. *Ground Water Manual*. US Department of the Interior, 480 pp.
2. AMERICAN SOCIETY OF CIVIL ENGINEERS, 1972. *Ground Water Management*. Manual of Engineering Practice, 40, 216 pp.
3. MENARD, H. W., 1974. *Geology, Resources and Society*. W. H. Freeman, San Francisco.
4. SASMAN, R. T. and SCHICHT, R. J., 1978. To mine or not to mine groundwater. *J. Amer. Water Works Assn*, **70**, 156–61.
5. THOMAS, H. E., 1951. *The Conservation of Ground Water*. McGraw-Hill, New York, 327 pp.
6. HELWEG, O. J., 1977. A non-structural approach to control salt water accumulation in ground water. *Ground Water*, **15**, 51–7.
7. CLENENDEN, F. B., 1955. Economic utilization of ground water and surface water storage reservoirs. Paper presented at the Amer. Soc. Civ. Engrs Meeting, San Diego, Ca.
8. FOWLER, L. C., 1964. Ground-water management for the nation's future — ground-water basin operation. *J. Hydraulics Divn. Amer. Soc. Civ. Engrs*, **90**, HY5, 51–7.
9. CONOVER, C. S., 1961. *Ground-water Resources — Development and Management*, US Geol. Surv. Circ. 442, 7 pp.
10. CALIFORNIA DEPARTMENT OF WATER RESOURCES, 1968. *Planned Irrigation of Ground Water Basins: Coastal Plain of Los Angeles County*. Bull. 104, Sacramento, 25 pp + appendices.
11. MUNDORFF, M. J. et al., 1976. *Hydrologic Evaluation of Salinity Control and*

Reclamation Projects in the Indus Plain, Pakistan — a Summary. US Geol. Surv. Water Supply Paper 1608–Q, 59 pp.
12. ACHISON, M. and LEWIS, D., 1985. Libya's man-made river. In: Newsweek, 10 June, 47.
13. STAATLICHES AMT FÜR WASSER-UND ABFALLWIRTSCHAFT DÜSSELDORF, 1980. Wasserwirtschaft und Braunkohle. KV Büro GmBH, Bochum.
14. SCHWARTZ, F. W. and CROWE, A. S., 1985. Simulation of changes in groundwater levels associated with strip mining. Bull. Geol. Soc. Amer., **96**, 253–62.
15. SHATA, A. A., 1983. Management problems of the major regional aquifers in North Africa and the Arabian peninsula. Intl Conf. Groundwater and Man, Sydney, Australia, pp. 263–71.
16. BURDON, D. J., 1975. Mechanisms for movement of fossil groundwaters. 3° Convegno Internazionale sulle Acque, Sotterrance, Palermo, Italy.
17. THORWEIHE, U., SCHNEIDER, M. and SONNTAG, C., 1984. New aspects of hydrogeology in southern Egypt. Berliner geowiss. Abh., (A), **50**, 209–16.
18. HEINL, M. and HÖLLANDER, R., 1984. Some aspects of a new groundwater model for the Nubian aquifer. Berliner geowiss. Abh., (A), **50**, 221–31.
19. DINÇER, T., PAYNE, B. R., MARTINEC, J., TONGIORGI, E. and FLORKOWSKI, T., 1968. An environmental isotope study of the snowmelt-runoff in a representative basin. IAEA Tech. Rets Ser., **91**, 190.
20. PAYNE, B. R., 1970. Water balance of Lake Chala and its relations to groundwater from tritium and stable isotope data. J. Hydrol., **11**, 47–58.
21. DINÇER, T., 1968. The use of oxygen-18 and deuterium concentrations in the water balance of lakes. Water Res. Res., **4**, 6, 1289–1306.
22. BEGEMANN, F. and LIBBY, W. F., 1957. Continental water balance, groundwater inventory and storage time, surface ocean mixing rates and worldwide water circulation patterns from cosmic ray and bomb tritium. Geochim. Cosmochim. Acta, **12**, 277–96.
23. BROWN, R. M., 1970. Distribution of hydrogen isotopes in Canadian waters. Isotope Hydrology, Symp. 9–13/3/70, IAEA, Vienna pp. 3–22.
24. ERIKSSON, E., 1963. Atmospheric tritium as a tool for the study of certain hydrologic aspects of river basins. Tellus, **15**, 3.
25. GAT, J. R., 1970. Environmental isotope balance of Lake Tiberias. Isotope Hydrology, Symp. 9–13/3/70, IAEA, Vienna, pp. 109–27.
26. BUDYKO, M. I., 1956. The Heat Balance of the Surface of the Earth. Leningrad, USSR. Translation by N. A. Stepanova, Washington, DC.

APPENDIX 1

Dictionary of Water Terms

Absolute drought: Usually applied to periods of 15 consecutive days or more characterized by less than 0·25 mm rainfall daily.

Absorption (water): The imbibing of water by a soil or rock, a quantity expressed in percentage terms of the original dry weight.

Acidification: Using acid (normally hydrochloric) to increase water supply from a borehole failing due to encrustation on screens and slotted pipes.

Acre: A unit of area equal to 4840 yd^2 (or 0·405 hectare).

Acre foot: That volume of water which will cover 1 acre to a depth of 1 ft. Equal to 1233·5 m^3.

Adit: A rectangular heading or tunnel either horizontal or inclined for tapping groundwater. Often driven from a shaft and may be lined or unlined.

Adsorbed water: That water which is retained in a mass of soil by physico-chemical forces.

Aeration, zone of: Subsurface between the surface and the water table divisible into a belt of soil water, an intermediate region and a lowermost capillary fringe.

Air-lift pump: A piece of equipment capable of lifting water in a well and comprising an air compressor at the surface and two pipes hanging down vertically with one inside the other. The smaller pipe delivers compressed air to the depth of occurrence of water where a nozzle discharges it into the free water and, by aeration, causes its density to drop. Thereafter, the water/air mixture is forced upwards by the head of groundwater.

Airline correction: A correction necessary in the measurement of water depth in order to determine true depth.

Apparent velocity of groundwater: The apparent rate of movement of groundwater in the zone of saturation is expressible thus: $V = Q/A$, where Q is the volume of water passing through a cross-section of area A in unit time.

Appropriated rights, water: An individual's rights to the exclusive usage of water based strictly upon priority of appropriation and the beneficial utilization of the water without limitation of the place of usage.

Aquiclude: A stratum of low porosity absorbing water slowly and not transmitting it freely enough to comprise useful supplies for a well.

Aquifer: A permeable deposit which can yield useful quantities of water when tapped by a well.

Aquifuge: An impervious rock devoid of interconnected fissures, voids or openings which cannot either absorb or transmit water.

Artesian aquifer (confined aquifer): An aquifer in which the water is under pressure and confined beneath an impermeable deposit.

Artesian head, negative: Used in regard to a well in which the hydrostatic pressure is negative and the free water level is below the existing water table.

Artesian head, positive: Used in regard to a well in which the hydrostatic pressure is positive and the free water level is above the existing water table.

Artesian slope: This refers to an artesian aquifer dipping beneath impermeable strata, groundwater being stored there under pressure.

Artesian well: A well tapping artesian water.

Artificial recharge: The augmentation of natural recharge, usually by spreading of water on the surface.

Attached groundwater: The part of groundwater which is retained on particle surfaces against the force of gravity during pumping or drainage.

Augers: Manually operated or power driven boring tools from about $1\frac{1}{2}$ to 24 in (40 to 610 mm) diameter.

Average velocity of groundwater: The mean distance covered by mass of groundwater per unit of time (equal to total volume of groundwater passing through unit cross-sectional area per unit of time divided by the porosity of the medium).

Backblowing: Improving water yield from boreholes, especially in fissured rocks, by use of compressed air which is either pumped into them or employed to pump in air until maximum pressure is attained and then released.

Basin: This is topographically either a river-drained area or low lying land encircled by hills. The geological meaning is different and given to an area in which stratified rock strata dip towards a central point, these strata possessing a centroclinal dip.

Basin recharge: The difference between precipitation and runoff plus other losses, i.e. that part of precipitation which resides as groundwater, surface storage and soil moisture.

Battery of wells: Several wells in a convenient radius which are connected to a main pump for water withdrawal.

Belt of phreatic fluctuation: That mass of rocks in the lithosphere in which the fluctuation of water table takes place.

Belt of soil water: The upper part of the zone of aeration containing soil moisture.

Bernoulli's Theorem: Relating to flow in conduits, this asserts that if a perfect, incompressible fluid is flowing in a steady stream, then, neglecting frictional and eddy current effects, the total energy is constant.

Blowing well: A water well from which is periodically blown a current of air.

Borehole: A hole drilled from the surface or from a subsurface excavation into the ground in order to obtain geological data or for the drainage or abstraction of water or for access to hydraulic works, etc. In the UK, depths for tapping water may reach 600 ft (180 m) or more. Diameters vary up to 40 in (1 m).

Borehole casing: A plain or perforated pipe of steel or some other material which is inserted into a borehole, often in loose ground.

Borehole log: A record, principally of the rock strata penetrated during the drilling of a borehole.

Bourdon pressure gauge: An instrument for measuring water pressure and porewater pressure.

Brackish: A word applied to water ranging from 1000 ppm up to the dissolved salt content of sea water.

Breathing well: A water well in which air is alternately blown out and sucked in, a phenomenon apparently related to barometric pressure.

Brine: A water having a dissolved salt content exceeding that of sea water.

Cable tool: A sharp chisel-edged bit utilized in drilling a deep well by lifting and dropping so as to break rock by impact, the fragments being removed by a bailer (a section of pipe with a foot-valve through which the cuttings enter).

Capillary action: A term applied to the movement of liquids due to capillary forces.

Capillary fringe: The belt of ground immediately above the water table, i.e. at the bottom of the zone of aeration and containing capillary water.

Carbonate hardness: Temporary hardness.

Catchment area: A land area from which precipitation drains into a reservoir, pond, lake or stream.

Cavitation: The formation of cavities during high-speed pumping, resulting in corrosion of metal parts because of the liberation of oxygen from the water.

Cavity well: Sometimes termed a boulder well, a water well drilled into a thick aquifer comprising sand, gravel and boulders.

Chalk: The SE England Chalk is the most famous and largest aquifer in the UK. Interestingly, the rock itself is impermeable, but water can flow along fissures and wells put down in fissured regions give high yields.

Chezy Formula: An empirical formula relating mean flow velocity, V, hydraulic mean depth, R, hydraulic gradient, S, and Chezy coefficient, C, thus: $V = C\sqrt{(RS)}$.

Coefficient of permeability: The rate of flow of water through unit cross-section of a medium under a hydraulic gradient of unity and at a specified temperature. Also known as coefficient of conductivity and coefficient of transmission.

Coefficient of viscosity: A quantitative expression of the friction between the molecules of a fluid when in motion. The capacity of a rock or soil to transmit water varies inversely with the coefficient of viscosity of the water.

Compensation water: Water which legally must be released from a reservoir in order to meet the needs of downstream users who received a water supply before a dam was constructed.

Cone of depression: This is the inverted conical depression in the water table round a well or borehole in which pumping is being effected. Also known as cone of exhaustion and cone of influence.

Confined water: This is a term for artesian water used in the USA.

Connate water: Original water in the interstices of a sedimentary rock which was not expelled during consolidation. Also known as fossil water.

Darcy's law: This is used to determine the velocity of percolation of water through natural materials of granular type.

Deep well: One exceeding 100 ft (30 m) in depth.

Depletion: Exhaustion of a water well caused by (a) excessive pumping by neighbouring wells, (b) pumping in excess of replenishment, (c) defective casing allowing leakage.

Dip: Maximum angle of inclination of any surface, which may be natural or artificial.

Drawdown: The lowering of the water table in and around a well or borehole by pumping.

Dug well: One which is excavated by manual labour.

Effective porosity: The ratio of the volume of water in a pervious mass previously saturated with water which can be drained by the force of gravity, to the total volume of the mass.

Effective velocity of groundwater: The volume of groundwater passing through unit cross-sectional area divided by effective porosity of the material. Also known as field, true or actual velocity.

Effluent: Flowing out, flow of sewage from a process plant.

Electro-osmosis: A method of lowering groundwater and especially applicable to silts. It accelerates natural drainage away from surface works and excavations.

Evapotranspiration: Loss of moisture from a soil by evaporation and plant transpiration.

Fault: A break in rock strata along which displacement of one side relative to the other has occurred parallel to the fracture.

Fault trap: A geological structure in which water in a porous deposit has been trapped by an impervious deposit thrown opposite it by a fault.

Field capacity: The maximal amount of water which can be held by a soil against free drainage.

Field coefficient of permeability: The coefficient of permeability at the temperature of the water.

Field moisture: Term used for adhesive water (pellicular water, *q.v.*) found above the water table.

Fixed groundwater: Water stored in rocks with fine voids.

Flow: The amount of water flowing in a pipe, aquifer, etc., expressed as volume per unit time.

Flowing well: A well in which the hydrostatic pressure of the water is sufficient to cause it to rise and flow out at the surface.

Flow line: Line shown in a flow net.

Flow net: A net of equipotential lines and flow lines intersecting at right angles.

Fluctuation of water table: The alternate upward and downward movements of the water table due to periods of intake and discharge of water in the zone of saturation.

Free groundwater: Groundwater which is not trapped or confined by an overlying impervious rock.

Fresh water: Water containing less than 1000 dissolved parts of salt per million parts of water.

Froude number: A dimensionless number expressing the ratio between influence of inertia and gravity in a fluid. It is the velocity squared divided by length times the acceleration due to gravity. In analysis of hydraulic models, the ratio should be similar in both model and full-scale plant.

Gravitational water: Water in soils and rocks above the water table.

Gravity groundwater: That water which would drain out of a rock in the zone of saturation assuming the zone and the capillary fringe moved downwards for a period, no water entered the area and none was lost except through the force of gravity. Water discharged from springs and that withdrawn from wells is gravity groundwater.

Gravity spring: Water discharged at the surface from permeable beds under the influence of gravity.

Groundwater: Water in the zone of saturation.

Groundwater balance (budget): An estimate of water resources usually applied to a groundwater basin or province. Recharge, storage and discharge are important factors in it.

Groundwater basin: A basin-shaped group of rocks containing groundwater and with geologic/hydraulic boundaries suitable for investigation and description. A basin of this type normally includes both the recharge and the discharge areas.

Groundwater dam: A subterranean impervious mass or a fault which prevents or at least impedes the lateral movement of groundwater.

Groundwater decrement: A decrease in groundwater storage by withdrawal from wells, spring flows, infiltration tunnels, etc.

Groundwater discharge: Discharge of water from the zone of saturation into bodies of surface water or on land.

Groundwater divide: The line of maximal elevation along a groundwater ridge where the water table slopes downwards in opposite directions.

Groundwater equation: The balance between water supplied to a basin and the quantity leaving the basin.

Groundwater flow: Part of streamflow derived from the zone of saturation through seepage or springs. Also the movement of groundwater in the aquifer.

Groundwater inventory: An estimate of amounts of water forming groundwater increment balanced against estimates of amounts forming groundwater decrement for a particular area or basin.

Groundwater lowering: Localized lowering of the water table so that excavations can be made in relatively dry conditions. It may be effected by wellpoints, *q.v.*

Groundwater mound: An elevation formed in a groundwater body by influent seepage.

Groundwater province: Any area wherein the groundwater conditions are everywhere similar.

Groundwater recession: Lowering of the water table in a basin or area.

Groundwater recharge (increment): The replenishment of water in the zone of saturation.

Groundwater runoff: Runoff which existed partly or wholly as groundwater since its last precipitation.

Groundwater storage: Estimate of the amount of water in the zone of saturation. That stage of the hydrologic cycle when water is leaving and entering groundwater storage.

Groundwater storage curve: A curve which shows the quantity of groundwater available for runoff at given rates of groundwater flow.

Groundwater tracers: These are chemical dyes or salts or compounds incorporating radioactive isotopes used to trace the source of water seeping into wells, shafts, tunnels or deep excavations. Radiotracers (as the last category is usually termed) are especially practical because they can be detected in very minute quantities.

Groundwater trench: A rather narrow depression in the water table and resulting from effluent seepage into a stream, channel or drainage ditch.

Head: The potential energy of water arising from its height above a given datum.

Headings (wells): Small adits or tunnels excavated into the water-bearing rock formations in order to increase the yield of a well.

Held water: Capillary water, water retained in the ground above the standing water level.

Hook gauge: A piece of equipment designed to measure the elevation of the free surface of a liquid and comprising a pointed hook attached to a vernier which slides along a graduated staff. The hook is lowered into, say, water and then raised until the upward point just cuts the water surface. It is capable of measuring water level with considerable accuracy.

Hydraulic: Refers to the flow of liquids, particularly water, through pipes or channels.

Hydraulic conductivity: A term occasionally used for coefficient of permeability.

Hydraulic discharge: Loss of groundwater by discharge through springs.

Hydraulic gradient: In a closed conduit, this is an imaginary line connecting the points to which water will rise in vertical open pipes extending upwards from the conduit. In an open channel, it is the free surface of flowing water.

Hydraulic models: A scale representation of a hydraulic structure which is geometrically similar at all solid–liquid boundaries. A resemblance to a prototype is desirable. The type of flow in both must also be similar.

Hydraulic permeability: The capacity of a rock or soil for transmitting water under pressure.

Hydraulic profile (aquifer): A vertical section of the piezometric surface from any given aquifer.

Hydrograph: A graph which shows level, velocity or discharge of water in a channel or conduit plotted against time.

Hydrograph, recession (normal recession curve): The curve obtained from specified lengths of hydrograph that represent discharge from channel storage or a natural valley after deducting base flow; the curve illustrating the decreasing rate of flow in a stream channel.

Hydrologic cycle: That series of transformations occurring in the circulation of surface waters to atmosphere, to ground as precipitation and back to surface and subsurface waters.

Hydrostatic pressure: The pressure at any given point in a liquid at rest; equals its density multiplied by the depth.

Impermeability factor (runoff coefficient): A factor enabling runoff to be calculated; the ratio of the direct runoff to the average rainfall over the entire drainage area for any storm.

Impermeable (impervious): Word used to describe a soil, rock or other substance permitting the passage of water at an extremely slow rate.

Induced recharge (aquifer): Recharge of an aquifer by inflow of stream water.

Infiltration: Slow movement of water through or into the interstices of a soil (see Figs. 1.1, 3.2).

Infiltration area (well): That area of water-bearing rocks penetrated in a well and which discharges water into it.

Infiltration capacity: The maximum infiltration rate of a soil or other porous material.

Infiltration coefficient: The ratio of infiltration to precipitation for a soil under specified conditions.

Infiltration rate: The rate at which water is absorbed by a soil or other porous material; varies with the infiltration capacity.

Influence basin (well): The basin-shaped depression in the water table around a well due to withdrawal of water by pumping.

Initial detention: That portion of rainfall not appearing as surface runoff or as infiltration during period of precipitation; it includes evaporation, interception by vegetation and depression storage (this last refers to the volume of water needed to fill all natural depressions in an area to their overflow levels).

Intake (well): Voids in a water-bearing rock through which water passes into a well.

Intake area (aquifer): The area of outcropping permeable rocks from which an aquifer is fed with surface waters.

Interception: That process by which precipitation is retained by foliage and vegetation prior to reaching the ground.

Intermediate belt: That portion of the zone of aeration lying between the capillary fringe and the belt of soil water, *q.v.*

Internal water: Subterranean water below the zone of saturation.

Interrupted water table: A water table with a difference in level near a fault or other obstruction to the lateral flow of water.

Interstices: Voids, *q.v.*

Interstitial water: Water contained within voids of a rock or soil.

Inverted capacity: The maximum rate at which an inverted well, *q.v.*, can remove surface or near-surface water by discharge through openings into deposits at its lower end.

Inverted well: A well in which the water flow is downwards; a recharge well.

Isobath (water table): A line on a plane connecting points at the same height or elevation above an aquifer or water table.

Isopiestic line: An imaginary line connecting points possessing the same static level; a contour of the piezometric surface of an aquifer.

Jetting: A hydraulic method of inserting wellpoints, *q.v.*, or piles into sandy material and utilized in situations in which a pile hammer could damage structure in the neighbourhood.

Juvenile water: Magmatic water or plutonic water, i.e. water derived from magmas or molten masses of igneous rock during their crystallization or from lava flows as steam.

Lysimeter: An instrument for measuring percolation of water through soils and determining soluble constituents removed.

Magmatic water: Juvenile water, *q.v.*

Main water table: The surface of the zone of saturation, *q.v.* Also termed the phreatic surface.

Maximum capacity (well): The maximum rate at which water can be withdrawn from a well. Expressed in various ways, e.g. gallons per minute.

Menard Pressure Permeameter: An instrument for measuring directly the permeability of rock in a borehole.

Meteoric water: Water falling as rain or derived from snow, hail or dew.

Micropores: Rock voids smaller than 0·005 mm. If porosity in a rock is mostly microporous, the passage of water through it is inhibited.

Net intake: The quantity of water reaching the zone of saturation.

Non-flowing artesian well: A well which has tapped water with sufficient pressure to cause it to rise in the well but not enough to reach and overflow at the surface.

Normal depletion curve: A curve showing the normal loss of water from groundwater storage:

Observation well: A well sunk to aid groundwater investigations.

Open-end well: A well with a pipe form of lining where the water enters near the bottom through a screen or other arrangement within the lining.

Outcrop: That part of a rock appearing at the surface.

Overdraft: Any draft from an aquifer in excess of safe yield.

Overpumping: That situation in a well when the water pumped out exceeds the rate of replenishment from outcropping parts of an aquifer.

Pellicular water: Sometimes called adhesive water, this comprises water which is retained in a mass of soil as a result of molecular attraction, coating the soil particles and occasionally migrating from one to another.

Pellicular zone: The portion of the ground below the surface to which evaporative effects can penetrate.

Perched water table: The upper surface of a small water body above a main water table and retained in its elevated position by an impervious stratum.

Permeability: Capacity of a rock or soil or other substance to transmit water.

Pervious rock: One permitting the easy passage of water.

pH: An expression for the acidity or alkalinity of a solution, actually the logarithm of the reciprocal of the hydrogen ion concentration. A one unit change in pH value equals a tenfold change in the hydrogen ion concentration; 7 is the neutral point. Values above this indicate alkalis and values below it indicate acids.

Phreatic: A word applied to groundwater and its concomitants. Thus, groundwater may be referred to as phreatic water.

Phreatic wave: An undulation in the water table which moves laterally away from a zone where a large intake of water occurs in the zone of saturation.

Piestic interval: The vertical distance between two isopiestic lines on a map.

Piezometer: An instrument for measuring pressure head; normally a small pipe tapped into the side of a closed or open conduit and connected to a gauge.

Piezometric surface (potentiometric surface): The imaginary surface to which water will rise under its full head from any given groundwater aquifer.

Plane of saturation: The water table.

Plutonic water: Juvenile water, magmatic water.

Pore pressure: The pressure of water and air in the interstices between the grains of a rock or soil mass.

Pores: Voids.

Porewater: Water occupying the interstices of rocks or soils.

Porewater pressure: The pressure of water in a saturated soil.

Porosity: The percentage ratio of the volume of voids to the total volume of a rock or soil sample. Thus
$$P = 100 \frac{W-D}{W-S}$$
where P is the percentage porosity, W is the saturated weight, D is the dry weight, S is the weight of saturated sample when suspended in water.

Potential gradient (groundwater): The rate of change in potential in a mass of groundwater. Where no direction is specified, that of maximum gradient is taken.

Potential yield: The maximum rate at which water may be extracted from an aquifer throughout the foreseeable future, ignoring recovery cost.

Precipitation: Total quantity of water falling as rain, hail, snow and expressed as millimetres or inches of rainfall over a specified period. Moisture deposited as dew.

Prescribed rights (water): Legal title to the use of water acquired by possession and use over a long period without any protest from other parties.

Pressure head: Describes the water pressure in a system, expressible as N/mm^2 or psi or as metres or feet head.

Primitive water: Water trapped in the interior of the Earth since its formation.

Pumping level (well): That level from which water has to be pumped after the cone of depression has been established in the local water table.

Pumping test: (1) Water yield: quantities and water levels are recorded during the test period. The test pumping rate is usually greater than that at which water will be needed and covers a period long enough to show whether the yield can be maintained. (2) Water quality: taking samples of water during the test to determine by chemical analysis the major constituents and organic purity. Such tests may extend over 2 weeks.

Radial well: A well in which a number of strainer pipes are driven laterally into a water-bearing deposit in radial fashion from a main sump.

Radius of influence (water well): The radius of the circular base of the cone of depression formed in the groundwater around a well when pumping is in progress.

Recharge of aquifer, induced: Flow of stream water into an aquifer.

Recharge well: An inverted well, i.e. one which conducts superficial water into an aquifer at shallow or moderate depth.

Rejuvenated water: Water of compaction and water released during metamorphism.

Resistivity method: A method of geophysical prospection in which direct measurements are made of the ratio of voltage to current when a current is forced to flow through the ground to be tested. The conductivity of a rock is governed by its water content and its salinity; where these values are high, conductivity is high and resistivity is low, and vice versa.

Reynolds number: A dimensionless number symbolized by R applied to fluid flow in a tube:

$$R = \frac{\text{mean velocity of flow} \times \text{pipe diameter}}{\text{kinematic viscosity}}$$

Riparian: Situated on or pertaining to the banks of a river or other water body.

Riparian rights: The right of a riparian owner to the use of a stream or other water bordering or flowing through his land.

Runoff: That part of precipitation flowing from a catchment area and finding its way into streams, lakes, etc. Includes direct runoff and groundwater runoff.

Salinity (groundwater): The content of totally dissolved solids (TDS) in the water; measured using the electrical conductivity method. For seawater, the content is approximately 3·5% (35 000 ppm).

Screened well: A cased well in which water enters through one or more screens and not through holes in the casing.

Seasonal recovery (groundwater): The replenishment of water in the zone of saturation during and following a wet season with rise in level of water table.

Seismic method: A method of geophysical prospection in which the velocities of transmission of shock waves through the ground under test are utilized. A shock wave is initiated by the firing of an explosive charge at a known point or sometimes by impacting a sledge hammer on a metal plate, the shot. Thereafter, the travel time for selected waves to arrive at receivers (geophones) is recorded. Wave velocities can vary from 600 m/s for loose sediments to 6000 m/s for granite.

Shallow well: A well less than 100 ft (30 m) deep.

Shot firing: Augmenting the supply of water in boreholes put down in deposits such as sandstone, chalk or limestone by detonating small charges of gelignite or dynamite in order to shatter and enlarge the fissures. Water increases of up to 300% are feasible with optimal siting of the explosives.

Soil: The upper layer of earth on which rain falls and in which plants grow. Composed of mineral particles, some organic material and water and ranging from fine clay to gravel or boulders. The normal classification is into:

Grade	Dominant grain size (mm)
Gravel	2 and above
Very coarse sand	2–1
Coarse sand	1–0·5
Medium sand	0·5–0·25
Fine sand	0·25–0·1
Silt	0·1–0·01
Clay	Less than 0·01

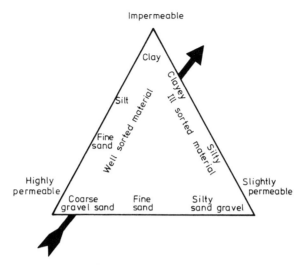

Fig. A.1. Classification of soils with respect to permeability. Arrow indicates direction of decrease in K.

In Europe, a category intermediate between fine sand and silt is recognized, namely schluff (German) or mo (Scandinavian). However, a classification based upon permeability is shown in the accompanying Fig. A.1.

Specific capacity (of a well): This is the rate at which water can be pumped from a well per unit of drawdown, q.v.

Specific retention: The ratio of the weight or volume of water which a soil will retain against the force of gravity, to its own weight or volume. It can be determined after soil has once been saturated.

Specific yield: The quantity of water which a unit volume of soil or rock will yield after being saturated and allowed to drain under specified conditions. Expressed as a percentage of volume.

Storage: The impounding of water in an aquifer or in a surface reservoir.

Storage coefficient (aquifer): The ratio of (a) the volume of water taken into or released from storage in a prism of aquifer of unit surface area and the total thickness of the aquifer, to (b) the volume of the prism of aquifer per unit change in the component of pressure head normal to that surface.

Storage of aquifer: The amount of water released from storage in an aquifer with a given lowering of head.

Strike: A line on a rock stratum at right angles to the full dip of that stratum.

Structure contours: Contours drawn on the upper surface of an aquifer and used in order to depict its form and depth.

Sub-artesian water: Artesian water which the pressure makes rise in a well, but not to overflow at the surface.

Submersible pump: Usually an electrical centrifugal pump capable of operating entirely submerged in water.

Subnormal pressure (water): A pressure converse to that in artesian water, i.e. related to confining beds in which water is being pressed downwards.

Subsurface water: All water which is subterranean, whether solid, liquid or gaseous.

Synthetic unit hydrograph: A unit hydrograph, *q.v.*, prepared for a drainage basin which is ungauged; its basis is the basin's known physical characteristics.

Tested capacity (well): The maximum rate of yield by pumping from a well without depleting its water supply. The capacity is determined by tests over a specified period.

Thermal spring: A spring which is fed by hot groundwater, usually mineralized.

Total head (pump): The sum of the delivery pressure head, static pressure head and friction head.

Total porosity: This term includes capillary porosity, i.e. the small voids holding water by capillarity, and non-capillary porosity, i.e. the large voids which will not hold water by capillarity.

Total runoff: Includes both surface runoff and shallow percolation.

Tracer: A dye or a salt or another substance such as radioactive chemical compound which is employed to track the movement of water.

Transmissibility: See transmissivity coefficient.

Transmission coefficient: See coefficient of permeability.

Transmissivity, transmissivity coefficient: The product of thickness of saturated portion of aquifer and the field coefficient of permeability.

Transpiration: The emission of water vapour by living plants into the atmosphere, almost always in daylight.

Unconfined groundwater: Groundwater which is not restrained in its movement by an impervious or confining bed above or below.

Underground water: Groundwater.

Unit hydrograph: In reference to a river, this is that hydrograph produced by an isolated storm of a given duration and of uniform intensity over the entire drainage area, which had an equivalent runoff of precisely 1 in (25·4 mm) of rain.

Unit weight of water: The weight of water per unit volume of water. Normally taken as 1 g/cm^3 or 62·4 lb/ft^3.

Vadose water: Water held in the zone of aeration.

Void: A pore or cavity between particles in a rock or soil mass which may be filled by air or water or both or by some other substance. An interstice.

Voids ratio: The ratio of the volume of voids to the volume of solids in a soil sample.

Water balance: The accounting for all water inputs and all water outputs within a system.

Water of capillarity: Porewater in rocks or soils above the water table.

Water table: The surface of the zone of saturation. Subject to fluctuation, it follows in a flatter form the profile of the land surface.

Water table contour plan: This shows contour lines on the water table.

Water table gradient: The inclination of the water table.

Water table level: That level at which the water table is encountered in borehole or well.

Water table profile: A vertical section of the water table in a specified direction constructed from a water table contour plan or from levels in wells or boreholes.

Water well: A well put down to provide a supply of water.

Wave of the water table: See phreatic wave.

Well: An excavation from the surface to obtain water ranging from shallow level to about 400 ft (120 m).

Well hydrograph: A graph showing fluctuations of water level in a well.

Well interference: That condition arising when the cone of depression of a well is affected by that of an adjacent well, resulting in a decrease in its yield of water. Such a diminution can be expressed as a lowering of the water table in feet or metres or alternatively as a reduction in yield.

Well log: A description of the rocks passed through in a well.

Wellpoint: A shallow well with a pump to drain a water-bearing soil around or along an excavation. A wellpoint strengthens ground and reduces the pumping necessary inside the excavation. Essentially, it comprises a tube of about 2 in (50 mm) diameter which is driven into the ground by jetting, *q.v.* The tube is fitted with a close mesh screen at the bottom and connected through a header pipe to a suction pump at the surface. On a major project, several wellpoints are sunk and connected to a common header pipe. Fine sands, of course, cannot be drained by wellpoints.

Well screen: A strainer inserted in a well when pumping water from a loose and gravelly deposit, functioning to exclude solid particles over a certain size but permitting fine sand to enter with the water and be removed.

Yield: Usually the economic yield of a well. Probable yield can be estimated if the permeability of the strata is known and a short pumping test is effected in order to give the different values of drawdown for successive increases in rate of pumping.

Zone of aeration: The ground above the main water table and extending to the surface. Comprising in ascending order the capillary fringe, *q.v.*, an intermediate belt, *q.v.* and the belt of soil water, *q.v.* Obviously variable in thickness.

Zone of capillarity: The ground above the water table containing water of capillarity, *q.v.*

Zone of saturation: The mass of water-bearing ground below the main water table and comprising solid rocks and incoherent materials.

APPENDIX 2

Dictionary of Isotope Terms

^{14}C *age, adjusted:* The age calculated from the radiocarbon content of a water sample and from an initial content of this radioisotope evaluated on the basis of a given correction method, usually on the basis of the $\delta(^{13}C)$ value of dissolved carbonate species.

^{14}C *age, apparent:* The age calculated from the radiocarbon content of a water sample assuming an initial content of this radioisotope equal to 100% of the modern (this condition is not attained for carbonate species dissolved in water). The apparent age is given by: $8270 \ln(100/C)$ years, where C is the ^{14}C content of the water sample relative to modern.

^{14}C *Modern:* The radiocarbon content of natural compounds of carbon is given as a percentage of the ^{14}C content of 'modern' carbon. This is accepted as equal to 95% of the ^{14}C content (in 1950) of an oxalic acid standard distributed by the National Bureau of Standards in Washington, DC, corresponding to the normal ^{14}C content of carbon dioxide in the atmosphere prior to 1890.

Cosmic radiation: Primary cosmic radiation has two constituents, one originating in the sun, the solar component, and the second originating in the stars, the galactic component. It comprises very high energy protons and alpha particles together with a small fraction of heavier nuclei. Secondary cosmic radiation is produced by interaction of the primary cosmic radiation with the components of the upper atmosphere. It comprises a variety of nuclear particles such as neutrons, protons, mesons, etc. Reactions of neutrons and protons with atmospheric components produce tritium and radiocarbon.

Counters: Instruments for detecting radioactive events and concentrations of associated substances. There are various types. The Geiger–Müller counter consists of a metallic cylinder with an axial wire which is positively charged to 1000 or 1500 V with respect to the cylinder. Alpha, beta or gamma particles which enter the cylinder promote ionization and discharge between the wire and the wall, this being detected and recorded electronically.

The proportional counter is commonly utilized in tritium and radiocarbon determinations and is basically similar to the Geiger–Müller, but with the potential of the wire adjusted so that current pulses produced by incoming radiation are related to the energies released. Hence, it is possible to distinguish radiations possessing different energies.

The scintillation counter uses the fact that when reacting with a radiation some substances emit a flash of light, which can be detected by a photomultiplier and recorded. For tritium and radiocarbon detection, liquid scintillation counters are employed, in which a liquid scintillates on interacting with beta particles mixed with the water sample.

Curie, Ci: The unit of radioactivity; it corresponds to the quantity of any radioisotope undergoing $3\cdot7 \times 10^{10}$ disintegrations per second. 1millicurie $(mCi) = 10^{-3}$ Ci; microcurie $(\mu Ci) = 10^{-6}$ Ci; the nanocurie $(nCi) = 10^{-9}$ Ci; picocurie $(pCi) = 10^{-12}$ Ci.

Delta, δ‰: This is defined by the expression:

$$\delta\text{\textperthousand} = \left(\frac{R_{\text{sample}}}{R_{\text{standard}}} - 1\right) \times 1000$$

where R is D/H or $^{18}O/^{16}O$ or $^{13}C/^{12}C$. Original standards used were SMOW for the isotopes of oxygen and hydrogen and PDB for carbon with the notations δD‰, $\delta^{18}O$‰, $\delta^{13}C$‰. Thus, a δ‰ of $+10$ or -10 means that the heavy isotope of the sample in question, i.e. D or ^{18}O or ^{13}C, is higher or lower by 10‰ than the standard content. Accuracies in determination are of the order of 1‰ for D/H and 0·1‰ for $^{18}O/^{16}O$ and $^{13}C/^{12}C$. In nature, relative variations of the D/H ratio in waters are usually up to eight-fold greater than those of $^{18}O/^{16}O$; therefore accuracies for hydrogen and oxygen are practically comparable.

Environmental isotopes: Isotopes of natural and artificial origins occurring in nature globally and in a manner which is not susceptible to control by mankind. Variations of these isotopes in natural waters are used by investigators in order to study significant aspects of surface water and groundwater, including origins, ages, flow rates, directions of flow, etc. When waters contain appropriate isotopes with concentrations and distributions that are known both spatially and in time, specific water masses can be labelled in order to trace their origins and provenances. As well as through the atmosphere, such environmental tracers may enter the hydrologic cycle by leaching, by exchange with organic matter or rock material in soils, or by deeper-seated exchange processes or radioactive decay processes.

Evaporation effect: With stable isotopes, this is given by:

$$\delta D \permil = a \delta^{18}O \permil + b$$

with $4 < a < 8$ and $b < 10$ fitting points which represent the isotopic compositions of waters which have undergone different degrees of free surface evaporation under similar environmental conditions from an original state in which all had the same isotopic composition.

In a $\delta^{18}O$–δD diagram, the evaporation lines will lie to the right of the meteoric water line; the point where evaporation and meteoric water lines meet represents the $\delta^{18}O$ and δD values of the water in question prior to evaporation.

Such evaporated waters are to be found among surface waters, especially in lakes. However, the isotopic composition of groundwater arising in part or entirely from surface waters which have undergone evaporation can deviate markedly from the meteoric water line and follow an evaporation line.

Half-life, $T_{1/2}$: The time required to reduce the concentration of a given radioactive isotope to half its initial value by radioactive decay.

$$T_{1/2} = \frac{\ln 2}{\lambda}$$

where λ is the decay constant characteristic of the radioisotope; see also radioactive law of decay.

Isotopes: Atoms of the same chemical element having different masses.

They have the same number of protons in the nucleus, but a different number of neutrons. Thus, for example, hydrogen has three naturally occurring isotopes with atomic masses of 1, 2 and 3, called protium, deuterium and tritium, respectively. Oxygen also has three naturally occurring isotopes of interest in hydrology and these have atomic masses of 16, 17 and 18 of which the compositions are as follows:

$^{16}_{8}O$, i.e. 8 protons and 8 neutrons in the nucleus.
$^{17}_{8}O$, i.e. 8 protons and 9 neutrons in the nucleus.
$^{18}_{8}O$, i.e. 8 protons and 10 neutrons in the nucleus.

The subscript at the left of the chemical symbol indicates the number of protons in the nucleus (the atomic number, z), the superscript indicates the mass number A (the sum of the number of protons, Z, together with the number of neutrons, N). Thus, $A = Z + N$.

In a neutral atom, the nuclear charge balances that of the surrounding electrons; where this is not the case, it becomes an ion.

The diameter of an atomic nucleus is of the order of 10^{-13} cm and the diameter of the atom about 10^{-8} cm (1 Angstrom unit, Å). The density of the atomic nucleus is approximately 10^{14} g/cm^3.

Stable isotopes have nuclei with stable configurations and their concentrations in a closed system do not alter with time unless they are produced by some radioactive element within the system.

Radioactive isotopes possess unstable nuclei and change into isotopes of other elements with time (radioactive decay). The rate of decay is expressed by the half-life or mean life, q.v. The concentration of radioactive isotopes in a closed system decreases exponentially with time unless there is production at the same time by other radioactive processes.

Isotopic exchange: Definable as the exchange of atoms between chemical compounds, a process permitting different isotopic compositions of coexisting compounds or phases to attain an equilibrium characteristic of the conditions of the system involved. In hydrology, the most important processes of isotopic exchange are:

(a) Water/rock, especially water/limestone. Here, the oxygen isotopic composition both of the water and the limestone can be affected, but the process is extremely slow at the temperatures normally prevailing in aquifers. It becomes significant only in geothermal waters where the rate of exchange is accelerated. Hydrogen is unimportant since its occurrence in rocks is very small indeed.

(b) Carbon isotopic exchange between dissolved bicarbonate and solid calcium carbonate in the aquifer can modify the ^{13}C and ^{14}C contents of the bicarbonate utilized for radiocarbon age determination. Again, this process is extremely slow at normal temperatures and the quantity of dissolved bicarbonate is very small, but nevertheless its isotopic composition can be affected and the $\delta^{13}C$ value displaced in a positive direction.

Isotopic fractionation: A difference in the distribution of isotopes of the same element in two chemical phases in mutual isotopic exchange. Isotopic fractionation takes place because the isotopes possess very slightly different physico-chemical characteristics due to their mass differences. Hence, isotopic fractionation is usually greater in light elements because the relative mass differences are greater.

Isotopic fractionation factor: The ratio between the isotopic proportions of an element in two different chemical phases, usually indicated by α. For instance, in the case of the equilibrium between water and water vapour at 0° and 25°C, the fractionation factors for the isotopes of oxygen and hydrogen are as follows:

$$\alpha\,^{18}O = \frac{^{18}O/^{16}O \text{ water}}{^{18}O/^{16}O \text{ vapour}} \quad 1 \cdot 0115\,(0°C),\ 1 \cdot 0092\,(25°C)$$

$$\alpha\,D = \frac{D/H \text{ water}}{D/H \text{ vapour}} \quad 1 \cdot 108\,(0°C),\ 1 \cdot 077\,(25°C)$$

The fractionation factors approach unity with increasing temperature and, when $\alpha = 1$, the isotopic ratios become equal in both the phases considered, therefore there is no isotopic fractionation. Fractionation may be expressed also as $\varepsilon\text{\textperthousand} = (\alpha - 1) \times 1000$.

Mass spectrometer: An instrument permitting determination of relative abundances of constituents possessing different masses in a given substance, e.g. isotopic species of a chemical compound.

Commonly, the investigated compound is ionized, often through impact by an electron beam, the ions then being accelerated through an electric field to a magnetic field normal to their path where they follow circular trajectories of which the radii depend upon the masses of the ions. The following expression is applicable:

$$r=\frac{1}{H}\sqrt{\frac{2Vm}{e}}$$

where r is the radius, H is the magnetic field intensity, V is the accelerating voltage, m is the mass of the ion and e is its electrical charge. Ions with different masses emerge from the magnetic field following different trajectories and can be collected separately in order that their relative concentrations can be measured.

In hydrology, mass spectrometers are employed in measuring the abundances of stable isotopes. The abundances of radioactive isotopes are too low to be so measured.

Mean life: The mean life, τ, of a radioisotope is given by:

$$\tau = \frac{\int_0^\infty C \, dt}{C_0} = \frac{1}{\lambda} = \frac{T_{1/2}}{\ln 2}$$

(see also radioactive law of decay).

Meteoric water line: For stable isotopes, in a $\delta^{18}O - \delta D$ diagram, this is the line of the equation $\delta D ‰ = 8 \, \delta^{18}O ‰ + 10$ which best fits the points representing the isotopic compositions of samples of precipitation from all over the world. In some areas, the line is displaced so that the intercept on the deuterium axis has a value other than 10; it then constitutes a regional meteoric line. Groundwater usually follows the same $\delta^{18}O$ and δD relationship as surface water except where the latter is substantially evaporated (as in a lake). Free surface water evaporation enriches the residual water in the heavy stable isotopes, but not in that relative proportion established by the above relationship.

PDB: Calcium carbonate obtained from the calcareous guard of a Cretaceous Belemnite from the Peedee Formation of South Carolina in the USA, belonging to the species *Belemnitella americana*. It had a carbon isotopic composition corresponding to a reasonable average of that of marine limestone. It possessed a ^{13}C abundance of 1·111%. PDB also constituted a reference standard for $^{18}O/^{16}O$ determinations in paleotemperature measurements. Unfortunately, it no longer exists and other standards, with a known relationship to it, have to be substituted, e.g. NBS-20 (Solenhofen limestone) and TS (marble) were distributed by the National Bureau of Standards in Washington, DC, and the International Atomic Energy Agency in Vienna respectively. PDB, SMOW and SLAP standards are discussed in detail in Section 12.2. However, it may be added that the IAEA Advisory Group in its meeting on stable isotope reference samples for geochemical and hydrological investigations which took place in Vienna on 19 to 21 September, 1983, presented a new carbonate standard, NBS-19. This is another replacement for PDB and has the following values in relation to it:

$$\delta^{18}O = -2\cdot20\%_{oo}$$
$$\delta^{13}C = +1\cdot95\%_{oo}$$

Copies of relevant reports may be obtained from R. Gonfiantini at the IAEA, P.O. Box 100, 1400 Vienna, Austria.

Radioactive law of decay: This is given by:

$$\frac{dC}{dt} = -\lambda C$$

where C is the concentration of radioactive nuclei in the system at a time t, dC/dt is the rate of decay and λ is the decay constant characteristic of each isotope. Integrating gives:

$$C = C_0 \exp(-\lambda t)$$

where C_0 is the concentration of radioactive nuclei present initially at time $t=0$. If C_0, λ and C, the present concentration, are known it is possible to evaluate the age of the system because:

$$t = \frac{1}{\lambda} \ln(C_0/C)$$

Radioactivity: Refers to that property of unstable isotopes of emitting radiation while undergoing transformation, i.e. while decaying, into isotopes of other elements by nuclear processes. The most common processes of radioactive decay are as follows.

(a) α-decay: an alpha particle with mass 4 and charge 2, i.e. a helium nucleus, is emitted by the nucleus of the parent isotope which is transformed thereby into an isotope of a different element with a mass number which is lower by 4 units and an atomic number which is lower by 2 units, e.g. $^{238}_{92}U \xrightarrow{\alpha} {}^{234}_{90}Th$.

(b) β-decay: an electron is emitted by the nucleus of the parent isotope which is thus transformed into an isotope of a different element having the same mass number and an atomic number which is higher by 1 unit, e.g. $^{3}_{1}H \xrightarrow{\beta^-} {}^{3}_{2}He$; $^{14}_{6}C \xrightarrow{\beta^-} {}^{14}_{7}N$.

(c) Electron capture (EC) decay: the nucleus of the parent isotope captures an orbital electron and is transformed into an isotope of a different element having the same mass number and the atomic number which is lower by 1 unit, e.g. $^{40}_{19}K \xrightarrow{EC} {}^{40}_{18}Ar$.

The product of decay is usually in an excited state and returns to the ground state by emitting energy in the form of X-rays or gamma rays (electromagnetic radiations).

SMOW (Standard Mean Ocean Water): A standard water with an isotopic composition close to that of mean ocean water. The abundance of the three main isotopic species of water in SMOW are $H_2^{16}O$, 99.73%; $HD^{16}O$, 0.031%; $H_2^{18}O$, 0.199%. SMOW was proposed because the oceans comprise the beginning and end of important hydrological circuits and contain about 97% of all water existing on the Earth's crust, this having a rather uniform isotopic composition. Variants such as V-SMOW and SLAP are discussed in Section 12.2.

Tritium unit (T.U.): Used to express the tritium concentration in samples of natural water. One tritium unit corresponds to a concentration of one tritium atom per 10^{18} atoms of hydrogen. Tritium units can be converted to $\mu Ci/ml$ thus:

$$1 \text{ T.U.} = 3.24 \times 10^{-9} \, \mu Ci/ml$$

APPENDIX 3

Analysis of Water Quality with Reference to its Chemistry[a]

Chemical species or property	Usual water content (ppm)	Effects		General characteristics
		Biological	Industrial	
Silica	<30	Nil	Nil	In waters with pH 6–8.5, insoluble
Calcium	10–100	Essential	Sometimes necessary	Loosens soils, maintains pH above 6
Iron	<1	Essential	Some industries have low iron tolerance	Solubility greater in water of pH 4.0
Magnesium		Essential	Sometimes necessary	Buffers undesirable toxics and loosens soils
Manganese		Traces are essential; larger amounts are toxic	Some industries have low Mn tolerance	Imparts objectionable taste to potable water
Aluminium	<0.1	Harmful to eyes at stated concn	Nil	Insoluble below pH of 9.0
Sodium, potassium		Essential	Nil	Very soluble, packs soils thereby preventing aeration and nitrogenation
Chlorine		Traces are essential	Bad for the dairy, ice and sugar industries	Corrosive

APPENDIX 3

Fluorine	0–4	Traces are essential	Nil	Corrosive and toxic as a gas
Sulphate		Traces are essential	Bad for the dairy, ice and sugar industries	Very soluble, usually a pollution measure in rain and forms strong acid
Nitrate		Carcinogenic	Bad in dyeing and fermenting industries	Reacts with organic compounds and promotes excess plant growth when in excess; excess reduces permeability of soils
pH (optimal range 5-8.5)		Not important	Only of the individual acids or bases	Alkalinity produced by hydroxides buffers harmful pH changes
Hardness		Nil	Undesirable	Ca and Mg are the main constituents
DS (dissolved solids)	50–3 000	More than 1 000 ppm bad for humans	Industry normally needs less than 1 000 ppm	More than 2 000 ppm are bad for irrigation
DO (dissolved oxygen)	4–8	Aquatic fauna need specific quantities	Insignificant	Measure of pollution — the lower the DO, the greater the pollution
Free CO_2		Small	Small	Fish are normally adapted to specific free CO_2 ranges
Turbidity		5 ppm or less are recommended for human consumption		Normal quantities do not affect stream life

(continued)

Chemical species or property	Usual water content (ppm)	Effects		General characteristics
		Biological	Industrial	
Temperature		Each life form has a specific range		Rise causes increase in organic growth and reduces palatability, i.e. warmer irrigation water promotes plant growth
Radioactivity		Variable	Damage to photographic processes	Excess radiation is deleterious to biological systems; the relevant public water supply standards are: gross beta, 1 000 pCi/litre; radium-226, 3 pCi/litre; strontium-90, 10 pCi/litre.

[a] Data derived from SKINNER, BRIAN J., 1969. *Earth Resources*. Prentice-Hall, Englewood Cliffs, N.J.

APPENDIX 4

Some Representative Porosities and Permeabilities for Various Geologic Materials[a]

Material	Representative porosities (% void space)	Approximate range of permeability (gal/day/ft^2; hydraulic gradient = 1)
Loose		
Clay	50–60	0.00001–0.001
Silt and glacial till	20–40	0.001–10
Alluvial sands	30–40	10–10 000
Alluvial gravels	25–35	10 000–10^6
Indurated		
Sedimentary		
Shale	5–15	0.0000001–0.0001
Siltstone	5–20	0.00001–0.1
Sandstone	5–25	0.001–100
Conglomerate	5–25	0.001–100
Limestone	0.1–10	0.0001–10
Igneous and metamorphic		
Volcanic (basalt)	0.001–50	0.0001–1
Weathered granite	0.001–10	0.00001–0.01
Fresh granite	0.0001–1	0.0000001–0.00001
Slate	0.001–1	0.0000001–0.0001
Schist	0.001–1	0.000001–0.001
Gneiss	0.0001–1	0.0000001–0.0001
Tuff	10–80	0.00001–1

[a]Data from: WALTZ, J. P., 1976. Ground water. In *Introduction to Physical Hydrology*, ed. Richard J. Chorley. Methuen, pp. 122–30.

APPENDIX 5

Groundwater in Law

Legal consequences may arise if something goes wrong after submission and implementation of reports on groundwater, and therefore geologists preparing such materials should be careful to include only conditions actually encountered during exploration. Thus, extrapolating borehole logs to bedrock is most inadvisable and it is important to attach dates of measurement to groundwater levels, in order to guard against substantial fluctuations which may cause the water table to rise prior to construction. Also, reference to water should be made in all logs because if it is not, a contractor may assume that there is none present. If he later encounters water, he could claim for additional costs incurred as a result. All these considerations are even more important in matters relating to land use. For instance, in 1966, San Antonio, Texas, entered into a land use mapping programme together with the US Soil Conservation Service and, as a result, soils maps and a soils handbook were produced. Various sites were designated as recreational, residential, etc., but it was found later that not enough data were supplied for subsurface conditions below 3 m. C. L. McGuiness in 1951 called for hydrologically sound statutes to avoid problems of this type.[1]

In legal practice, a right to take possession of water deriving from a natural source of supply and apply it to a beneficial use constitutes a water right.

In the USA, two laws govern water: a common law of riparian rights and a law of prior appropriation. As regards groundwater, the first of these laws refers to the ownership of land which overlies a water-bearing formation, the term 'land-ownership right' being applicable.

The concept of land ownership right originated in the United Kingdom in the early part of the nineteenth century. In 1942, W. A. Hutchins listed pragmatic arguments that favoured it.[2] These are as follows:

(a) The origin of the relevant waters is unknown, therefore not subject to law.
(b) The recognition of correlative rights, i.e. the rights of adjacent landowners, would interfere with important public projects.

The English rule of *unlimited* use proved inappropriate in the USA, particularly in the west. From a law suit in 1862, that of *Basset v. Salisbury Manufacturing Company*, 43 N.H. 569, 82 Am. Dec. 179, the principle that right of use of percolating water will be *limited* by the corresponding use of a neighbour was established. Essentially, it means that the rights of others are recognized and, later, this concept was termed the American rule of reasonable use. In 1903, there was a modification called the California rule, which stated that the rights of landowners over a common groundwater basin are co-equal. This means that no landowner can use more than his or her share if, by so doing, the rights of others are affected adversely. In conditions of inadequate supply, reason must be exercised and reasonable shares allocated.

The land ownership approach involves a spatial position. The matter of prior appropriation involves a chronological position, i.e. the case where earlier rights override later ones. If adopted, it would mean that a landowner would not necessarily enjoy an inherent right to water from sources on or near his or her land; rather, the rights to such sources would be grounded on the chronological priority of beneficent utilization and could be lost if such beneficial utilization ceases. The origin of this approach was in California where miners used streamwater in placer mining on public land. The Californian Supreme Court recognized prior appropriation in 1855 and Congressional Acts in 1866 and 1870 recognized the right of appropriating water on public land and even extended protection to land which later became private. In 1877, the Desert Land Act also recognized it and, in fact, all the western states in the USA acknowledge it to varying degrees. During drought, later appropriators must cease using water in reverse order of priority. There are no rights for non-beneficial use of water.

The prior appropriation doctrine gained strength over riparian rights because it was believed to have fewer defects, and that it could promote vested rights; see for instance the views of the National Planning Resources Board in 1943.[3]

Some US states allow prescriptive rights to claimants; this entails loss of rights by prior appropriators or landowners if they use water adversely, under certain legally specified conditions.

Earlier legal interpretation arose because groundwater was not understood; for instance in the suit of *Roath* versus *Driscoll* 20 Conn 533 in 1850 it was referred to as moving through factors beyond apprehension.

However, even then, and of course in later times, attempts were made to differentiate between water courses and percolating waters. The former were defined as natural surface streams or subterranean water flowing along channels from definite points of origin. Percolating waters are those moving downwards, but not constituting part of an underground water stream. Groundwater law should apply to water in the saturated zone.

The concept of states' rights complicates the situation in the USA because there, it is agreed that the local control of groundwater ought to be in the hands of the individual state or states concerned, and not in those of the Federal Government. This results in inconsistency in law between states and can cause difficulty in controlling waters shared jointly between states. Climate plays an important rôle. In the 31 eastern states, there is usually a surplus of water and precipitation is normally sufficient to provide an adequate water supply. Thus, it is not surprising that riparian rights are favoured. On the other hand, the Rocky Mountain states are arid with insufficient rainfall; consequently, water must be imported except in the actual mountains. Prior appropriation is preferred. In the Pacific coast states and those of the Great Plains (Montana to Texas), both doctrines occur, probably because the former contain both arid and humid regions whilst the latter have a semi-arid climate intermediate in type between those of the arid west and the humid east of the continent.

The phenomenon of overdraft is widespread, and controlling groundwater resources in such circumstances is so complex legally that it is often handled by the courts, state administrative agencies or landowner associations. It may be minimized by reducing pumping and restricting the number of years through which this is permitted to continue, essentially rationing the groundwater available for use. A famous instance may be given relating to the Raymond Basin (San Gabriel Valley) in Southern California. The case commenced because an auxiliary water supply, Metropolitan Water District water conveyed by conduit from the Colorado River, was available. One user was the City of Pasadena which initiated an action in 1938 for adjudication of all subterranean water rights in the basin after city engineers decided that the draft greatly exceeded the recharge to the Raymond Basin. No less than 31 parties became involved to the suit and, in 1939, after a petition brought by one of these, the court referred the matter to the State Division of Water Resources so as to obtain a technical report on the existing physical situation. This stated in 1943 that recharge was about 70% of the draft and a final settlement was given by the Californian Supreme Court in 1949. This ruled that the pumping by all concerned was to be proportionately reduced so as to ensure that the total annual pumpage would not exceed the safe yield. A watermaster was appointed to monitor this. Those parties possessing other sources of water, e.g. the City of Pasadena, were allowed to sell annually all or part of the water from the basin to which they held rights to other parties without access to other sources. The necessary reduction in usage was carried out quite amicably and water sold could be pumped by the purchaser from his own wells. Later, a law was enacted to encourage development of substitute water supplies other than groundwater, a sensible idea in areas liable to the problems of overdraft.

A complication arises where surface water is used to supplement groundwater in order to overcome overdraft. The title to the use of such water can be based on different grounds from those which apply to natural groundwater itself. It may be invested in the overlying landowner or several such landowners, or even a state in the USA. Where no claim of ownership is made by whoever was responsible for injecting such surface water, the owner of overlying land may claim it. The supply introduced is regarded then as abandoned water which mixes with groundwater. However, the operation of injecting such recharge water is expensive and involves physical construction. Where this is the case, individual landowners may not be able to pay, even in association. Then the state usually provides finance and retains title to the supplementary water, as has happened in several American states. A Federal agency of the US Department of the Interior, the Bureau of Reclamation, can and does reserve rights to all return flows and wastage from project water delivered to contracting regions. Any unappropriated groundwater belongs to the state where it is. Where a state acquires surface water and introduces it underground on lands not owned by the state itself, it is considered as acting in the public interest as a non-landowning body.

Among general requirements for optimum development of groundwater are the following.

(a) Drillers must be licensed and must report on prescribed forms, giving all facts relating to installation, deepening or re-perforating of a well.

(b) Geological logs must be kept and refer to factors such as depth, stratal lithological details, structure of the subterranean beds, etc.
(c) Artesian wells with flowing water must be capped for conservation.
(d) Abandoned wells must be sealed to prevent accidents or pollution.
(e) Water pumped for air conditioning or to cool must be re-cycled into the aquifer using recharge wells. Otherwise, it is disposed of through sewers, when a sewer tax may become payable.
(f) Pollutants disposal is strictly regulated because their introduction can degrade the groundwater system, e.g. through addition of brine.

In the United Kingdom, a Water Resources Act became law in 1963. This provided for establishment of River Authorities and a Water Resources Board as well as providing for the transfer to these River Authorities of functions previously exercised by River Boards and other bodies. The term 'water resources' was defined as water for the time being contained in any source of supply in (an) area; this source of supply includes inland waters and any water in underground strata. No distinction was made between water in the unsaturated zone and groundwater. The following observation was made: water for the time being contained in a well, borehole or similar work, including any adit or passage constructed in order to facilitate collection of water in the well, borehole, etc., as well as water for the time being contained in any excavation into such underground strata where the level of the water is a function of the entry of water from these same strata, i.e. depends wholly or mainly upon such an entry taking place, constitute components of it.

Conservation, augmentation or re-distribution of water are primary purposes of the Water Resources Board in the Act.

Subsequently, the Water Act of 1973 aimed at a national policy for water, including the same purposes listed above, (a)–(f), and the present Water Authorities were established on a regional basis.

In the Federal Republic of Germany, the situation is reminiscent of that in the USA because the individual states have a large degree of autonomy. Consequently, in North Rhine Westfalen, statutes were enacted in respect of the development of brown coal resources and interference with groundwater conditions west of Cologne which were described in Section 17.8.6. This brown coal plan was a part of an overall land development programme and of course, it entailed open-cast mining. This was correctly envisaged as having widespread consequences, and appropriate actions to ameliorate adverse results were incorporated into the plan; some of them were put into effect. Legally, the plan came under Art. 1 Nr. 22 of laws relating to changes in the affected area made on 20 November, 1979 (GV NW S. 730/SGV NW 230) and 28 November, 1979 (GV NW S. 878/SGV NW 230).

REFERENCES

1. McGuiness, C. L., 1951. *Water Law with Special Reference to Groundwater*. US Geol. Surv. Circ. 117, 30 pp.
2. Hutchins, W. A., 1942. *Selected Problems in the Law of Water Rights in the West*. US Department of Agriculture, Misc. Publ. 418, Washington, DC, 513 pp.
3. National Planning Resources Board, 1943. *State Water Law in the Development of the West*. Water Resources Committee, Sub-committee on State Water Law. Washington, DC.

APPENDIX 6

Organizations Connected with Ground Matters

Advisory Commission of Inter-Governmental Relations, 1701 Pennsylvania Avenue, NW, Washington, DC 20575, USA.

American Geological Institute, 4220 King Street, Alexandria, Virginia 22302, USA.

American Public Power Association, 919 18th Street, NW, Washington, DC 20006, USA.

American Society of Civil Engineers, 345 East 47th Street, New York, NY 10017, USA.

American Society of Limnology and Oceanography, Oregon State University, Corvallis, Oregon 97331, USA.

American Water Resources Association, St Antony Falls, Hydrology Laboratory, Minneapolis, Minnesota 55414, USA.

American Water Works Association, 6666 West Quincy Avenue, Denver, Colorado 80235, USA.

Australian Atomic Energy Commission, Research Establishment, Lucas Heights Research Establishment, Sutherland, NSW 2232, Australia.

Australian Water Resources Association, Department of National Development, Canberra, Australia.

British Geological Survey, Geological Museum, Exhibition Road, London SW7, UK.

British Non-ferrous Metals Research Association, Euston Street, London NW1, UK.

British Waterworks Association, 34 Park Street, London W1Y 3PF, UK.

Bureau of Mineral Resources, Geology and Geophysics, Canberra, ACT 2601, Australia.

CSIRO, Institute of Earth Resources, North Ryde, NSW 2113, Australia.

Department of Hydrology and Water Resources, University of Arizona, Tucson, Arizona 85721, USA.

Federal Institute of Hydrology, Koblenz, Federal Republic of Germany.

Federal Power Commission, 414 G Street, Washington, DC 20426, USA.

FAO (Food and Agriculture Organization), Via delle Terme di Caracalla, 00100 Rome, Italy.

Geological Society of London, Burlington House, Piccadilly, London W1V 0JU, UK.

Geophysical Isotope Laboratory, University of Copenhagen, Haraldsgade 6, DK 2200 Copenhagen N, Denmark.

Institution of Civil Engineers, Great George Street, London SW1P 3AA, UK.

Institute of Environmental Physics, University of Heidelberg, Im Neuenheimer Feld 6900, Heidelberg, Federal Republic of Germany.

Institute of Hydrology, Wallingford, Berks, UK.

Institute of Hydrodynamics and Hydraulic Engineering (ISVA), Technical University of Denmark, Building 115, DK 2800 Lyngby, Denmark.

Institution of Water Engineers, 11 Pall Mall, London SW1Y 5LU, UK.

International Association of Engineering Geology, c/o Laboratoire Central des Ponts et Chaussées, 75732 Paris, Cedex 15, France.

International Association of Hydrogeologists, P.O. Box 9090, 6800 GX Arnhem, The Netherlands.

International Atomic Energy Agency, P.O. Box 100, 1400 Vienna, Austria.

Jet Propulsion Laboratory, California Institute of Technology, Pasadena, California 91109, USA.

National Water Well Association, 500 West Wilson Bridge Road, Suite 135, Worthington, Ohio 43085, USA.

New Mexico Institute of Mining and Technology, Socorro, New Mexico 87801, USA.

Nuclear Structure Research Laboratory, University of Rochester, NY 14627, USA.

Remote Sensing Research Unit, Department of Geological Sciences, University of California, Santa Barbara, California 93106, USA.

Research School of Earth Sciences, Australian National University, Canberra, ACT 2600, Australia.

UNESCO, 7 Place de Fontenoy, 75700 Paris, France.

UNICEF, United Nations Plaza, New York, NY 10017, USA.

United Nations, United Nations Plaza, New York, NY 10017, USA.

APPENDIX 6

United States Geological Survey, Reston, Virginia 22092, USA.

US Water Resources Council, 1205 Vermont Avenue, NW, Washington, DC 20005, USA.

Water Research Association, Ferry Lane, Medmenham, Marlow, Bucks, UK.

APPENDIX 7

Measurements

TABLE A.7.1
METRIC UNITS

Measurement	Unit	Formula
Basic		
Length	Metre, m	
Mass	Kilogram, kg	
Time	Second, s	
Electrical current	Ampere, A	
Thermodynamic temperature	Kelvin, K	
Derived		
Acceleration	Metre per second squared	m/s^2
Area	Square metre	m^2
Density	Kilogram per cubic metre	kg/m^3
Electrical		
capacitance	Farad, F	$A\,s/V$
conductance	Siemens, S	A/V
field strength	Volt per metre	V/m
inductance	Henry, H	W/A
potential difference	Volt, V	W/A
resistance	Ohm, Ω	V/A
Electromotive force	Volt, V	W/A
Energy	Joule, J	$N\,m$
Force	Newton, N	$kg\,m/s^2$
Power	Watt, W	J/s
Pressure	Pascal, Pa	N/m^2
Quantity of electricity	Coulomb, C	$A\,s$
Radioactivity	Disintegration per second	(Disintegration)/s
Specific heat	Joule per kilogram-kelvin	$J/kg\,K$
Stress	Pascal, Pa	N/m^2
Thermal conductivity	Watt per kelvin-metre	$W/m\,K$
Velocity	Metres per second	m/s
Viscosity, dynamic	Pascal-second	$Pa\,s$
Viscosity, kinematic	Square metre per second	m^2/s
Voltage	Volt, V	W/A
Volume	Cubic metre	m^3
Work	Joule, J	$N\,m$

TABLE A.7.2
METRIC EQUIVALENTS

Measurement	Metric unit	Equivalents
Length	1 metre (m)	10^3 mm, 10^2 cm, 10^{-3} km
Area	1 hectare (ha)	10^4 m^2, 10^{-2} km^2
Volume	1 litre (litre)	10^{-3} m^3
Mass	1 kilogram (kg)	10^3 g

TABLE A.7.3
CGS/IMPERIAL EQUIVALENTS

Measurement	Equivalents
Length	1 cm = 0·393 7 in, 2·54 cm = 1 in, 1 m = 3·281 ft (39·37 in), 1 km = 0·621 4 mile, 1·61 km = 1 mile.
Area	1 cm^2 = 0·155 in^2, 1 in^2 = 6·45 cm^2, 1 m^2 = 10·76 ft^2, 1 ha = 2·471 acre, 1 km^2 = 0·386 1 mile.
Volume	1 cm^3 = 0·061 02 in^3, 1 litre = 0·264 2 gal (0·035 31 ft^3), 1 m^3 = 264·2 gal, 35·31 ft^3.
Mass	1 g = 2·205 × 10^{-3} lb, 1 kg = 2·205 lb, 9·842 × 10^{-4} long ton.
Flow rate	1 litre/s = 15·85 gpm, 0·022 82 mgd, 0·035 31 cfs, 1 m^3/s = 1·585 × 10^4 gpm, 22·82 mgd, 35·31 cfs, 1 m^3/d = 0·183 4 gpm, 2·642 × 10^{-4} mgd, 4·087 × 10^{-4} cfs.
Velocity	1 m/s = 3·281 ft/s, 2·237 mi/h, 1 km/h = 0·911 3 ft/s, 0·621 4 m/h.
Temperature	°C = K − 273·15, (°F − 32)/1·8.
Pressure	1 Pa = 9·872 × 10^{-6} atm, 1 × 10^{-5} bar, 0·01 mb, 10 dyn/cm^2, 3·346 × 10^{-4} ft H$_2$O(4°C), 2·953 × 10^{-4} in Hg (0°C), 0·102 kg(force)/m^2, 0·020 89 lb(force)/ft^2.
Heat	1 J = 0·238 8 cal = 10^7 erg 1 J/m^2 = 8·806 × 10^{-5} BTU/ft^2, 2·39 × 10^{-5} cal/cm^2. 1 J/kg = 4·299 × 10^{-4} BTU/lb(mass), 2·388 × 10^{-4} cal/g.
Hydraulic conductivity	1 m/day = 24·54 gpd/ft^2, 1 cm/day = 2·121 × 10^4 gpd/ft^2. 1 m/day = 1·198 darcy (water at 20°C). 1 cm/day = 1 035 darcy (water at 20°C).
Water quality	1 mg/litre = 1 ppm, 0·058 4 grain/gal. Equivalent weight of ion = atomic weight of ion/valency of ion. meq/litre of ion = mg/litre of ion/equivalent weight of ion. 1 meq/l = 1 me/litre, 1 epm. 1 μS/cm = 1 μmho/cm, 0·65 mg/litre, 0·1 meq of cations/litre (the last two are approximations for most natural waters in the range 100–5 000 μS/cm at 25°C.
Viscosity	1 Pa s = 1 × 10^3 cS, 10 P, 0·020 89 lb(force)s/ft^2. 1 m^2/s = 1 × 10^6 cS, 10·76 ft^2/s.
Force	1 N = 1 × 10^5 dyne, 0·102 kg (force), 0·224 8 lb (force).
Power	1 W = 9·478 × 10^{-4} BTU/s, 0·238 8 cal/s, 0·737 6 ft lb (force)/s
Energy:	1 J = 9·478 × 10^{-4} BTU, 0·238 8 cal, 0·737 6 ft lb (force). 2·778 × 10^{-7} kWh (745·7 W = 1 horsepower).

TABLE A.7.4
RELEVANT PHYSICAL PROPERTIES

Property	Value
Water	
Density, ρ, at 10°C	$1\,000\ \text{kg/m}^3$ ($1.94\ \text{slugs/ft}^3$)
Specific weight, γ,	
at 10°C	$9.807 \times 10^3\ \text{N/m}^3$ ($62.4\ \text{lb/ft}^3$)
Specific heat at 20°C	$4.168\,6\ \text{kJ/kg K}$
Latent heat of fusion	$334\ \text{kJ/kg}$
Latent heat of vaporization	$2\,135\ \text{kJ/kg}$
Thermal conductivity at 10°C	$0.419/\text{Wm K}$ at 0°C
Dynamic	
viscosity, μ,	
at 10°C	$1.3 \times 10^{-3}\ \text{Pa s}$ ($2.73 \times 10^{-5}\ \text{lb s/ft}^2$)
at 20°C	$1.0 \times 10^{-3}\ \text{Pa s}$ ($2.05 \times 10^{-5}\ \text{lb s/ft}^2$)
Kinematic	
viscosity, v,	
at 10°C	$1.3 \times 10^{-6}\ \text{m}^2/\text{s}$ ($1.41 \times 10^{-5}\ \text{ft}^2/\text{s}$)
at 20°C	$1.0 \times 10^{-6}\ \text{m}^2/\text{s}$ ($1.06 \times 10^{-5}\ \text{ft}^2/\text{s}$)
Atmosphere	
Gravitational acceleration,	
g, std, free fall	$9.807\ \text{m/s}^2$ ($32.2\ \text{ft/s}^2$)
Pressure, p, std	$1.013 \times 10^5\ \text{Pa}$ ($14.7\ \text{psia}$)

TABLE A.7.5
ADDITIONAL DATA ON QUANTITIES

Atmospheric pressure

On weather charts, this is recorded in millibars (mb), where

$100\ \text{mb} = 1\ \text{bar} = 750.062$ standard millimetres of mercury (mm Hg) $= 100\,000\ \text{Pa}$

and the standard millimetre of mercury is based upon mercury density $13\,595.1\ \text{kg/m}^3$, $g = 9.806\,65\ \text{m/s}^2$.

Standard atmospheric pressure (1 atm) $= 101\,325\ \text{Pa} = 14.7\ \text{psia}$.

Density of half-saturated air at 20°C and 760 mm Hg $= 1.199\ \text{kg/m}^3$.

The air density increases in proportion to increased pressure and decreases in inverse proportion to absolute temperature. At 20°C, it decreases with humidity by approximately $0.000\,1\ \text{kg/m}^3$ for each 1% increase in relative humidity, and it decreases with altitude approximately thus:

Altitude (m)	900	1 800	3 000	4 300	5 500
Relative density	0.9	0.8	0.7	0.6	0.5

Gravity

Acceleration due to gravity varies with latitude and slightly also with altitude; thus at the equator $g=9.78$ m/s^2, in London $g=9.81$ m/s^2, and at the poles, $g=9.83$ m/s^2. The values arise from a combination of the Earth's attraction, expressed as G, the gravitational constant, and the centrifugal force caused by the rotation of the planet (equivalent to approximately 0.03 m/s^2 at the equator). Standard acceleration (standard gravity) $= 9.816\,65$ m/s^2 (ca 32.174 ft/s^2).

Thermometry

The thermometric scales are based upon the Kelvin thermodynamic scale of temperature.

In practice, the International Practical Scale of Temperature 1948 (IPST) is used. This relates to various fixed points such as the following.

Boiling point of oxygen	$-182.970°$C.
Melting point of ice	$0°$C.
Boiling point of water	$100°$C.
Boiling point of sulphur	$444.6°$C.
Freezing point of silver	$960.8°$C.
Freezing point of gold	$1\,063.0°$C.

Above $1\,063.0°$C, temperatures are defined in terms of Planck's law of radiation.

On the thermodynamic scale of temperature which commences at 0 K at the absolute zero of temperature, the melting point of ice is 273·15 K and the triple point of the equilibrium state (ice, water, water vapour) is 273·16 K.

Atomic physics

Avogadro's number	$6.022\,52 \times 10^{23}$ atoms/mol.
Loschmidt's number	$2.687\,13 \times 10^{19}$ atoms/m^3 at STP.
Faraday's constant of electrolysis	$96\,487$ C/mol.
Charge of electron	$1.602\,07 \times 10^{-19}$ C.
Mass of electron	$9.108\,5 \times 10^{-31}$ kg.
Ratio of mass to charge of electron, m/e	$1.758\,9 \times 10^{11}$ C/kg.
Planck's constant	$6.625\,2 \times 10^{-34}$ Js.
Rydberg constant (for hydrogen)	$10\,967.758$ mm^{-1}.
Mass of proton	$1\,836.12 \times$ mass of electron.
Mass of neutron	$1\,836.6 \times$ mass of electron.
Electron volt (eV)	$1.602\,19 \times 10^{-19}$ J.
Unified atomic mass unit	$1.660\,53 \times 10^{-27}$ kg.

General

Boltzmann's constant	$1.380\,42 \times 10^{-23}$ J/K.
Stefan–Boltzmann constant	5.69×10^{-8} Wm^{-2} K^{-4}.
Wien's constant	$0.002\,89$ mK.

Mathematical constants:

π	3·141 592 7
$\log_{10}\pi$	0·497 149 9
e	2·718 281 8
$\log_e 10$	2·302 585 1

Velocity of sound in water at 20°C = 1 484 m/s.
Velocity of sound in dry air at 0°C = 331·46 m/s.

Earth

Radius for a sphere of equal volume	6 371 km.
Polar radius	6 356·6 km.
Equatorial radius	6 378·1 km.
Surface area	$5·101 \times 10^{14}$ m².
Land area	$1·49 \times 10^{14}$ m².
Ocean area	$3·61 \times 10^{14}$ m².
Volume	$1·083 \times 10^{21}$ m³.
Mass	$5·976 \times 10^{24}$ kg.
Mean density	5 517 kg/m³.
Density at 5 000 m depth	11 500 kg/m³.

Author Index

Numbers in italic type indicate those pages on which references are given in full.

Aaronson, M. A., 331, *335*
Abbe, C., 26, *47*
Achison, M., 347, *360*
Ackermann, W. C., 18, 20, *24*
Adderley, E. E., 2, *23*
Ad Hoc Panel on Hydrology, 2, 20, *23*
Airey, P. L., 90, *102*, 268, *274*
Alembert, J. d', 12
Allison, G. B., 259, *272*
Alter, J. C., 26, *47*
Althaus, E., 14, *23*, 59, *76*
American Public Health Association, 200, *213*
American Society of Civil Engineers, 283, *287*, 338, *359*
American Water Works Association, 138, *150*, 203, 206, *213*
Appel, C. A., 332, *335*
Aristotle, 10
Armstead, H. C. H., 166, 168, *186*
Arnorsson, S., 173, *186*
Aronson, D. A., 238, *243*
Austin, C. L., 184, *187*

Bachmat, Y., 332, *335*
Back, W., 188, *212*, 266, *273*
Baertschi, P., 249, *272*
Baffa, J. J., 239, *243*
Baker, E., 191, 193, *212*
Balmer, G. G., 121, *129*

Balzer, R., 263, *273*
Bardsley, W. E., 332, *335*
Barnes, D. F., 317, *321*
Baturic-Rubcic, J., 329, *334*
Bavel, C. H. M. van, 100, *103*
Baydon-Ghyben, W., 15, 215, *226*
Bear, J., 58, *76*, 92, *103*, 323, 324, 328, *333*, *334*
Beaumont, P., 9, *23*
Beer, J., 263, *273*
Begemann, F., 357, *360*
Belidor, B. F., 12
Bell, R. T., 158, *164*
Bennett, T. W., 149, *151*
Bentley, H., 90, *102*, 268, *274*
Berg, G., 14, *23*
Bernhard, A. P., 147, *151*
Bernoulli, D., 11, 12
Bernoulli, J., 11, 12
Bertrand, A. R., 40, *48*
Bianchi, W. C., 235, *243*
Bierschenk, W. H., 128, *130*
Bigeleisen, J., 248, *271*
Bishop, A. W., 314, *321*
Biswas, A. K., 9, *23*
Black, R. F., 318, *321*
Blavoux, B., 267, *274*
Bleasdale, A., 27, *47*
Blom, R., 278, *287*
Boast, C. W., 86, *102*
Bonani, G., 263, *273*

Borowczyk, M., 270, *274*
Bottinga, Y., 171, *186*, 248, *271*
Bottomley, D. J., 212, *213*
Bouchardeau, A., 40, *48*
Boulton, N. S., 117, *129*
Boussinesq, J., 15
Bouw, T., 276, *287*
Bouwer, H., 99, 100, *103*
Bowen, E. G., 2, *23*
Bowen, R., 214, *226*
Bradley, E., 266, *274*
Braithwaite, F., 215, *226*
Brantly, J. E., 16, *24*
Bredehoeft, J. D., 159, *164*, 332, *335*
Breed, C. S., 277, *287*
Brogan, G. E., 277, *287*
Brown, F. F., 234, *243*
Brown, R. M., 358, *360*
Browne, P. R. L., 172, *186*
Browning, L. A., 71, *77*
Brucker, R. W., 64, *76*
Bruington, A. E., 223, *226*
Brutsaert, W. F., 331, *334*
Bryan, K., 73, *77*
Budyko, M. I., 358, *360*
Bugg, S. F., 283, *287*
Burdon, D. J., 355, *360*
Busch, K.-F., 6, *23*
Buswell, A. M., 2, *23*

California Department of Water Resources, 331, *334*, 346, *359*
California State Water Resources Control Board, 210, *213*
Calf, G. E., 90, *102*, 268, *274*
Campana, M. E., 332, *335*
Campbell, M. D., 137, *150*
Carroll, D., 69, *76*
Cartwright, K., 72, *77*
Casagrande, A., 51
Case, C. M., 119, *129*
Castelli, 11
Cedergren, E. R., 91, *103*
Celati, R., 184, *187*
Chadwick, D. G., 286, 287, *287*
Chamberlain, T. C., 15

Chow, V. T., 113, 115, *128*
Chun, R. Y. D., 228, 229, *243*
Clarke, P. F., 20, *24*
Clarke, W. E., 156, *164*
Clausen, H. B., 265, *273*
Clayton, R. N., 225, *226*
Clenenden, F. B., 344, 345, *359*
Clodius, S., 6, *23*
Cluff, L. S., 277, *287*
Collins, M. A., 219, *226*, 324, *333*
Collins, W. D., 208, *213*
Conover, C. S., 346, *359*
Conrad, G., 266, *273*
Cook, K. L., 282, *287*
Cooley, R. L., 119, *129*
Cooper, H. H., Jr, 113, *128*, 153, *164*
Cortecci, G., 172, *186*, 255, *272*
Craig, H., 248, 254, *271*, *272*
Cressey, G. B., 9, *23*
Crowe, A. S., 333, *335*, 354, *360*

Dagan, G., 92, *103*, 331, *334*
D'Amore, F., 264, *273*
Dansgaard, W., 255, *272*
Darcy, H., 15, 78, *102*
Darton, N. H., 15
Daubrée, G. A., 15
Davis, G. H., 3, *23*, 258, 266, *272*, *274*, 302, *304*
Davis, S. N., 90, *102*, 189, 190, *212*, 268, *274*
Debrine, B. E., 328, *334*
Demaster, D. J., 264, *273*
Denny, K. J., 82, *102*
Descartes, R., 10
DeWiest, R. J. M., 189, 190, *212*
Dinçer, T., 258, 265, *272*, *273*, 357, *360*
Downing, R. A., 241, *244*
Drabbe, J., 215, *226*
Drower, M. S., 7, *23*
Dupuit, J., 15, 105, 106, 109, *128*
Durfer, C. N., 191, 193, *212*

Eagon, J. B., Jr, 152, *164*
Edworthy, K. J., 241, *244*

Ehlig, C., 117, *129*
Elachi, C., 278, *287*
Ellis, A. J., 168, 183, *186*, *187*
Ellis, T. G., 16
Elmore, D., 90, *102*, 268, *274*
Epstein, S., 249, *271*
Erickson, C. R., 148, 149, *151*
Eriksson, E., 191, *212*, 358, *360*
Everett, L. C., 211, *213*

Feronsky, V. I., 295, *304*
Ferrara, G. C., 264, *273*
Ferris, J. E., 122, 123, *129*
Feth, J. H., 214, *226*
Feulner, A. J., 149, *151*
Florkowski, T., 357, *360*
Fontes, J.-C., 265, 266, 267, *273*, *274*
Forsberg, H. G., 269, *274*
Forscheimer, P., 15, 322
Förstel, H., 254, *272*
Fournier, R. O., 167, 169, 173, 174, 175, 179, 181, 183, 184, *186*, *187*
Fowler, L. C., 344, *359*
Foxworthy, B. L., 234, *243*
Franke, O. L., 160, *164*
Freeman, B. N., 149, *151*
Freeze, R. A., 154, 161, *164*, *165*, 326, 331, *334*
Friedman, I., 225, *226*, 249, 251, *271*, *272*
Frisi, P., 12, *23*
Fritz, P., 332, *335*
Frontinus, S. J., 9, 11, *23*

Gaillard, B., 269, *274*
Galilei, G., 11, 12
Gambolati, G., 161, *165*, 331, *334*
Gat, J. R., 358, *360*
Geithoff, D., 265, *273*
Gelhar, L. W., 219, *226*
Geyh, M. A., 303, *304*
Giefer, G. J., 21, *24*
Gilkeson, R. H., 72, *77*
Gilliland, J. A., 157, *164*
Glover, R. E., 121, *129*

Gonfiantini, R., 248, 249, 254, 265, 266, *271*, *272*, *273*
Gorder, Z. A., 147, *151*
Gordon, L. I., 254, *272*
Gould, W. D., 71, *77*
Gove, H., 90, *102*, 268, *274*
Grabmaier, K., 276, *287*
Graf, D. L., 225, *226*
Graham, R., 276, *287*
Graham, W. G., 234, *243*
Grandelement, G., 269, *274*
Gray, W. G., 331, *335*
Green, D. W., 331, *334*
Gregg, D. O., 158, *164*
Gregory, K. J., 2, *23*
Grolier, M. J., 278, *287*
Gruner, E., 312, *321*
Guizerix, J., 269, *274*

Haas, J. L., Jr, 178, 180, *187*
Habermehl, M. A., 90, *102*, 268, *274*
Hagemann, R., 249, *271*
Halepaska, J. C., 117, *129*
Halevy, E., 84, 90, *102*, 269, 270, 271, *274*
Hall, E. T., 264, *273*
Halley, E., 10, 11
Hansen, B. L., 317, *321*
Hansen, V. E., 325, 326, *334*
Hanshaw, B. B., 188, *212*, 266, *273*
Hanson, J. M., 332, *335*
Hantush, M. S., 119, 120, 126, *129*, *130*, 237, *243*
Harleman, D. R. F., 331, *334*
Harmeson, R. H., 231, 232, *243*
Haynes, C. V., 278, *287*
Hazen, A., 15
Hazzaa, I. B., 271, *274*
Heaton, T. H. E., 71, *77*
Heberden, W., 26, *47*
Heinl, M., 356, *360*
Hele-Shaw, H. S., 324, *333*
Helweg, O. J., 343, *359*
Hem, J. D., 195, 207, *212*
Hendry, M. S., 71, *77*
Heraclitus, 1
Herschel, C., 11, *23*

Herzberg, A., 15
Herzberg, B., 215, 226
Hill, P. G., 178, 179, 186
Hirsch, P., 14, 23
Holland, D. J., 27, 47
Holländer, R., 356, 360
Hölting, B., 15, 16, 24, 38, 48
Holzer, T. J., 164, 165
Homer, 10
Honda, M., 265, 273
Horton, R. E., 39, 42, 48
Howard, E., 276, 287
Howard, K. W. F., 70, 77
Hubbert, M. K., 96, 98, 103, 217, 226
Hughes, M. W., 259, 272
Huisman, L., 134, 150, 215, 226
Hutchins, W. A., 396, 401
Hützen, H., 254, 272
Huygens, C., 10, 12

Ineson, J., 157, 164
Irmay, S., 99, 103
Issar, A., 303, 304
Issawi, B., 278, 287

Jackson, R. D., 100, 103
Jacob, C. E., 15, 113, 119, 126, 128, 129, 130, 157, 158, 164, 237, 243
Jansa, V., 240, 244
Jantsch, K., 264, 273
Javandel, I., 326, 334
Jensen, L., 286, 287, 287
Johnson, A. I., 50, 51, 54, 62, 76, 89, 102
Johnson, G. E., 316, 321
Johnson, T. M., 72, 77
Johnston, R. H., 153, 164
Jones, J. F., 153, 164
Judd, W. R., 315, 321
Jung, K. D., 14, 23, 59, 76

Kantrowitz, I. H., 208, 213
Kasameyer, P. W., 332, 335
Kaufman, M. I., 269, 274
Kaufman, W. J., 269, 274

Kazmann, R. G., 4, 8, 23, 91, 103, 239, 242, 244
Keenan, J. H., 178, 179, 186
Keilhack, K., 15
Kennard, M. F., 314, 321
Kepler, J., 10
Keyes, F. G., 178, 179, 186
Keys, W. S., 290, 292, 294, 301, 304
King, F. H., 15
Kirby, M. E., 131, 150
Kirkham, D., 86, 102
Knutsson, G., 269, 274
Koch, E., 239, 244
Konikow, L. F., 331, **334**
Koschmieder, H., 26, 47
Kreitler, C. W., 71, 77
Krumbein, W. C., 52, 82, 102
Kruseman, G. P., 119, 129
Krynine, D., 315, 321
Kumar, R., 96, 103, 332, 335
Kunji, B., 276, 287

LaFleur, R. G., vi, vii
Lakshinarayana, V., 126, 129
Lal, D., 257, 260, 265, 272, 273
Lamarck, J., 15
Lee, C. K., 266, 274
Legget, R. F., 307, 311, 316, 320
LeGrand, H. E., 64, 76, 211, 213, 222, 226
Lehr, J. H., 137, 150
Lenda, A., 271, 274
Leonardo da Vinci, 9
Letolle, R., 267, 274
Leverett, F., 15
Lewis, D., 347, 360
Libby, W. F., 357, 360
Lindner, M., 265, 273
Lizzi, F., 306, 320
Lloyd, J. W., 283, 287
Lloyd, R. M., 175, 186
Loewengart, S., 191, 212
Logan, J., 197
Lohman, S. W., 57, 73, 76, 77
Longinelli, A., 254, 266, 272
Lonnquist, C. G., 329, 334
Lorenz, R., 276, 287

AUTHOR INDEX

Luckner, L., 6, *23*
Lusczynski, N. J., 217, *226*

McCarty, P. L., 197, 198, *212*
McCauley, J. F., 278, *287*
MacCrary, L. M., 290, 292, 294, 301, *304*
McCready, R. G. L., 71, *77*
McGuiness, C. L., 396, *401*
McKee, J. E., 237, *243*
McKenzie, W. F., 176, *186*
McMichael, F. C., 237, *243*
Maddeus, W. O., 331, *335*
Mahon, W. A. J., 167, 168, 183, *186*, *187*
Mairhofer, J., 270, *274*
Manger, G. E., 50, *76*
Mariotte, E., 10, 11, 38
Martel, E., 15
Martinec, J., 357, *360*
Masch, F. D., 82, *102*
Matlock, W. G., 134, *150*
Matthess, G., 14, *23*, 59, *76*
Mayeda, T. K., 225, *226*, 249, *271*
Mead, D. W., 16, 17, *24*
Meents, W. F., 225, *226*
Meinzer, O. E., 2, 11, 15, 17, *23*, 73, 77, 156, *164*
Mekel, J. F. M., 278, *287*
Menard, H. W., 20, *24*, 341, *359*
Mercado, S., 182, *187*
Meyer, A. F., 17, *24*
Meyer, C. F., 242, *244*
Michel, G., 303, *304*
Miller, D. W., 208, 209, *213*
Mitchell Perry, H., Jr, 211, *213*
Mobasheri, F., 100, *103*
Mogg, S. L., 149, *151*
Monk, G. D., 82, *102*
Mooney, H. M., 282, *287*
Moore, J. G., 178, 179, *186*
Morgante, S., 266, *273*
Morris, D. A., 50, 51, 54, *76*, 89, *102*
Moser, H., 270, 271, *274*
Mosetti, F., 266, *273*
Muffler, L. J. P., 180, *186*
Muller, S. W., 318, *321*

Mundorff, M. J., 347, *359*
Münnich, K. O., 259, 260, *272*

Nace, R. L., 20, *24*
Nassif, S. H., 39, 42, *48*
National Planning Resources Board, 397, *401*
Nativ, R., 303, *304*
Neff, E. L., 27, *47*
Neuman, S. P., 126, *129*
New York State Water Resources Control Board, 236, *243*
Nief, G., 249, *271*
Nightingale, H. I., 235, *243*
Nipher, F. E., 26, *47*
Nir, A., 270, *274*
Noble, D. G., 133, *150*
Norris, S. E., 152, *164*
Norton, W. H., 15
Noto, P., 184, *187*, 255, *272*
Nuti, S., 255, 264, *272*, *273*

Oeschger, H., 263, *273*
Olive, P., 267, *274*
O'Neil, J. R., 251, *272*
O'Neil, P. G., 331, *335*
Orellana, E., 282, *287*
Orlob, G. T., 269, *274*
Osgood, J. O., 209, *213*

Paces, T., 169, *186*
Palissy, B., 9, 10
Panichi, C., 167, 169, 172, 184, *186*, *187*
Papadopoulos, I. S., 126, *130*
Payne, B. R., 257, 265, 266, *272*, *273*, *274*, 357, *360*
Pearson, F. J., Jr, 266, *273*
Pekdeger, A., 14, *23*, 59, *76*
Penman, H. L., 31, 33, 35, *47*
Pennink, J., 15
Perrault, P., 10, 11
Peterson, F. L., 149, *151*
Phillips, F., 90, *102*, 268, *274*
Phillips, J., 26, *47*

Pinder, G. F., 153, *164*, 331, *335*
Piper, A. M., 198, *213*
Plato, 10
Pliny, 10
Pluhowski, E. J., 208, *213*
Poland, J. F., 162, 163, *165*
Potter, R. W., II, 175, 181, *186*
Powell, J. W., 16
Prickett, T. A., 118, *129*, 322, 329, *333*, *334*
Prinz, E., 15
Probst, J. L., 70, *76*

Rades-Rohkohl, E., 14, *23*
Ragone, S. E., 239, *244*
Rajagopalan, S. P., 126, *129*
Rao, D. B., 124, *129*
Rast, N., 59, *76*
Rauert, W., 59, *76*
Reardon, E. J., 332, *335*
Reeder, H. O., 234, *243*
Revelle, R., 224, *226*
Richards, H. J., 259, *272*
Richards, L. A., 203, *213*
Richter, R. C., 228, 229, *243*
Ridder, N. A. de, 119, *129*
Ripple, C. D., 29, *47*
Robinson, E. S., 158, *164*
Robinson, K. E., 312, *321*
Roche, M. A., 265, *273*
Rodebush, W. H., 2, *23*
Rodier, J., 40, *48*
Roether, W., 259, *272*
Rorabaugh, M. I., 126, 127, *130*, 153, *164*
Roth, E., 249, *271*
Rowe, P. C., 259, *272*
Rubin, M., 266, *273*
Ruby, P., 269, *274*
Rutledge, J., 303, *304*

Sabroux, J. C., 264, *273*
Samuels, A., 264, *273*
Sarton, G., 7, *23*
Sasman, R. T., 341, *359*
Sawyer, C. N., 197, 198, *212*

Sceva, J. E., 234, *243*
Schaber, G. G., 278, *287*
Schicht, R. J., 341, *359*
Schiedig, A., 315, *321*
Schneider, M., 356, *360*
Schoeller, H., 199, *213*
Schudel, P., 46, *48*
Schulz, H. D., 59, *76*
Schwartz, F. W., 333, *335*, 354, *360*
Schweitzer, S., 81, *102*
Schwille, F., 14, *23*, 67, 68, *76*
Seaburn, G. E., 238, *243*
Seares, F. D., 223, *226*
Seneca, 10
Sevenhuysen, R. J., 326, *334*
Shahbazi, M., 100, *103*
Shamir, U. Y., 331, *334*
Shata, A. A., 355, *360*
Sheahan, N. T., 127, *130*
Sherman, L. K., 42, *48*
Shestakov, V. M., 328, *334*
Shimp, N. F., 225, *226*
Silvey, W. D., 234, *243*
Simpson, E. S., 332, *335*
Singh, J., 96, *103*, 332, *335*
Singh, K. P., 154, *164*
Singh, R., 96, *103*, 332, *335*
Skinner, B. J., 394
Slichter, C. S., 15
Smith, D. B., 259, *272*
Smith, J., 90, *102*, 268, *274*
Smith, S. M., 20, *24*
Sondhi, S. K., 96, *103*, 332, *335*
Sonntag, C., 356, *360*
Sophocleous, M. A., 96, *103*, 332, *335*
Sor, K., 40, *48*
Speedstar Division, Koehring Co., 134, *150*
Squarci, P., 184, *187*
Staatliches Amt für Wasser- und Abfallwirtschäft Düsseldorf, 354, *360*
Stauffer, B., 263, *273*
Stewart, D. M., 289, *304*
Stiff, H. A., Jr, 198, *213*
Streltsova, T. D., 117, *129*
Stringfield, V. T., 64, *76*, 222, *226*
Suess, H. E., 38, 257, 260, *272*

AUTHOR INDEX

Suter, M., 231, *243*, 263, *273*
Swarzenski, W. V., 217, *226*

Taffi, L., 184, *187*
Talma, A. S., 71, *77*
Task Group on Artificial Ground Water Recharge, 229, *243*
Tedd, D. W., 62, *76*
Tenu, A., 255, *272*
Thales, 10
Thatcher, L. L., 257, *272*
Theis, C. V., 15, 112, 113, 116, *128*
Thiem, A., 15
Thiem, G., 15, 107, *128*
Thiessen, A. H., 28
Thomas, H. E., 342, *359*
Thomas, R. G., 126, *129*, 324, *333*
Thompson, J. M., 184, *187*
Thornthwaite, C. W., 35, 36, *47*
Thorweihe, U., 356, *360*
Todd, D. K., 15, 21, *24*, 69, 70, *76*, 98, *103*, 121, *129*, 193, 210, *212*, 215, 219, *226*, 229, 230, 241, *243*, *244*, 269, *274*, 286, *287*, 328, 330, *334*
Togliatti, V., 265, *273*
Tongiorgi, E., 265, 266, *273*, 357, *360*
Torgerson, T., 90, *102*, 268, *274*
Torricelli, E., 12
Toth, J. A., 96, *103*
Trescott, P. C., 330, *334*
Truesdell, A. H., 167, 169, 173, 176, 179, 180, *186*
Turcan, A. N., Jr, 293, *304*
Turkevich, A., 264, *273*
Tyson, H. H., Jr, 330, *334*

UN Department of Economic and Social Affairs, 62, *76*
Urey, H. C., 1, 247, 250
US Bureau of Reclamation, 118, 123, *129*, 136, 140, 141, *150*, 337, *359*
US Department of Agriculture, 54, *76*
US Environmental Protection Agency, 134, *150*, 201, 202, 203, *213*
US Public Health Service, 144, 146, *151*

Vandenberg, A., 332, *335*
Van Nostrand, R. G., 282, *287*
Van Zuylen, L., 276, *287*
Vaughan, P. R., 314, *321*
Vecchioli, J., 239, *243*
Verfel, J., 315, *321*
Visocky, A. P., 153, *164*
Visser, J., 276, *287*
Vitruvious, 10
Vogel, J. C., 71, *77*, 261, *272*
Vorhis, R. C., 160, *165*
Vries, J. J. de, 332, *335*

Walton, W. C., 127, *130*, 139, *150*
Waltz, J. P., 394
Waring, G. A., 74, *77*
Watt, S. B., 132, *150*
Wearn, P. L., 259, *272*
Weber, E. M., 330, *334*
Weiss, L. L., 26, *47*
Welchert, W. T., 149, *151*
Wenzel, L. K., 114, *129*
White, D. E., 168, 179, 180, *186*, 188, *212*, 266, *273*
White, W. B., 64, *76*
Wilcox, L. V., 206, *213*
Williams, J. R., 75, *77*
Wilson, E. M., 30, 39, 42, *47*, *48*
Wilson, W. T., 26, *47*
Witherspoon, P. A., 126, *129*, 326, *334*
Wölfi, W., 263, *273*
Wood, W. E., 132, *150*
Working Group on Nuclear Techniques in Hydrology, 295, *304*

Young, C. P., 28, *47*
Younker, L. W., 332, *335*

Zee, C. H., 326, *334*
Zellhofer, O., 270, 271, *274*
Zimmerman, U., 259, *272*
Zohdy, A. A., 286, *287*
Zuber, A., 270, 271, *274*

Subject Index

Acoustic logging, 301
Adsorption/desorption reactions, 60
Adsorption mechanism, 61
Advection, 67
Aeration zone, 55, 56, 59
Agriculture, 70, 209
Airfield Classification, 51
Alaska, 75
Alberta, Canada, 71
Alberta Basin, 225
Algae, 242
Aliphatic chlorohydrocarbons, 67
Aliphatic hydrocarbons, 67
Alkali soils, 205
Alpha particle emitters, 297
Aluminium, 212
Amsterdam, 228, 241
Anchorage, Alaska, 160
Animal hazards, 241
Animal wastes, 210
Anions, 198
Antecedent precipitation index, 43, 44
Apparent resistivity, 282
Aquifers, 14, 17, 49, 62, 341
  anisotropic, 87–9
  artesian, 66
  classification of, 65
  coastal, 222, 342
  confined, 66, 72, 73, 106, 107, 108, 120, 122, 125, 156, 161, 242, 342

Aquifers—*contd.*
  deep, 214
  fresh water, 215
  interconnections between, 266
  isotropic homogeneous, 66, 87, 98, 106, 108
  layered, 219
  leaky, 66, 117, 118, 154
  non-homogeneous, 331
  perched, 65
  permeable, 150
  phreatic, 302
  pressure, 66
  shallow, 96, 166
  sloping, 126
  steam-bearing, 185
  storage in, 72
  two-layered, 126
  unconfined, 65, 66, 100, 106, 108–11, 117–18, 122, 157, 259, 275, 284
Arabian Peninsula, 355
Argon, 252
Arizona, 162, 164
Arsenic, 212
Artificial recharge, 227–42
  activities in Europe, 240–1
  basin method, 229, 235
  ditch and furrow method, 231
  induction, 239
  Long Island, New York, 238

SUBJECT INDEX

Artificial recharge—*contd*
  major problems in, 241–2
  methods, 227–38
  mounds, 237
  pit method, 231
  project objectives, 228
  stream channel method, 229–30
  unplanned, 235
  well method, 232
Atmospheric pressure, 152, 157, 159
Atterberg limits, 53
Auguers, 132–3
  hole methodology, 86
Australia, 90, 160, 267
Austria, 302

Back Bay, 102
Bacteria, 14, 61, 242
  active mobility of, 59
  lateral dispersion of, 59
  transport, 60
Bahamas, 221
Barbados, 221
Barometric changes, 156
Barometric efficiency, 156–8
Base flow, 153, 154
Beacon Hill, 101
Belemnitella americana, 250
Bentonite, 72
Bermuda, 221
Bessemer process, 16
Bingham bodies, 52
Biochemical oxygen demand (BOD), 208–9, 236
Biological analyses, 194, 200
Birmingham, UK, 101, 211
Black Sand, 178
Blaney–Criddle method, 339
Boreholes, 84–6, 131, 288
Boron, 206
Boston, Massachusetts, 101
Boston Harbour, 102
Boulder Dam, 315
Box Tunnel, 305
Bracciano Lake, 265
Broadlands, New Zealand, 183

Building foundations, 306–8
Bunter Sandstone, 241

Cache La Poudre Valley, 339
Calcium, 169, 190
California, 162, 184, 215, 227–30, 234, 332, 397
Caliper logging, 300
Canada, 75, 212
Capillary fringe, 56
Capillary rise, 56–7
Capillary water, 56
Capillary zone, 56, 57, 100
Carbon, 246, 247, 259–64
Carbon dioxide, 4, 69, 169, 171, 179, 188, 251
Carbonic acid, 4
Carcinogens, 71
Casing logging, 301
Catchment wetness index (CWI), 44
Cations, 198
Cayman Islands, 283
Cerro Prieto, 182, 183, 184
Chad Lake, 265
Chala Lake, 265, 357
Chalcedony, 173
Chemical analysis, 194, 198
Chemical equivalence, 194
  convergence factors, 195
Chemical geothermometers
  qualitative, 170
  quantitative, 167–70
Chemical oxygen demand (COD), 208–9
Chemical thermometry, 167
Chloride–bicarbonate ratio, 224
Chlorination, 242
Chlorine, 267
Chlorohydrocarbons, 67–9
Chow method, 115
Circulation cells, 185
Clipstone, Nottinghamshire, 241
Clogging coefficient, 140
Coal mining, 349
Coal strip mining, 333
Colorado, USA, 339
Conductive liquid technique, 327

Conjunctive usage, 344
Connate water, 38
Connecticut, 215
Construction hazards, 305–21
Continental Intercalaire, 266
Cooper–Jacob method, 114, 123, 128
Coriolis force, 2
Coso Springs, 184
Cretaceous Edwards aquifer, 71
Crustal uplift, 163–4
Cucamonga Creek, 230

Dakota, 156
Dams, 16, 18, 230, 312–16
Darcy's law, 78–87, 89, 94, 96, 98, 104, 322, 330
DBCP (dibromochloropropane), 210
Dead Sea, 303
Denitrification, 70, 71
Depth-to-water, 289
Deserts, 100
Desorption velocity, 60
Deuterium, 1, 246, 248, 249, 254
De-watering, 162
DIALOG, 22
Digital computers, 329–33
Dispersion coefficients, 271
Dispersivity, 271
Dissolved constituents, 190–1
Dowsing, 288
Drainage, 153, 203, 278, 307
Drawdown, 114, 115, 121, 125
  curve, 106–11, 117, 127
Drilling, 16
  bits, 136, 137
  methods, 135
  mud, 136–7
  rig, 134, 135
Drilling-time log, 131
Drinking water, 200–2
Drought, 154, 157
Dupuit equation, 105, 111
Dyes, 83, 323

Earth tides, 158
Earthquakes, 160

EDTA (ethylenediaminetetraacetic acid), 269
Egypt, 160, 278, 356
Electrical conductance, 197
Electrical log, 292–4
Electrical resistivity, 280–1
England, 157, 241
Environmental effects, 152–65
Environmental isotope logging, 302–3
Escherichia coli, 59, 60
Euphrates River, 6–8
Evaporation, 3, 11, 17, 29, 30, 34, 35, 37, 154, 155, 225
Evapotranspiration, 3, 35, 44, 154–6, 164
Extraction of groundwater, 336–7

Far West Rand, 162
Federal Republic of Germany, 101, 215, 240, 254, 348–54, 400
Fertilizers, 70, 210
Field capacity, 58
Filtration, 60, 210
  efficiency, 60, 61, 232
  velocity, 75
Finite-difference method, 329
Finite-element method, 331, 333
Fire hazard, 241
Flashing, 183, 184
Flooding, 8, 101, 153
Florida, 215
Floridan aquifer, 90
Flow
  base, 106
  between two fixed water bodies with constant heads, 105
  direction, 85–6, 89–101
  equation, 96, 329
  net, 91, 92, 122
  pattern, 96
  radial, 108, 109
  rate, 89–101, 170
  steady, 104–11
    radial, 106, 107
  testing, 182
  unidirectional, 105

SUBJECT INDEX

Flow—*contd.*
  unsteady, 112–17
    radial, 117–20
    velocity, 105
Fluid-conductivity logging, 300
Fluid-velocity logging, 301
Fluoride, 191
Formation loss, 127
Fractionation
  factors, 247–50, 254
  process, 247, 250, 251
France, 101
Fresh water/saline water interface, 219–20
Fresno, California, 235
Freundlich isotherm, 59
Frost
  crack, 59
  heave, 41
  layers, 158
Funarole, 75

Gambier Plain, 259
Gamma–gamma logging, 296
Gamma radiation, 298–300
Gases in groundwater, 69
Geiger–Müller detector, 85
GEOARCHIVE, 22
Geochemical cycle, 3–5
Geochemical indicators, 167, 172
Geological investigations, 288–9
Geological log, 288
Geophysical investigations,
  above-surface methods, 275–9
  on-surface methods, 279–87
  subsurface methods, 288–304
  surface methods, 275–7
Geophysical log, 290
GEOREF, 22
Geostatic pressure, 161
Geothermal fields, 180–1
Geothermal fluid, 172
Geothermal gradient, 166
Geothermal waters, 166–87
Geyser Hill, 178, 179
Ghyben–Herzberg relation, 215–18
Gilbert island, 221

Girdle of Fire, 166
Glasgow, 211
Göteborg, 240
Grand Cayman Islands, 71
Grand Erg Occidental, 266
Gravel pack material, 141
Gravity method, 285
Great Artesian Basin, 90, 267
Great Plains, 398
Greenland, 75
Grevenbroich, 349–52
Groundwater
  basins, 73, 336, 338–40
  contours, 92
  flow. *See* Flow
  head, 153
  history, in, 6–22
  hydraulics, 104–30
  levels, 93, 152–4
    recovery of, 115
  movement, 78–103
    vertical, 101–2
  occurrence, 62–75
  quality, 188–213
  storage, 340
  surface, 93
Guam, 221
Gulf Coast, 225

Hague, The, 228, 241
Halogenated hydrocarbons, 67
Hardness, 197
  classification, 198
Haschendorf, 303
Haute Savoir, France, 267
Hawaii, 74, 215, 219, 221
Heat
  index, 36
  inertia, 37
  storage, 242
Hedke method, 339
High Plains, New Mexico/Texas, USA, 345
Homo erectus, 278
Honolulu, 219
Hot-water reactions, 168
Hudson River, 308

Hydraulic conductivity, 51, 86–9, 92, 94, 99, 105, 107, 109, 119, 139, 154, 218, 220, 221, 270, 327
Hydraulic engineering, 12
Hydrocompaction, 162
Hydrodynamic dispersion, 67
Hydrogen, 1–2, 171, 196, 247, 252
  atoms, 1
  bonds, 1
Hydrogen sulphide, 69, 171
Hydrogeology, 15
Hydrologic balance, 159
Hydrologic (continuity) equation, 5, 29
Hydrologic cycle, 2–6, 9, 38
Hydrologic equilibrium equation, 338, 340
Hydrological data, 339
Hydrology, 1–24
  history, in, 6–22
  isotopic, 17
Hydrometeorology, 25
Hydroseisms, 160
Hydrostatic balance, 215, 216
Hydrostatic pressure, 180
Hydrothermal reactions, 172–6
Hyperthermal geothermal fields, 185
Hysteresis effects, 98

Iceland, 75, 173
IFOV (instantaneous field of view), 277
Illinois, 225
  State Water Survey, 232
Indus River Valley, 346
Infiltration, 38–47, 68, 342
  capacity, 39
  index, 42
  investigating, 41
  rate, 40
Infiltrometers, 41
International Hydrological Decade, 323, 356
Ionic concentration, 153
Iran, 9
Iron, 190
Irrigation, 9, 162, 203–6, 209, 231, 236, 342, 343, 347

Isotope(s), 17, 246
  artificial, 268–71
  environmental, 90, 245–68, 302
  hydrology, 245–74
  radioactive, 18, 19, 224–5, 246, 255, 295
  ratio, 246, 252
  stable, 246
  terminology, 384–91
  thermometers, 171–2
  vapour pressures, 247
Isotopic abundances, 248, 251, 252
Isotopic exchange, 255
Isotopic thermometry, 167
Israel, 191, 215, 303
Italy, 184, 266

Japan, 75, 170, 215
Juvenile water, 38

Karst, 64, 222
Kenilworth, 212
Kenya, 265
Kilsby tunnel, 306
Korea, 266

Land ownership right, 396
LANDSAT, 277
Laplace equation, 98, 104, 325, 326
Larderello, 172, 184
Lea River, 241
Lead, 212
Leamington, 212
Legal aspects, 396–401
Leiden, 228, 241
Libya, 347, 356
Lichfield, 212
Limestone, 64, 280
Liverpool, 215
London, 215, 241
Long Beach, California, 224
Long Island, 160, 208, 237–9, 306, 311
Long Valley, California, 176
Los Alamos, 242
Los Angeles, 230, 234, 236
  Coastal Plain, 346

Louisiana, 293
Lower Beacon Hill, 102
Lowry–Johnson method, 340

Magmatic water, 38
Magnesium, 190
Magnetic fields, 288
Magnetic methods, 288
Main River, 240
Management of groundwater, 336–60
Manchester, 211
Marianas island, 221
Mediterranean Sea, 74
Membrane models, 325
Mersey River, 101
Meteorological data, 25–38
Meteorological fluctuations, 156–8
Methamoglobinemia, 71
Methane, 69, 171
Method of images, 120–4
Mexico, 184
Michigan, 225
Micro-organisms, 59–61
Microseisms, 160
Milwaukee, Wisconsin, 307
Mineral oil products, 67
Mining, 349
  yield, 340
Mittendorf, 303
Models, 322–35
  blotting paper, 326
  digital computer, 329–33
  electrical analogue, 327–9
  groundwater characteristics, 322
  hydrothermal, 332
  membrane, 325
  Moiré pattern, 326
  non-electrical analogue, 323–7
  numerical, 332
  porous media, 322–3
  sand, 322
  thermal, 326
  viscous fluid, 324
Modry Dul, 357
Moiré pattern, 326
Monitoring, 72
Most probable number (MPN), 200

Mudpots, 75
Multibarriers, 72
Multispectral scanning, 277

Natural gamma logging, 295
Nebraska, USA, 316
Negative pressure, 99
Netherlands, 215, 228, 240, 316
Neutron
  gamma logging, 298–300
  logging, 297–8
  probe, 46
New Brunswick, 212
New York, 215, 310
New York City, 101
New Zealand, 75, 168, 170, 183
Niers catchment area, 354
Nieuwe Maas, 316
Nile River, 6–7
Nitrate, 70–1, 189
Nitrogen, 3, 69, 71, 171
Nitrosamines, 71
North Africa, 355
Nova Scotia, 212

Oceanic islands, 220–2
Oceanic tides, 158
Ohm's law, 322
Oil in groundwater, 68
Orange County, 234
Organo-chlorine insecticides, 67
Overdraft, 342
Oxygen, 1, 2, 69, 246, 247, 249, 251, 252, 254

Pacific Belt, 166
Particle size, 54
Pasadena, 399
Pattern diagrams, 198
Pearl Harbor, 219
Peirera pan, 37
Penetration factor, 125
Penman
  method, 339
  theory, 31

SUBJECT INDEX

Peoria, Illinois, 231
Pepoli Museum, Trapani, Italy, 306
Percolation, 38–47, 342
Perennial yield, 341–3
Permafrost, 75, 149, 317–20
Permeability, 49, 50, 395
  intrinsic, 51
Permeameter
  constant head, 82
  falling head, 83
Pesticides, 210
pH values, 58, 61, 168, 172
Phi-scale, 52
Phosphorus, 271
Photography, 275–7, 279
Photomaps, 276
Phreatic surface, 55
Phreatic water, 55
Physical analysis, 194, 199
Piezometric head, 119, 224
Piezometric surface, 66, 106, 119, 160, 161, 217
Point dilution method, 84, 90
Poland, 162
Pollutants, 69, 211–12, 236
Pollution, 57, 66–70, 72, 160, 200, 208–11, 283
Polychlorobiphenyls, 67
Population centres, 159–60
Porosity, 49, 50, 62, 270, 283, 395
Porous media models, 322–3
Potability standards, 211
Potential evapotranspiration, 35–7
Potential flow theory, 218
Potentiometric surface, 66
Precipitation, 3, 17, 25–7, 44, 188
Pump chamber casing, 138
Pumping, 14, 101, 106, 142–6, 161, 162
  test, 86, 115–18, 123, 127, 340, 349
Pumps, 15

Qanats, 9, 10
Quality
  analyses, 194–203
  classification, 206
Quartz, 167, 168, 173
  geothermometer, 173

Radar, 277, 278
Radiation, 31, 32
Radioactive decay, 255–6
Radioactive (nuclear) logging, 295–300
Radioactive waste disposal, 71–2
Radiocarbon, 259–64
Radioisotopes. *See* Isotopes
Radionuclides, 255, 256
Radiotracers, 86
Radon, 264
Rain gauges, 26, 29
Raindrops, 40, 41
Rainfall, 25, 27, 28, 157
Rayleigh waves, 160
Raymond Basin, 398
Recession curves, 153, 154
Recovery procedures, 69
Reservoir(s), 337
  engineering, 182–4
  temperature estimates, 172–6
Resistivity log, 292–4
Resistance–capacitance network, 328
Resistance networks, 329
Reverse-circulation rotary technique, 137
Rhine River, 227, 240
Rio Hondo, 236
Rising water, 152
River Authorities, 400
River Boards, 400
Rocky Mountains, 398
Rotary-percussion method, 137
Rotterdam, 316
Ruhr, 240

SAAR, 277, 278
Safe yield, 338, 340–3
St. Helier Hospital, Surrey, UK, 308
Salem Heights, 234
Saline contamination, 224–5
Saline water, 203, 214–26
  intrusion control, 222–4
Salinity, 188–94, 204, 343
Salton Sea, 332
San Antonio, Texas, 396
San Gabriel River, 236
Sandstones, 65

SUBJECT INDEX

Santa Cruz River Basin, 164
Satellite imagery, 276, 339
Saturated zone, 55, 56, 68–9
Saturation
  vapour pressure, 33, 34
  zone, 55, 56
Scandinavia, 75
Scavenger wells, 220
Schlumberger array, 281, 282
Schoeller diagram, 199
Scintillation detector, 85
Scunthorpe, 212
Sea-level response, 157
Seawater, 225
  intrusion, 238
Second Elizabeth River Tunnel, 311
Seine River, 10
Seismic reflection, 283
Seismic refraction, 283, 285
Seismic shock, 284
Seismic velocity, 283
Seismic waves, 283, 284
Seismicity, 160
Self-potentials, 294
Selset reservoir, 315
Serratia marcescens, 60
Settlement, 310–12, 315
Severn Trent Water Authority, 101
Sewage systems, 208–9, 238, 342
Sewer construction, 306
Shock waves, 284–5
Silica, 167–9, 173, 189
  geothermometer, 179
Silicon, 264–5
Sinai–Negev paleowater, 303
Sinkholes, 162
SLAR, 277
SMOW, 250, 254
Snake River, 234
Sodium
  absorption ratio (SAR), 206
  content, 205
Sodium–potassium
  geothermometer, 173, 183
  ratio, 168, 169, 182
Sodium–potassium–calcium
  geothermometer, 173, 175
  molal concentrations, 169
  ratio, 168, 169

Soil(s), 204–5
  classification of, 49, 51
  cover, 45
  moisture, 43–5, 340
  texture, 54
Solar energy, 101
Solar radiation, 29, 207
South Africa, 160, 162
South Dakota, 160
Space shuttle, 278
Specific capacity, 126–8
Specific yield, 62
Spontaneous potential logging, 294
Spreading techniques, 229
Springs, 176
  artesian, 73
  discharge, 73
  fissure, 73
  gravity, 73
  perennial, 74
  periodic, 74
  submarine, 74
  thermal, 74, 75, 168, 170, 181–2
  volcanic, 73
Sprinkler tests, 41
Steady-state drawdown, 111
Storage coefficient, 72, 73, 97, 117, 122
Storativity, 72
Strasbourg Cathedral, 307
Streamflow, 153
  hydrograph analysis, 154
Streams, 121, 353
Subsidence, 161–4
Subterranean geochemical variations, 207
Subterranean mixing and boiling phenomena, 177–80
Sudan, 278
Surface tension, 68
Surging, 142
Sutton Coldfield, 212

Taurus Mountains, 265
Television logging, 301
Temperature
  effects, 207–8
  evaluation, 168

Temperature—*contd.*
　criteria, 172
　fluctuations, 157
　logging, 300
　measurement, 167, 184
Tetrachloroethylene, 69
Texas, 71, 162, 215
Theis equation, 113, 114, 120
Thermography, 279
Thiem equation, 107
Threshold saturation, 99
Tidal efficiency, 158
Tidal fluctuations, 158–9
Tigris River, 6
Tracer tests, 83–4
Transmissibility coefficient, 270
Transmissivity, 107, 109, 110, 116, 119, 270
Transpiration, 154, 155
Transport equation, 58
1,1,1-Trichloroethane, 69
Trilinear diagram, 198
Tritium, 246, 257–9, 266
Tunisia, 266
Tunnelling, 305
Tunnels, 101, 309–10
Turkey, 265, 357
Type curve, 114

Unified Soil Classification, 52
Uniformity coefficient, 54
United Kingdom, 211, 241
Uranium, 247
USA, 64, 75, 208, 214, 215, 227, 311
USSR, 75, 157, 348

Vadose water, 46, 55, 56
Van der Waals force, 61
Vapour pressure, 33
Vegetation, 342
Venice, 161
Vienna Basin, 302
Viruses, 59, 61
Void ratio, 49
Voids, 49–77, 162
Volcanic rocks, 64, 73

Wairakei, 183
Wallingford soil moisture probe, 45
Warwick, 212
Wastewater, 159, 160, 234, 236–7
Water
　Act 1973: 400
　aspects of, 1–6
　balance, 356–9
　　global, 358–9
　divining, 288
　heavy, 1
　isotopic species, 2
　level, 157, 333
　molecule, 1
　quality, 342, 392–4
　resources, 400
　Resources Act 1963, 400
　Resources Board, 400
　spreading, 228
　table 44, 55, 100, 105, 106, 117
　witching, 288
WATERF, 332
WATERQ, 332
Watershed movement, 353
WATSTORE, 21
Weed growth, 241
Well(s), 13–17, 106–11, 120–2, 131–51, 156, 159, 162, 168, 181–2, 221, 234, 236
　casing, 138–9
　cementing, 138
　chemicals, 143
　collector, 150
　construction technology, 132–42
　deep, 134, 146
　development of, 142–3
　driven, 133
　efficiency, 128
　function, 113
　hot storage, 242
　incrustation, 149
　interfering, 124–6
　jetted, 133
　large-diameter, 126
　lining, 132
　losses, 126–8
　non-vertical, 149–50
　partially penetrating, 124–6

Well(s)—*contd.*
 protection, 147–9
 recovery, 147
 restoration, 147–9
 rock, 149
 screens, 138, 140, 149
 shallow, 134, 146
 yield and drawdown, 143
Wenner array, 281, 282
Wentworth scale, 52
West St. Paul, Minnesota, 234
Whittier Narrows, 236
Wickrathberg benchmark, 353

Winchester Cathedral, 306
Wind, 29, 157
Winter rain
 acceptance potential, 45
 classification, 45
Wisbech, 212

Yellow River, 9
Yellowstone National Park, 75, 123, 168, 176, 178, 181

Zarand, Iran, 9